PHYSICS RESEARCH AND TECHNOLOGY

UV RADIATION

PROPERTIES, EFFECTS, AND APPLICATIONS

PHYSICS RESEARCH AND TECHNOLOGY

Additional books in this series can be found on Nova's website under the Series tab.

Additional e-books in this series can be found on Nova's website under the e-book tab.

PHYSICS RESEARCH AND TECHNOLOGY

UV RADIATION

PROPERTIES, EFFECTS, AND APPLICATIONS

JAMES A. RADOSEVICH, PH.D.
EDITOR

nova publishers
New York

For permission to use material from this book please contact us:
Telephone 631-231-7269; Fax 631-231-8175
Web Site: http://www.novapublishers.com

Library of Congress Cataloging-in-Publication Data

ISBN: 978-1-63321-090-5

Published by Nova Science Publishers, Inc. † New York

CONTENTS

PREFACE

Ultraviolent (UV) radiation represents a comparatively small part of the total electromagnetic spectrum, yet this portion of the spectrum is responsible for an extremely wide range of effects in many different settings. In biological systems, it can produce good and bad outcomes. We use UV radiation to drive chemical reactions that are desired, and we know that it also drives environmental reactions that are harmful to living organisms. We use UV radiation to interrogate molecules and to analyze cells. Some animals view our world in the UV radiation range, while astronomers use UV to study the universe; in both cases, what is observed is very different from that observed using other parts of the electromagnetic spectrum. This volume gives the reader just the very beginning of the broad range of properties, applications, and effects of UV radiation and is intended to stimulate the reader to explore more of this fascinating part of the electromagnetic spectrum.

Chapter 1 - Sun light or the visible spectrum is consisted of colors from violet to red, which is the only part of the electromagnetic spectrum that a typical human eye can see. On the other hand, ultraviolet (UV) radiation comes in handy for some animals to carry out various vital tasks like finding their mates and food. The discovery of UV radiation was by chance at first, nonetheless, comprehensive understanding of its properties, relations with electromagnetic waves, and effects on human and living organisms, was a gradual process that involved contributions from chemists, physicists, engineers, and biologists. The ample scientific knowledge and evidence the authors have gathered thus far regarding its properties have facilitated the beneficial applications of UV radiation. UV radiation is ever-present in our life; its applications ranging from medicine to military, industry to disease control, personal defense to atmospheric sciences are increasing rapidly as technology improves. While too much of it might be dangerous, we cannot and we should not entirely avoid UV radiation as less than the specific exposure levels may create enormous health afflictions including chronic diseases and skeletal disorders due to vitamin D deficiency. Therefore, it is imperative to have a systematic effort for gathering our resources and revisiting our knowledge for evaluating, understanding and properly communicating the adverse health effect caused by excess exposure to UV radiation, and for increasing public awareness while exploring innovative and safe means to increase its beneficial applications.

Chapter 2 - Vegetables play a particularly important role in human diet and their stabilization after harvesting and during sub-sequent storage is critical. Among several long-term preservation processes, freezing is widely recognized as a proven and efficient method for vegetables.

Most vegetables require a short heat treatment (blanching) prior to freezing to reduce microbiota and inactivate enzymes responsible for deteriorative reactions. However, detrimental effects occur during this process, such as alterations of plant tissues and consequent texture changes, solubilisation of nutrients in the blanching medium, thermal degradation of nutrients, weight loss and color changes, resulting in quality losses. Therefore, a rigorous balance between enzyme inactivation (correlated to degradation) and minimization of quality losses must be achieved in order to produce the highest final quality of frozen product.

The increasing consumers demand for high-quality food standards, have stimulated research on alternative and/or mild processing technologies that extend shelf-life without the detrimental effects caused by severe heating. From an industrial perspective, thermal processes are extremely energy consuming and have high environmental impact. Non-thermal processes, or their combination with less severe heat treatments applied before low-temperature storage, are emergent challenges with potential applications in the food industry. The application of ultraviolet irradiation is an example of such processes.

Ultraviolet radiation (UV) occupies a wide band in the non-ionizing region of the electromagnetic spectrum, with a wavelength between 100 and 400 nm. Application of UV-C radiation in the germicidal range (240-260 nm) as a non-thermal food processing technology, by pulses or continuously to food right after thermal processing of frozen vegetables, may improve the process' quality by reducing heat. It is generally claimed that the main parameters influencing sensory and quality changes on food are the dose of UV radiation applied (measured in intensity of flux in J/m^2), treatment time, and the distance between the UV source and product. UV-C pre-treatments present a high potential to retain the vegetable quality after freezing process and storage.

Chapter 3 - Global assessments of ultraviolet radiation (UVR) by the end of the 20th century indicated worldwide increases in UVR. While with the implement of the Montreal Protocol ozone-depleting substances are starting to decrease, studies in coral reef platforms have shown a 1% increase in UVR during the past two decades. Despite the exponential light attenuation in the water column, reef areas not exposed to a high sediment influx and other particulate and dissolved matter are continually exposed to high levels of Photosynthetically Active Radiation (PAR) and UVR. In the tropics, UVR is also enhanced due to the low solar zenith angle, shorter atmospheric path length and natural thinning of the ozone layer. Nonetheless, coral reefs and other associated marine ecosystems (i.e., mangroves and seagrasses) have thrived for millennia due to the evolution and development of UV-protecting mechanisms. In the case of reef corals, such mechanisms exist in the form of mycosporine-like amino acids (MAAs) thanks to their symbiotic relationship with unicellular photoautotrophic dinoflagellates known as zooxanthellae. Mangroves and seagrasses have a wide range of flavonoids which absorb in the UV region of the electromagnetic spectrum. Nonetheless, these organisms are still susceptible to physiological damages caused by UVR including DNA and protein damage, production of reactive oxygen species, and photo-oxidation of chlorophyll and other photosynthetic pigments. This chapter presents a summary of a series of experiments performed by researchers of the Bio-optical Oceanography Laboratory of the Department of Marine Sciences of the University of Puerto Rico aimed to study the effects of actual, lowered and enhanced levels of UVR on the shallow-water reef branching corals *Acropora cervicornis* and *Porites furcata*, the turtle grass *Thalassia testudinum* and the red mangrove *Rhizophora mangle* in La Parguera, Puerto Rico between 1997-2006.

Our results show that slight increases in UVR can terminate the coral's reproductive process, reduce linear extension rates, skeletal density and photosynthetic pigments production, and increase MAAs concentrations while not necessarily affecting zooxanthellae densities within the coral tissue. Both species of macrophytes responded to ambient and increased levels of UV by changes in their photosynthetic and photoprotective pigments.

Also, mangroves exposed to full solar radiation showed changes in leaf optical properties. Some of the responses observed in tropical coastal marine organisms indicate that under actual or enhanced UV scenarios most of the available energy is diverted into adaptive

Chapter 4 - Melatonin is a biogenic amine with hormonal nature found in animals. One of the most studied physiological functions is its role as a regulator of the sleep-wake cycles. In addition, melatonin is a potent antioxidant with demonstrated protective properties in various cellular dysfunctions and diseases. Its exceptional anti-oxidative qualities are mainly due to a cascade of reactions whereby generated-intermediates prolong and accentuate its anti-oxidant properties. In these reactions, light, especially UV radiation, seems to play a decisive role.

Melatonin was discovered in plants in 1995, and since then studies have focused on the identification and quantification of melatonin levels in many plant species and, also on the possible physiological roles that this indoleamine could have in plants. One of the aspects that has aroused much interest is their possible role as a regulator in light-dark cycles, as in animals, and as an indicator of light stress (at high PAR) and under UV radiation. In this work, the most significant data related to the effect that light, in natural and artificial conditions, can have on the regulation of endogenous levels of melatonin in plant tissues are presented. The authors discuss the physiological consequences that this natural anti-oxidant molecule has in stressed plants.

Chapter 5 - "The Hallmarks of Cancer" were first documented as six traits that cells acquire during their progression to malignancy: a) limitless proliferation potential, b) sustained angiogenesis, c) apoptosis resistance, d) self-sufficiency in growth signals, e) insensitivity to anti-growth signals and f) ability to tissue invasion and metastasize to distinct organs. Indeed, radiation resistance of tumor cells might be explained in part by a variety of DNA damage in oncogenes and/or onco-suppressors and metabolic reprogramming that are involved in signaling pathways related to cell survival. Herein, the authors focus our analysis on environmental conditions that play key role in enhanced cell proliferation, evasion from apoptosis and acquiring radiation resistance such as hypoxia and oxidative stress. The text will be divided into three main parts. The first will be a comprehensive introduction of role of hypoxia and oxidative stress in tumor cell radio-sensitization. The second part will address about how UV-radiation induces mutagenesis with emphasis on dysfunctions of oncogenes and onco-suppressors. The third part will cover the reprogramming of energy metabolism that accompanies cell transformation focusing on those aspects that entail UV-induced tumor microenvironment (TME). This preamble was included in order to facilitate the understanding of the ongoing debate concerning role of glycolytic pathway and mitochondria in the process of malignant cell transformation. Finally, the authors will explore how they might combine treatments to kill UV-resistant cancer cells.

Chapter 6 - Both arsenic and bioenergetics modulate cellular responses to UV radiation. Exposure of our skin to UV radiation leads to DNA damage. Cellular responses to DNA damage are crucial for the stability of the genome and prevention of skin cancer. Non-melanoma skin cancer is the most common cancer in humans. The main risk factor is exposure to UV radiation, although on a global level arsenic exposure, usually through contamination of drinking water, is also a significant cause. UV and arsenic act as co-carcinogens in the skin. Arsenic has a range of cellular effects that are not fully understood. It increases oxidative stress and also modulates the cell's response to DNA damage. The cell's response to UV-induced DNA damage is also highly dependent on bioenergetics. Cellular ATP and NAD+ are depleted following exposure to UV radiation, subsequently compromising repair pathways and leading to inefficient DNA repair. Nicotinamide, the amide form of vitamin B3, replenishes ATP and NAD+. It ensures that the cell has adequate energy following UV radiation, maximising the capacity for DNA repair. In summary, arsenic and bioenergetics are two clinically relevant factors that alter the cells capabilities to respond appropriately to DNA damage and can ultimately affect genomic stability and skin carcinogenesis.

Chapter 7 - Human skin is constantly challenged by environmental stimuli, like microorganisms, chemicals and sunlight-contained ultraviolet radiation (UVR). Skin is composed of a stratified epithelium -epidermis- and the underlying connective and adipose tissue -dermis and hypodermis respectively. There are several structures contained in the dermis like: 1) blood vessels, 2) lymphatic vessels, 3) sebaceous glands, 4) sweat glands, 5) hair follicles and, 6) sensitive nerve terminals. Among the different cells belonging to this important tissue, keratinocytes are the main epidermal cell type (almost 95% of all epidermal cells). As a consequence, these cells are more frequently exposed to the harmful effects of UVR (especially those induced by UVB), since this radiation has a low penetration capacity.

UVR affects keratinocytes by producing direct DNA damage (leading to cell cycle arrest and apoptosis) and also alterations in the cellular metabolism, including mitochondrial dysfunction. However, direct cellular alterations are not the unique harmful effect produced on keratinocytes. These cells are capable of producing large quantities of different cytokines, growth factors, and lipid-derived inflammatory mediators in response to UVR exposure, leading to both local inflammation and systemic immunosuppression. Taken together, all of these effects can induce abnormal transformation of epidermal cells. Malignant transformed keratinocytes lead to the development of the most common kind of human tumor: non-melanoma skin cancer (being basal cell carcinoma -BCC- and squamous cell carcinoma -SCC- the more frequent types).

Cutaneous photosensitivity (CP) is a condition where there is an abnormal skin reaction to UVR. The mechanisms involved in CP development are mostly related to autoimmunity, and there has been great controversy about whether anti-nuclear antibodies produce photosensitive skin lesions, specially, anti-Ro/SS-A auto-antibodies.

In their laboratory, during the last decade, the authors have been studying UVR effects on the skin and immune system using two different models: a human keratinocyte cell line (HaCaT cells) and a immunocompetitive hairless mice strain (SKH:1). Using these models the authors have investigated cellular alterations, mitochondrial dysfunction and, acute vs chronic inflammation; as well as SCC development and treatment. Treatment strategies included a non-steroidal anti-inflammatory drug, naproxen, and a probiotic-isolated molecule, lipoteichoic acid from *Lactobacillus rhamnosus*. The authors have also studied anti-Ro/SS-A auto-antibodies in both models, and their relationship with CP.

Chapter 8 - Ultraviolet radiation is known to induce DNA damage via the production of thymine-thymine dimers. This insult to DNA integrity leads to the development of tumors primarily due to mutations in oncogenes. One of the natural agents playing a dual role in tumor biology is nitric oxide. It is a diffusible, ubiquitous secondary messenger, produced biologically by nitric oxide synthase (NOS) family of enzymes. Over-expression of NOS is responsible for maintaining tumor vasculature and facilitating neoplastic transformation. In the appropriate microenvironment and at higher concentration for extended duration, NO exposure initiates inflammation leading to mutations and clinical presentation of cancers.

In previous clinical studies, the authors and others have reported that patients presenting with or converting to, tumors with high levels of nitric oxide, tended to do more poorly than those which had low NO expressing tumors. The authors also reported for the first time that human head and neck cancer cells over-express nitric oxide and can produce high levels of nitric oxide *in vivo*. Nitrogen oxides have been shown to interact with DNA leading to carcinogenic adducts. Thus to better understand the role nitric oxide plays in tumor progression and to simulate the conditions observed in clinic, they produced cell lines that grow in comparatively high nitric oxide environments. This was done by gradually exposing parent cells to increasing NO levels, resulting in a new set of 'HNO' cell lines. Interestingly, the HNO cell lines show much more aggressive growth behavior and express characteristics present in tumor stem cells. They show better UV resistance compared to parent cell lines. Further, the cellular protective mechanisms and antiapoptic pathways are induced in HNO cells along with deregulation of cell cycle. In this review, the authors will discuss various mechanisms that lead to development of UV resistance in high NO cells.

Chapter 9 - The importance of ultra violet (UV) vision was overlooked by humans for years probably because the authors lack the ability to detect UV light. In fact, the ability of organisms to detect and respond to UV light was considered a fluke or insignificant. However, as investigators have explored the evolution of the visual system, it has become clear that UV vision is widespread taxonomically in vertebrates and invertebrates and that it plays an important role in many visually guided behaviors.

The following chapter delves into some of the salient features of UV vision and provides an overview on current thoughts on how and why UV vision has evolved in complex organisms.

Chapter 10 - Ultraviolet (UV) radiation is frequently applied to disinfect water intended for human consumption and wastewater. Due to the greater chemical contamination of water, UV radiation is increasingly proposed as a technology to remove organic micropollutants, underlining its high efficacy to eliminate certain pesticides and pharmaceuticals from water. Significant advances have recently been made in our understanding of the photo-chemical processes undergone by organic contaminants and pharmaceuticals in aqueous medium. However, fewer data are available on their photochemical transformation. The objective of this chapter is to summarize the efficacy of ultraviolet (UV) radiation in the direct or indirect photodegradation of some emergent micropollutants. For this purpose, i) a kinetic study was performed, determining the quantum yield of the process; and ii) the influence of the different operational variables was analyzed [initial concentration of pollutant, pH, presence of natural organic matter compounds, radicals promoters addition ($K_2S_2O_8$, H_2O_2, activated carbon, activated carbon/TiO_2) and chemical composition of water], and iii) the time course of total organic carbon (TOC) concentration and toxicity during micropolluntant photodegradation was studied. The very low quantum yields obtained for the compounds studied are responsible for the low efficacy of the quantum process during direct photon absorption in micropollutant phototransformation. The R_{254} values obtained show that the dose habitually used for water dis-infection is not sufficient to remove this type of compounds; therefore, higher doses of UV irradiation or longer exposure times are required for their removal.

The concentration of organic micropolluntants has a major effect on their photodegradation rate. The study of the influence of pH on the values of parameters ε (molar absorption coefficient) and k'E (phodegradation rate constant) showed no general trend in the behavior of emergent pollutants as a function of the solution pH. The components of natural organic matter, gallic acid (GAL), tannic acid (TAN), and humic acid (HUM), may act as promoters and/or inhibitors of $OH^{\bullet}$radicals via photoproduct ion of H_2O_2. It is interesting to note that the addition of radical promoters such as H_2O_2 or $K_2S_2O_8$ markedly increased the effectiveness of UV radiation through the generation of $HO^{\bullet}$ or $SO_4^{\bullet-}$ radicals, respectively.

Regardless of the system considered, the results obtained showed that the micro-pollutant degradation rate was higher with lower concentrations. The solution pH had a major effect on micropollutant degradation with the UV/H_2O_2, and $UV/K_2S_2O_8$ systems. The presence of activated carbon during the pharmacecutical photodegradation process markedly increases the removal rate. The results obtained indicate that activated carbon exerts the greatest synergic effect on diatrizoic acid removal by the UV/AC system, with a synergic contribution >53 % at one minute of treatment. Moreover, the presence of activated carbons with a high carboxyl groups content enhances 2,4-D photodegradation by the UV/TiO_2 system.

Carboxyl groups in the graphene planes of the activated carbon participate in the additional generation of $OH^{\bullet}$ radicals by interacting with the electrons produced by the UV/TiO_2 system.

Chapter 11 - Ultraviolet (UV) light pervades the Universe. Many astronomical sources emit electromagnetic radiation in the wavelength range from 20 to 320 nanometers among them, the brightest objects in the Universe. UV radiation is efficiently scattered by dust grains and interacts strongly with gas in space due to the atomic structure of matter. For this reason, the UV range is ideally suited to detect subtle astronomical structures and thin gas clouds.

From the ground, the UV night sky is dark. The Earth atmosphere blocks the penetration of UV photons from space. Solar radiation with wavelengths below 280 nanometers does not reach the surface of the Earth.

UV space missions and astronomical satellites show an amazing view of the Universe. This chapter is an introduction to the processes and sources of UV radiation in space. The characteristics of the UV radiation from the Sun, planets, comets in the Solar System, from stars and interstellar clouds in the Galaxy and from the Universe, at large, are outlined.

Chapter 12 - Flow cytometry and cell sorting are well-established technologies to identify, analyze and/or sort different populations of cells in basic biological research and clinical diagnosis in pathology. This technology makes use of the principles of light scattering by particles crossing a beam of light, and excitation and fluorescence emission of fluorochromes attached to specific molecules or expressed by cells. All flow cytometers are able to interrogate cells and generate data, but only cell sorters can sort cells in a second stage where electrostatic cell sorting takes place.

Flow cytometers and cell sorters use combined systems of fluidics, optics, electronics, and computers. The fluidics system in a typical bench top flow cytometer is used to introduce and focus the particles of interest into the light source (laser) for interrogation. The excitation optics consists of a laser and lenses to shape and focus the beam. The numbers of lasers available in flow cytometry vary. Argon gas is commonly used as an excitation laser. The energized light produced by these lasers emit at 488nm. Many different fluorochromes absorb at 488nm and emit light of longer wavelength in the green (535 nm), orange (585 nm) and red spectrum (600 nm). However, ultraviolet (UV) sensitive or far red fluorochromes require excitation lasers (UV or red diode) that emit in the UV (300-400 nm) or far red (630 nm) range. Typical bench top systems have two lasers (argon and red diode). Addition of a Krypton laser provides additional color capabilities in the violet range. All of these lasers will be broadly addressed in this chapter.

This chapter will begin with a discussion of basic principles of flow cytometry and cell sorting, including a technical description of factors, intensively focused on UV lasers, which contribute to the performance of these instruments. The remaining sections will focus on clinical- and research-based applications of flow cytometry and cell sorting.

Chapter 13 - Recently, ultraviolet (UV) radiation has found two promising applications in analytical mass spectrometry. UV light may be used to generate ions through photoionization of molecules of interest. The development of commercial atmospheric pressure photoionization (APPI) sources has triggered a growing use of this method. The large success of APPI comes from is compatibility with separative techniques and its large tolerance to salts, additives, and organic solvents. UV has also attracted attention as an activation method in tandem mass spectrometry. Indeed, photoabsorption in the UV range is a straightforward means of activation of ions. Although no commercial setups exist presently, UV activation of ions appears as an extremely promising method.

Chapter 14 - Irradiation of natural fiber and biomaterials with photo initiator by ultraviolet (UV) radiation leads to the formation of very reactive intermediates in the forms of excited states, ions and free radicals which are almost instantaneously used up in several reaction pathways and consequently result in the arrangement or formation of new bonds structures. These reactions lead to grafting, crosslinking, curing and degradation of biomaterials. Use of non-ionizing radiation (UV) has been employed successfully for significant physical and chemical changes as well as changes in surface structure and surface energy of the fibers. It is possible to replace synthetic polymer by irradiated biomaterials having better mechanical and swelling property. UV radiation induced photo curing provides many advantages, such as: low energy consumption, ambient temperature operation, no emission of solvents, reduces reaction time, enhanced product quality, thus reducing costs. Radiation processing offers a clean and additive-free method for preparation of value-added novel materials based on renewable, non-toxic and biodegradable natural polymers. Effect of UV radiation stimulate grafting, crosslinking, curing and degradation on various biomaterials like chitosan, calcium alginate, gelatin and natural fiber will be discussed in this chapter.

In: UV Radiation: Properties, Effects, and Applications ISBN: 978-1-63321-090-5
Editor: James A. Radosevich

Chapter 1

A BRIEF INTRODUCTION TO ELECTROMAGNETIC RADIATION AND UV LIGHT

Bulent Aydogan*[*], *Ph.D.
The University of Chicago, Department of Radiation
and Cellular Oncology, Chicago, IL, US

ABSTRACT

Sun light or the visible spectrum is consisted of colors from violet to red, which is the only part of the electromagnetic spectrum that a typical human eye can see. On the other hand, ultraviolet (UV) radiation comes in handy for some animals to carry out various vital tasks like finding their mates and food. The discovery of UV radiation was by chance at first, nonetheless, comprehensive understanding of its properties, relations with electromagnetic waves, and effects on human and living organisms, was a gradual process that involved contributions from chemists, physicists, engineers, and biologists. The ample scientific knowledge and evidence we have gathered thus far regarding its properties have facilitated the beneficial applications of UV radiation. UV radiation is ever-present in our life; its applications ranging from medicine to military, industry to disease control, personal defense to atmospheric sciences are increasing rapidly as technology improves. While too much of it might be dangerous, we cannot and we should not entirely avoid UV radiation as less than the specific exposure levels may create enormous health afflictions including chronic diseases and skeletal disorders due to vitamin D deficiency. Therefore, it is imperative to have a systematic effort for gathering our resources and revisiting our knowledge for evaluating, understanding and properly communicating the adverse health effect caused by excess exposure to UV radiation, and for increasing public awareness while exploring innovative and safe means to increase its beneficial applications.

Keywords: Ultraviolet lights, electromagnetic waves, melanoma, skin cancer, applications of ultraviolet light, UV

---

[*] baydogan@radonc.uchicago.edu.

ABBREVIATIONS

EM	Electromagnetic
IR	Infrared
nm	Nanometer
UV	Ultraviolet

1.0. INTRODUCTION

Alike beauty, color is in the eye of the beholder. Our eyes perceive the sunlight as a white light. Nevertheless, sun light or the visible spectrum is not comprised of only white light as Isaac Newton discovered in 1666. [2] The visible spectrum consists of colors from violet to red, which is the only part of the electromagnetic (EM) spectrum that a typical human eye can see. Some animal species, on the other hand, see a totally different world. Some birds, bees, butterflies, reindeers, certain fishes, reptiles, and insects can distinguish light that are beyond violet, which is called Ultraviolet (UV) light to which we human are simply blind. These animals make use of UV lights for performing many essential tasks. Butterflies are believed to have the widest visual range of any animals; they use UV markings to find healthier mates. [3] Mother birds of some species distinguish their hungry chicks by how UV shiny they are, i.e., the amount of UV light they reflect. [4] UV light is used as a honey guide, by bees. The Arctic is UV-rich because of scattering of the shorter wavelengths and their reflection from snow and ice. Scientist recently showed that arctic reindeers do not block all UV light and that their retina responds to these wavelengths. [5] They rely on UV light to spot lichens that they eat. They can also easily spot the UV-absorbent urine of predators among the UV-reflective snow. It is truly amazing that many insects use UV wavelength emissions from celestial objects to help navigate their flight. [6] This is also the basis for how insect light traps or zappers attract mosquitoes as well.

Being the mysterious and unseen, UV light has been the center of many investigations and experiments since its discovery in early 19th century. By the first quarter of the 20th century, only 120 years after its discovery, we had ample knowledge about UV light and their properties, which helped us to turn our focus to its applications. [7] Potential uses in industry, military, medicine, atmospheric sciences and commercial applications have been explored since then. Some of the uses include, but are not limited to, antique inspection, art forgery, curing chemicals like glue to harden quickly, finding fluorescent dye leaks, glass and glass repair inspection, ID and document verification, forgery detection, mineral lighting and identification, animal and animal stains detecting, observing and recording the UV light from astronomical objects, UV coating to protect the underlying material from its harmful effects or to last longer, hand stamp screening, and reading invisible fluorescent inks. In the health field, it is used for the treatment of some diseases like chronic obstructive bronchitis, psoriasis and jaundice, detecting and killing bacteria, and sterilization and disinfection. In the military, it has a wide range of uses from camouflage to detection, surveillance to communication, disinfection of water, food, and air. The list is long and very remarkable; a literature search using Mendeley [8], a free reference search tool and organizer, found 804,099 publications and articles as of August 2013. It is very clear that, it is impossible to summarize everything related to UV radiation in just a few pages. Here we will first present a brief history of important events related to the discoveries and experiments related to the study of EM waves and UV light. This will be followed by a short summary of our current knowledge on the properties of UV light leaving the main stage to the experts in the field who will be presenting our current knowledge and recent research efforts regarding the properties, applications, potential hazard, and protection methods from UV lights in the subsequent chapters.

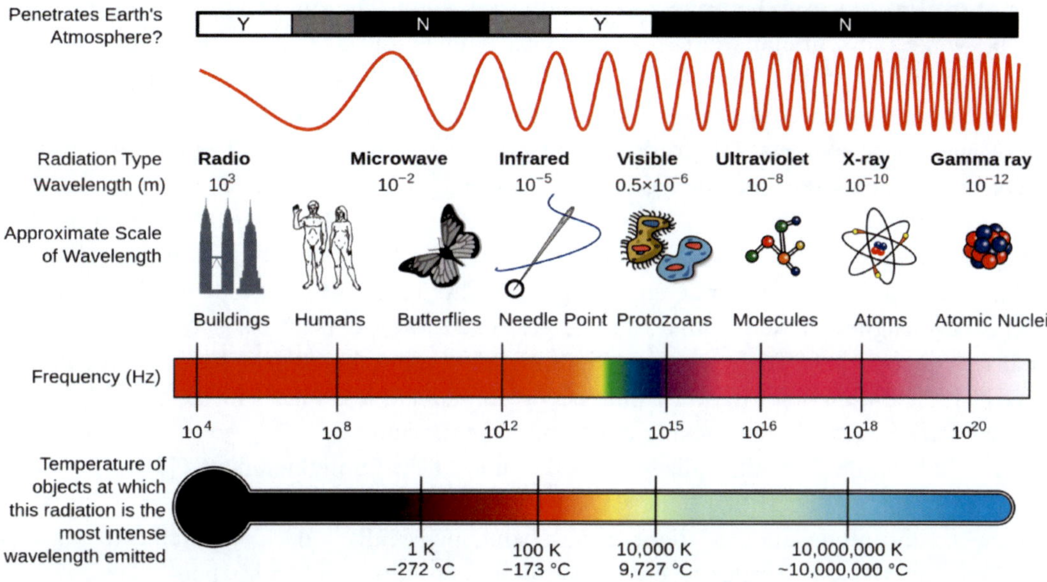

Figure 1. Electromagnetic spectrum [1].

1.1. History of EM Waves and UV Radiation

EM waves are a self-propagating transverse oscillating wave of both electric and magnetic fields together. [9] EM radiation, on the other hand, is a particular form EM field produced by moving charges. They carry energy and travel at the speed of light. They differ in wavelength, frequency and quantum energy, which determine their characteristics including, visibility, interaction with matter, hazard, and potential applications and use. As shown in Figure 1, x-rays and gamma rays are very short in wavelength and are very high in frequency, and as such they are very energetic EM radiations; they are harder to stop or block and therefore have a higher penetration power than the rest of the EM spectrum. They can cause severe damage by depositing their energies in living organism through several mechanisms, but mainly by setting bound electrons free from their orbits. They are known as ionizing radiation and can be harmful to our body while they have many applications including imaging in industry and medicine. Ironically they cause cancer and in the meantime are used to diagnose and treat cancer. Radio waves being on the other end of the spectrum possess very large wavelengths up to 60 miles and very low energy; a common comparison would be large buildings when describing their approximate scale of wavelength. Different frequencies of radio waves have different propagation characteristics in the atmosphere;

longer waves may cover a large part of our Earth uniformly whereas shorter waves can reflect off the ionosphere and travel around the world. This becomes very useful in communication; we use different frequencies of radio waves according the desired application. The portion of the EM spectrum that is visible to human eye is called visible spectrum that ranges between 380 to about 740 nanometers (nm). This part of the EM spectrum falls between the invisible infrared (IR) light with longer wavelengths and the invisible UV light with shorter wavelengths. UV rays are near the violet end of the light spectrum and are comprised of the portion of the EM spectrum between x-rays and visible light. The sun is our natural source of UV light, while the artificial sources include black lights, curing lamps, germicidal lamps, mercury vapor lamps, halogen lights, high-intensity discharge lamps, fluorescent and incandescent sources, and some types of lasers such as nitrogen lasers.

Stars and other sources of light do not all give off the same kind of emissions. These different rays are classified according to their wavelengths as shown in Figure 1. The EM spectrum is quite wide. There is a 10^{25} difference between the smallest size wavelength of gamma rays and the largest size wavelength of radio waves [10] Nearly all the rays emitted by the Sun are placed into just one small region of this large spectrum. Nearly 70% of the different wavelengths emitted by the Sun are restricted to a very narrow spectrum between 0.3 and 1.5 microns. [11] There are three kinds of light within this narrow band: visible light, infrared light, and ultraviolet light. Despite occupying a band less than 1 in 10^{25}, visible light represent 41% of all the lights reaching our Earth from the Sun.

Sun light was nothing but ordinary until the time of the 'Enlightenment'. [10] Christian Huygens was first to suggest that light was also a wave similar to sound waves in 1678. [12] He described the propagation of light waves being physically very comparable to waves successively spreading as circles when a stone is thrown into water in his book 'Treatise on Light' published in 1690. [13] Sir Isaac Newton, an English physicist and mathematician who is commonly regarded as one of the most influential scientists of all time and a key figure in the industrial and scientific revolution, had a pronounced influence on our current knowledge in optics and photophysics. [14] He is the first to challenge the theory that the sunlight is one entity, that is, it cannot be broken down into smaller constituents. He accidently noticed a chromatic aberration in a telescope creating a spectrum of colors, which later helped him to invent his own telescope known as "Newtonian telescope" to circumvent this problem. [10, 15] In 1666, Newton observed that the spectrum of colors exiting a prism is oblong, even when the light ray entering the prism is circular, which is to say, the prism refracts different colors by different angles. [16] Later he demonstrated using a prism that sunlight could be broken into its constituents, i.e., a spectrum of colors, leading him to conclude that color is a property intrinsic to light. [12] Newton was rewarded with a professorship based on this remarkable work, which nulled the theory that white light was just a single entity. He taught his first ever lecture, and also published his first ever paper in which he included his thoughts, experiments, and newly acquired knowledge on the components of light and other theories he has on light. [14] This hallmark manuscript appeared in the Philosophical Transactions of the Royal Society where he was accepted to be a member after donating a reflecting telescope. [17] Many praised this significant accomplishment, while his paper created controversy in the scientific community; scientists like Robert Hooke and even Huygens were skeptical and refuted Newton's theory. [14, 15] They argued what was seen by Newton was nothing but the perception of our eyes.

In 1800, William Herschel, a German-born British astronomer discovered IR radiation when he noted that filters of different colors seemed to pass different amounts of heat. [18] He hypothesized that the colors themselves might be of varying temperatures. He directed sunlight through a glass prism to create a spectrum and then measured the temperature of each color. He noticed that all of the colors had temperatures different than the visible light, which was his control. Herschel later measured the temperature beyond the red portion of the spectrum and performed additional experiments on what he called "calorific rays" (derived from the Latin word for heat) beyond the red portion of the spectrum. [10] He found that they were reflected, refracted, absorbed and transmitted in a way similar to visible light. What Herschel had discovered indeed was a form of light (or radiation) beyond red light, known as infrared radiation.

Understanding of EM waves and its spectrum led to discovery of UV radiation as well. Most of our current knowledge regarding UV radiation and its properties based on a series of unexpected observations in 17th and 18th century and vigilant experiments with specially developed and improved instrumentation in 19th and 20th centuries. [7] The first ever known observation was in early 17th century when sunlight changed the color of silver nitrate salts to black. [19] In 1777, Carl Wilhelm Scheele accidently observed that paper soaked in a silver chloride solution was darkened by sunlight. He further experimented with different parts of the light spectrum as "darkening" sunlight was directed through a prism onto silver chloride soaked paper. Being a chemist, he noted the release of metallic silver in a particular region of the light spectrum, but somehow he was not able to coin that it was indeed the UV light region. The opening of the 19th century marked the beginning of the most important discoveries regarding UV lights. In 1801, Wilhelm Ritter, famous for his work on galvanism, but widely accepted to be the originator of the modern electrochemistry, made his breakthrough observation that invisible rays beyond violet end of the spectrum were more effective in darkening paper soaked in silver chloride solution, which escaped the attention of Scheele. [20] He called them initially 'deoxidizing rays' which was replaced by 'chemical rays' as initial studies right after his discovery was directed on their ability to stimulate chemical reactions. [21] Niepce and Daguerre demonstrated in a series of experiments that silver iodide showed increased light sensitivity. [22] Their discovery was the basis of photography and later for x-ray films.

Discovery of UV light was followed by many invention and theories that changed our understandings of EM waves and the nature of light and established a new framework for studies of light and radiation. Thomas Young was the first to measure the wavelengths of different colors of light, in 1802. [23] Young showed that light behaved as waves by means of a diffraction experiment. He was the first to propose that different colors our eyes perceived were created by different wavelengths of light. One of the most important advances in photophysics was the invention of devices for measuring and quantifying radiation. [7] Chemical rays were shown to darken a metallic plate in 1815 by Planche. [22] Early devices were all based on his discovery that metals blackened when they were exposed to UV light. They, however, differed in how they quantified the amount of radiation measured. The first thermopile device used a stack of tightly packed metal plates to amplify the photoelectric signal. On the other hand, the bolometer invented by Langley in 1878, made use of the fact that blackened metal will have reduced electrical signal proportional to the amount UV light. The UV radiation below 200 nm was discovered in 1893 by the German physicist Victor Schumann and was named as vacuum UV because it was strongly absorbed by air. [11,23]

The invention of spectrometer led to the discovery of different elements and to the better understanding of atomic structure. It was determined that atoms of different elements absorb and emit different wavelengths of light and as such they have distinct spectra or atomic spectroscopy. [24] This allowed for the identification and quantitation and elemental composition of a sample by observing their emission spectra. In 1860, Robert Bunsen with Gustav Kirchhoff discovered two alkali metals, cesium and rubidium, with the aid of the spectroscope they had invented the year before. [25]

Atomic absorption lines are observed in the solar spectrum and referred to as Fraunhofer lines after their discoverer who mapped over 500 bands of sunlight including some in UV region in 1814. [23] A comprehensive explanation of the hydrogen spectrum was an early success of quantum mechanics and explained the Lamb shift, observed in the hydrogen spectrum, which led to the development of quantum electrodynamics. These earlier studies led to modern implementations of atomic spectroscopy for studying visible and UV transitions including arc (flame) emission spectroscopy, inductively coupled plasma atomic emission spectroscopy, glow discharge spectroscopy, and microwave induced plasma spectroscopy. [26, 27]

In 1845, Michael Faraday provided the first evidence that light was related to electromagnetism. He later theorized that light was a high-frequency EM vibration, which could propagate even in the absence of a medium such as the ether. [28] It was Faraday's inspiring work that motivated James Clerk Maxwell to study light and EM radiation. [9] Maxwell's theory of the EM field is accepted to be one of the main pillars of modern theoretical physics and played a key role in the formulation and the development of Einstein's special theory of relativity. [9] He suggested that light and sound possessed wave-like properties and are different part of the same energy spectrum that were generated by the interaction of magnetic and electric fields in 1865. This was the first time that the name 'EM wave' was coined. In 1873, he published 'A Treatise on Electricity and Magnetism', which contained a full mathematical description of the behavior of electric and magnetic fields, still known as Maxwell's equations. [29] Maxwell's speculation, however, did not imply that the light could be generated electromagnetically; in fact, he was silent about EM waves and their generation and detection. [9] It took almost a quarter of a century before Hertz discovered EM waves and his brilliant experiments confirmed Maxwell's theory. Distinguished from Maxwell and Hughes, Heinrich Hertz was the first to irrefutably prove the existence of EM waves using rigorous scientific methods and self-engineered transmitting and receiving antennas in 1886. [9]

Theoretical and empirical understanding of the UV light and EM radiation in general was intensified after Maxwell's hallmark theory and Hertz game changing experimental proof. It was the start of the 20th century when Planck suggested that radiation composed of tiny packets of energy called quanta which was later in 1905 described as massless particles released from atoms and molecules due to absorption of light, as defined by Einstein. Bohr, who helped develop quantum mechanics, proposed in 1913 that it was electron, indeed, to absorb the light and re-emit. [30] Finally Schrodinger, an Austrian physicist who developed a number of fundamental results in the field of quantum theory formulated the wave equation, which is known as Schrodinger equation. His famous paper 'Quantization as an Eigen value Problem', which was published in Annalen der Physik, has been universally celebrated as one of the most important achievements of the twentieth century and created a revolution in quantum mechanics, electromagnetism, and photophysics, indeed of all physics and chemistry. [31] He later received Nobel Prize for Physics in 1933 for the formulation of the Schrödinger equation of electromagnetic waves. [30]

1.2. The UV Spectrum and Its Properties

The portion of the EM spectrum between x-rays and visible light, i.e., between 100 and 400 nm (12.4–3.1 eV) is known as UV radiation. After atmospheric filtering, only about 3% of the total energy of sunlight at the zenith is UV radiation and this fraction decreases at other sun angles. [32] A large fraction of UV radiation, including all that reaches the surface of the Earth, is classified as non-ionizing radiation. However, the higher energy region of the UV spectrum from wavelengths about 100 nm to 280 nm, which is known as UVC radiation, is ionizing, i.e., have enough kinetic energy to liberate an electron from an atom or molecule, thereby ionizing it. Fortunately, these wavelengths are absorbed by nitrogen and dioxygen thus have an extremely short path length in air. In human body, UVC radiation is stopped in the outer layers of the epidermis. Accidental overexposure to UVC radiation is mainly from artificial sources and may cause corneal burns, commonly termed as welders' flash, snow blindness, and severe sunburn to the face. While UVC radiation injury usually clears up in a day or two, it can be extremely painful. [33]

UVB rays are essential for the synthesis of vitamin D; they are known to be responsible for triggering the production of vitamin D in our skin, but are only active in specific times of the year and of the day. [32] Nonetheless, it is the most destructive form of UV radiation since it has high enough energy, between 4.4 and 12.4 eV (315-280nm), not only to reach the Earth, but also to cause photochemical damage to cellular DNA. It is dangerous enough to cause erythema, cataracts, and skin cancer; nevertheless, a thin glass or proper clothing can easily stop it. Most solar UVB radiation is blocked by ozone in the atmosphere. Recent reductions in atmospheric ozone have emerged awareness for the possibility of increased rate of skin cancer.

UVA radiation is the most commonly encountered type of UV spectrum. It is the UVA exposure that cause tanning followed by erythema if the exposure is excessive due to initial pigment-darkening effect. Atmospheric ozone absorbs very little of this part of the UV spectrum. Overexposure to UVA radiation has been associated with toughening of the skin, suppression of the immune system, and cataract formation. UVA light is often called black light. Most phototherapy and tanning booths use UVA lamps. This is the type of UV light that is found indoors; windows absorb UVB light thus eliminating any claimed health benefit of the sunlight allowing only UVA light to penetrate in our offices and homes, which may enhance the risk of melanoma.

The photochemical effects of UV radiation can be exacerbated by chemical agents including birth control pills, tetracycline, sulphathizole, cyclamates, antidepressants, coal tar distillates found in antidandruff shampoos, lime oil, and some cosmetics. [33] Proper clothing, polycarbonate, glass, and acrylics may provide UV radiation protection. For instance, plastic diffusers are used in office lighting to block UV radiation. Sunscreens absorb or reflect UV radiation, thereby offering limited protection against UV radiation exposure and may help protect against sunburn and possibly cancer. Nevertheless, sunscreens are controversial and claimed to increase cancer risk and to reduce the health benefit of UVB radiations. [34] Some sunscreens only protect against UVB radiation, and not against the more dangerous UVA component of the spectrum. Incomplete protection against the full UV radiation spectrum, combined with increased time spent in the sun, can lead to an increase in the risks of developing malignant melanoma, a rare but more deadly form of skin cancer. [35]

Being invisible and not producing an immediate reaction, inadvertent UV radiation overexposure may instigate severe health problems in oblivious victims. Reported UV radiation accident scenarios often involve work near UV radiation sources with protective coverings removed, cracked, or fallen off. [33] Human exposure to excess UV radiation has significant public health implications. Skin cancer and malignant melanoma are among the most severe health effects, but a series of other health effects have been identified and reported. [36]

Prevention focuses on protecting the skin and other organs from UV radiation. On the other hand, UVB radiation exposure is necessary for much desired health benefit, mainly the bone health. Recent studies provide evidence that low Vitamin D levels are likely to be associated with other chronic diseases and some cancer types. In addition to skeletal disorders, calcium and vitamin D deficits increase the risk of malignancies, particularly of colon, breast and prostate gland, of chronic inflammatory and autoimmune diseases such as insulin-dependent diabetes mellitus, inflammatory bowel disease and multiple sclerosis, as well as of metabolic disorders including metabolic syndrome and hypertension. [37]

There is an urgent need for public health policy on UV radiation for preventing the disease burden associated with excessive UV exposure while promoting and ensuring its health benefit. Exposure guidelines for UV radiation have been established by the American Conference of Governmental Industrial Hygienists, [38] and by the International Commission on Non-Ionizing Radiation Protection. [39] Handheld meters to measure UV radiation are commercially available, but expert advice is recommended to ensure selecting the correct detector and diffuser for the UV wavelengths emitted by the source. [33]

CONCLUSION

We have an ample scientific knowledge regarding the properties of UV radiation and its beneficial and harmful effects. While it has many useful applications from medicine to military, industry to disease control, personal defense to atmospheric sciences, oblivious exposures to certain levels may be very dangerous. Therefore well-defined regulations and control of UV radiation hazards are needed more than ever to prevent accidental overexposures and to limit its harmful effect. It is an exposure, however, we cannot and we should not entirely avoid, since zero exposure would create a huge burden of skeletal and chronic diseases due to vitamin D deficiency. [36, 37] Therefore, it is imperative to have a systematic effort for evaluating and properly communicating the adverse health effect caused by excess exposure to UV and for increasing public awareness, while improving its beneficial and innovative applications to make our life better and safer. It is our sincere hope that this book will contribute largely towards this aim.

REFERENCES

[1] Electromagnetic spectrum. (2013) Available at: http://en.wikipedia.org/wiki/File:EM_Spectrum_Properties_edit.svg. Accessed August 15, 2013.

[2] Westfall RS. *Never at Rest: A Biography of Isaac Newton*. Cambridge: Cambridge University Press; 1983:530 – 1. Available at: http://article.wn.com/view/2013/07/17/Sir_Isaac_Newton_Granthams_blue_plaque_tributes/#/related_news. Accessed September 6, 2013.

[3] Morehouse NI, Rutowski RL. (2010)In the eyes of the beholders: Female choice and avian predation risk associated with an exaggerated male butterfly color. *The American naturalist*. 176:768–784.

[4] Reers H, Jacot A. (2011)The effect of hunger on the acoustic individuality in begging calls of a colonially breeding weaver bird. *BMC Ecology*. 11:3.

[5] Hogg C, Neveu M, Stokkan K-A, et al. (2011) Arctic reindeer extend their visual range into the UV. *Journal of Experimental Biology*. 214:2014–2019.

[6] Möller R. (2002) Insects could exploit UV-green contrast for Landmark navigation. *Journal of Theoretical Biology*. 214:619–631.

[7] Hockberger PE. (2002) A History of UV Photobiology for Humans, Animals and Microorganisms. *Photochemistry and Photobiology*. 76(6):561. Available at: http://www.bioone.org/doi/abs/10.1562/0031-8655(2002)076<0561:AHOUPF>2.0.CO;2. Accessed September 5, 2013.

[8] The Mendeley Support Team. (2011) Mendeley, A free web based reference search tool and organizer. *Mendeley Desktop*. 2011:1–16. Available at: http://www.mendeley.com.

[9] Sengupta DL, Sarkar TK. (2003) Maxwell, Hertz, the Maxwellians and the early history of electromagnetic waves. *IEEE Antennas and Propagation Magazine*. 1:14–17.

[10] Puff B. (2012) A Light History of Electromagnetic Waves: Waves and Particles before Wave-Particle Duality. *Colgate Academic Review*. Vol. 4, Article 14 Available at: http://commons.colgate. edu/car/vol4/iss1/14.

[11] Beeson S, Mayer JW. *Patterns of light: chasing the spectrum from Aristotle to LEDs*. New York: Springer Available at: http://en.wikipedia.org/wiki/UV#Fire_ detection. Accessed September 5, 2013.

[12] Parker AR. (2005) A geological history of reflecting optics. *Journal of the Royal Society, Interface / the Royal Society*. 2(2):1–17. Available at: http://www.ncbi.nlm.nih.gov/pmc/articles/PMC1578258. Accessed August 22, 2013.

[13] Huygens C. (2005) *Treatise on light.* Chicago: University of Chicago Press. Available at: http://www.gutenberg.org/files/14725/14725-h/14725-h.htm.

[14] O'Connor EF, Robertson JJ. Sir Isaac Newton. Available at: http://www-history.mcs.standrews.ac.uk/Biographies/Newton.html. Accessed September 5, 2013.

[15] Duarte FJ. (2000) Newton, Prisms, and the "Opticals" of Tunable Lasers. *Optics and Photonics News*. 11(5):24. Available at: http://en.wikipedia.org/wiki/Isaac_Newton. Accessed September 5, 2013.

[16] Whittaker ET. (1953) *A History of the Theories of Aether and Electricity.* Vol 2. Dublin: Dublin University Press.

[17] Christianson GE. (1996) *Isaac Newton and the scientific revolution*. Oxford University Press. Available at: http://books.google.com/books?id=O61ypNXvNkUC&pg=PA74. Accessed September 6, 2013.
Jones SH. (1933) The Herschel Chronicle: the Life-Story of William Herschel and his Sister Caroline Herschel : Abstract : Nature. *Nature*. 132:656–658. Available at: http://www.nature.com/nature/journal/v132/n3339/abs/132656a0.html. Accessed September 5, 2013.
[18] Abbott D. (1983) *Dictionary of Scientists: Chemists*. New York: Peter Bedrick Books;125–128.
[19] Barth J. (1987)Johann Wilhelm Ritter (1776-1810) and the discovery of UV irradiation 185 years ago. *Der Hautarzt Zeitschrift fur Dermatologie Venerologie und verwandte Gebiete*. 38:301–303.
[20] Frercks J, Weber H, Wiesenfeldt G. (2009) Reception and discovery: the nature of Johann Wilhelm Ritter's invisible rays. *Studies In History and Philosophy of Science Part A*. 40:143–156.
[21] Cole FJ. (1952) Dictionary of Scientists. *British medical journal*. 1:310–311.
[22] Bernard Cohen, Henry Crew, Joseph von Fraunhofer DWBB. (1981) *The Wave theory, light and spectra*. Ayer Publishing.
[23] I. Bernard Cohen, Henry Crew, Joseph von Fraunhofer, De Witt Bristol Brace, I. Bernard Cohen HC. (1981) *The Wave theory, light and spectra*. Ayer Publishing. Available at: http://en.wikipedia.org/wiki/Joseph_von_Fraunhofer. Accessed September 5, 2013.
[24] American Chemical Society, Society AC. Professor Robert W. Bunsen. (1900) *The Journal of the American Chemical Society*. 23:89–107. Available at: http://books.google.com. Accessed September 5, 2013.
[25] John CD. (1800-1930) Lines of light: the sources of dispersive spectroscopy. CRC Press; 1995:30–32.
[26] Misra P and Dubinskii MA. (2002) *Ultraviolet spectroscopy and UV lasers*. New York: Marcel Dekker Inc. 30:1-31. Available at: http://books.google.com/. Accessed September 5, 2013.
[27] Bence Jones, Henry, Bence Jones H. (1970) *The Life and Letters of Faraday*. Philadelphia: J. B. Lippincott and Company. Available at: http://books.google.com. Accessed September 5, 2013.
[28] A treatise on electricity and magnetism : Maxwell, James Clerk, (1831-1879). Free Download & Streaming : Internet Archive. Available at: http://archive.org/details /electricandmagne01maxwrich. Accessed September 5, 2013.
[29] Evans J, Thorndike AS. (2007) *Quantum mechanics at the crossroads: new perspectives from history, philosophy and physics*. Springer. Available at: http://books.google.com/books. Accessed September 6, 2013.
[30] Moore WJ. (1992)*Schrödinger, life and thought*. Cambridge University Press. Available at: http://books.google.com/books. Accessed September 5, 2013.
[31] NREL: Renewable Resource Data Center Home Page. Available at: http://www.nrel.gov/rredc/. Accessed September 5, 2013.
[32] Zeman G. UV Radiation. Available at: http://hps.org/hpspublications /articles/uv.html. Accessed September 5, 2013.

[33] Vainio H, Bianchini F. (2000) Cancer-preventive effects of sunscreens are uncertain. *Scandinavian journal of work, environment & health*. 26(6):529–31. Available at: http://www.sjweh.fi/show_abstract.php?abstract_id=578. Accessed September 5, 2013.

[34] Garland CF, Garland FC, Gorham ED. (1992) Could sunscreens increase melanoma risk? *American Journal of Public Health*. 82(4):614–615. Available at: http://www.ncbi.nlm.nih.gov/pmc/articles/PMC1694089. Accessed September 5, 2013.

[35] Lucas R, McMichael T, Smith W, Armstrong B. (2006) *Solar UV radiation: Global burden of disease from solar UV radiation*. (Annette Prüss-Üstün, Hajo Zeeb, Colin Mathers MR, ed.). Geneva: World Health Organization. Available at: http://www.who.int/uv/publications/solaradgbd/en/. Accessed September 5, 2013.

[36] Peterlik M, Cross HS. (2005) Vitamin D and calcium deficits predispose for multiple chronic diseases. *European journal of clinical investigation*. 35(5):290–304. Available at: http://www.ncbi.nlm.nih.gov/pubmed/15860041. Accessed August 11, 2013.

[37] ACGIH - Industrial Hygiene, Environmental, Occupational Health & Safety Resource. Available at: http://www.acgih.org/. Accessed September 5, 2013.

[38] ICNIRP: International Commission on Non-Ionizing Radiation Protection. Available at: http://www.icnirp.de/. Accessed September 5, 2013.

In: UV Radiation: Properties, Effects, and Applications
Editor: James A. Radosevich
ISBN: 978-1-63321-090-5

Chapter 2

EFFECTS OF UV PRE-TREATMENTS ON FROZEN VEGETABLES QUALITY

***Filipa I. G. Neves*[1], *Margarida C. Vieira*[2] *and Cristina L. M. Silva*[1*]**

[1]CBQF – Centro de Biotecnologia e Química Fina – Laboratório Associado, Escola Superior de Biotecnologia, Universidade Católica Portuguesa/Porto, Portugal
[2]CIQA – Centro de Investigação em Química do Algarve, Instituto Superior de Engenharia, Universidade do Algarve, Campus da Penha, Portugal

ABSTRACT

Vegetables play a particularly important role in human diet and their stabilization after harvesting and during sub-sequent storage is critical. Among several long-term preservation processes, freezing is widely recognized as a proven and efficient method for vegetables.

Most vegetables require a short heat treatment (blanching) prior to freezing to reduce microbiota and inactivate enzymes responsible for deteriorative reactions. However, detrimental effects occur during this process, such as alterations of plant tissues and consequent texture changes, solubilisation of nutrients in the blanching medium, thermal degradation of nutrients, weight loss and color changes, resulting in quality losses. Therefore, a rigorous balance between enzyme inactivation (correlated to degradation) and minimization of quality losses must be achieved in order to produce the highest final quality of frozen product.

The increasing consumers demand for high-quality food standards, have stimulated research on alternative and/or mild processing technologies that extend shelf-life without the detrimental effects caused by severe heating. From an industrial perspective, thermal processes are extremely energy consuming and have high environmental impact. Non-thermal processes, or their combination with less severe heat treatments applied before low-temperature storage, are emergent challenges with potential applications in the food industry. The application of ultraviolet irradiation is an example of such processes.

* Corresponding address: Escola Superior de Biotecnologia – UCP, Rua Dr. António Bernardino Almeida, 4200-072 Porto, Portugal. Tel.: +351 225 580 058; fax: +351 225 090 351; Skype contact: crislui; E-mail addresses clsilva@porto.ucp.pt (C.L.M. Silva).

Ultraviolet radiation (UV) occupies a wide band in the non-ionizing region of the electromagnetic spectrum, with a wavelength between 100 and 400 nm. Application of UV-C radiation in the germicidal range (240-260 nm) as a non-thermal food processing technology, by pulses or continuously to food right after thermal processing of frozen vegetables, may improve the process' quality by reducing heat. It is generally claimed that the main parameters influencing sensory and quality changes on food are the dose of UV radiation applied (measured in intensity of flux in J/m^2), treatment time, and the distance between the UV source and product. UV-C pre-treatments present a high potential to retain the vegetable quality after freezing process and storage.

Keywords: Vegetables; food industry; safety and quality preservation; freezing; thermal and non-thermal processes; blanching; UV-C radiation

ABBREVIATIONS

EU	European Union
FDA	Food and Drug Administration
HACCP	Hazard Analysis Critical Control Point
QA	Quality assurance
QC	Quality control
UV	Ultraviolet radiation

2.0. INTRODUCTION

Fresh fruits and vegetables consumption provides health benefits due to their nutritional values. However, they are perishable foods with extremely rapid deterioration hence their stabilization after harvesting and during sub-sequent storage is critical impairing their economic value. Ideally, preservation of foods should involve technologies that prevent undesirable microbial growth; retard degradation of quality attributes; and minimize nutrient losses. Preservation by freezing is one of the most important methods for retaining quality at its most of agricultural products over long-term storage. In particular, freshness of raw fruits and vegetables can be retained for long periods, extending well beyond the normal season of most horticultural crops [1, 2]. Frozen vegetables and potatoes form a significant proportion of the market in terms of frozen food consumption [3]. However, freezing must not be assumed as a magic technology, which prevents all changes. Produce still undergo deterioration during the freezing process though at a much lower rate.

The final quality of frozen vegetables depends upon a large number of factors, such as product type and variety, degree of ripening, initial quality, harvesting methods, elapsed time between harvesting and processing, pre-freezing treatments, freezing procedures, packaging and storage time and temperature [4]. All these factors, and their interactions, combined with possible deficiencies during the frozen distribution chain, point out the difficulty to predict the final product's quality [5, 6].

A flow chart of frozen fruits and vegetables, from the preparation of raw material to the consumer is shown in Figure 1.

The increasing consumer demand of high quality standards stimulated the search for new and gentle processing technologies to reduce enzyme activity and microbial load on products [7, 10] extending shelf-life without the negative effects caused by all the preparatory operations. A mild thermal treatment started to be developed in the late 20's with several improvements since then, characterized by using temperatures ranging from 70 to 100 °C lasting from seconds to a few minutes, denominated by "blanching" once it would avoid post-processing canning, drying or freezing darkness due to enzymatic action [11].

Non-thermal methods, such as UV radiation, have emerged as attractive alternatives to conventional thermal processing, and constitute challenging technologies aiming at reducing pernicious effects of thermal treatments, by preserving quality and nutritional attributes of fruit and vegetables while yielding safe and less-perishable products [12, 7, 10]. This chapter presents the importance and influence of UV pre-treatments on the physical, chemical and sensorial attributes of frozen horticultural products.

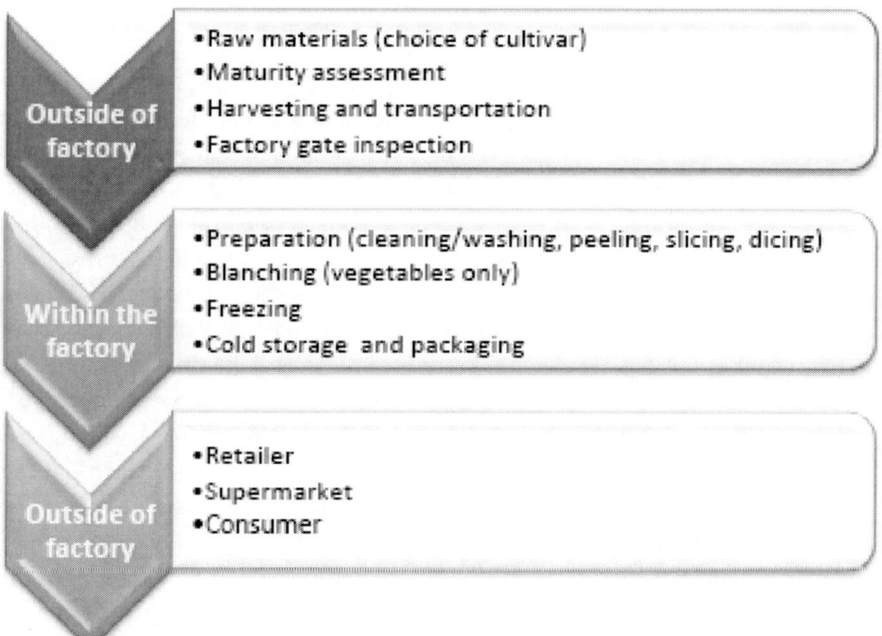

Figure 1. A general flow chart of frozen fruits and vegetables [3].

2.1. Safety and Quality Of Vegetables

The food processing industry requires a consistent supply of high-quality materials. Accordingly, a review on factors that influence the fundamental aspects and main stages of the freezing process and storage should be appraised, emphasizing their impact on safety and quality characteristics of vegetables, as well as the most relevant traditional and innovative technologies for vegetables decontamination, such as ultraviolet radiation.

2.1.1. Safety

Fruits and vegetables are often consumed uncooked after removal of the external protective cover leaving exposed more sensitive tissues that might bruise in contact with other surfaces and entities promoting product contamination. The inner tissue of fresh vegetables has long been considered as sterile. However, some studies have reported that microorganisms can be found in the inner tissue of vegetables. The husking of corn, the removal of peas and beans from the pod or the snipping and cutting of beans, are examples of processes that cause contamination of the inner vegetable tissue [13, 14]. This renders the original micro flora of the vegetable relatively unimportant. In general, fresh vegetables present great sanitary problems. Their rich nutrient composition and high water content make them attractive substrates for microorganisms activity and growth. Post-harvest contamination in processing plants is dangerous if convenient sanitizing practices are not implemented [15, 16]. The nutritional composition of vegetables depends upon several factors, such as genetic and/or environmental ones. It is also influenced by cultural practices, stages of maturity and postharvest handling.

Though vegetables' surface often contains a great diversity of microbial flora such as bacteria (some pathogenic), yeast and molds, parasites and viruses that may affect the quality of raw materials and of the products derived from them, the native micro flora that are initially observed on whole fruit or vegetable surfaces are soil inhabitants, members of a very large and diverse community of microorganisms that collectively are responsible for maintaining a dynamic ecological balance within most agricultural systems. Vectors for disseminating these contaminants include soil particles, airborne spores, irrigation water, and other agents such as animals, birds and insects. Most bacteria and fungi that arrive on the developing crop plant either are completely benign to the crop's health or, in many instances, provide a natural biological barrier to infestation by the subset of microorganisms responsible for crop damage. There is no relationship between the exterior look of vegetables and the presence of pathogenic bacteria on their surface. Several of these microorganisms are involved in the body's defense; most of them do not have any direct impact on human´s health, while some are responsible for particular diseases [17, 18]. The ones associated to outbreaks are included in Table I. Some of these pathogens are naturally present in the environment and one of the main sources of contamination is fecal deposits (human and/or animal).

Every effort should be made to minimize the risk of infection or intoxication associated with fruits and vegetables. Application of good hygienic practices during production, transport and processing, combined with the Hazard Analysis Critical Control Point (HACCP) system, will certainly minimize the contamination of fruits and vegetables and reduce the risk of illness associated with these foods [19]. Where the possibility of contamination cannot be excluded, the application of the most effective decontamination processes should be considered, as well as the germicidal effect of radiation. Since ultraviolet radiation hardly penetrates the tissue of the food products, and no residues remain on the surface it can be used as disinfection pre-treatment. Often other chemical agents or techniques are also used in combination with UV-C to reach the required level of disinfection in food products [20]. Nevertheless, the effectiveness of this treatment's application on microbial inactivation depends on several factors, such as type of product, target microorganism, initial microbial load level and physiological state of the bacterial cells, and time/doses of radiation applied.

Chemical contaminants, such as heavy metals and persistent organic pollutants (e.g. polycyclic aromatic hydrocarbons), are frequently found in fresh fruits and vegetables. They derive from the application of insecticides, fungicides, herbicides, rodenticides and also fertilizers. Many pesticides, due to their powerful neurotoxic chemicals, attack and disable portions of the nervous system. Compounds containing Pb, As, Cu, Hg and Zn are frequently used on annual crops and fruits trees [21]. Some studies revealed the presence of dangerous amounts of pesticides residing in fresh fruits and vegetables, which emphasizes the need for regular monotorization, especially for imported crops [22, 23].

Concerning fertilizers, they can be separated in two categories: organic (composed of decayed plant/animal matter) and inorganic (composed of simple chemicals and minerals). They may contain significant concentrations of trace metals including Cd and Zn (in phosphate fertilizers) and a wide range of metal and organic contaminants (in composts and sewage sludge) [21].

Table I. Pathogens associated to fruit and vegetables contamination and human illness [24, 25, 13, 14]

Organism	*Species*	*Transmission source*	*Food vehicle*
Bacteria	*Clostridium botulinum*	Soil, lakes, streams, coastal waters, intestinal tract of mammals, decaying vegetation, reptiles	Mushrooms, green pepper, coleslaw, cabbage, garlic, potato, carrot
	Escherichia coli	Animal feces, especially cattle, chicken, deer, sheep, goats, and pigs; cross contamination from raw meat	Berries, melon, lettuce, spinach, cantaloupe, alfalfa sprouts, cabbage, celery, watercress
	Salmonella spp.	Animal and human feces; cross contamination from raw meat, poultry or eggs	Berries, strawberries, melons, raw tomatoes, alfalfa sprouts, cabbage, celery, artichokes, endives, fennel, lettuce
	Shigella spp.	Human feces	Raw vegetables, lettuce, green onions, celery
	Listeria monocytogenes	Soil, food processing environments, surfaces	Lettuce, cabbage, carrots, chicory, potatoes, radish, bean products
Parasite	*Cryptosporidium* spp.	Water, animal and human feces	Vegetable salads
	Cyclospora spp.	The route of transmission is unclear; water is often referred	Raspberries, blackberries, lettuce, basil and other fresh produce
Virus	Hepatitis A	Water, human feces and urine	Lettuce, melons, strawberry, watercress, tomatoes, green onions, fresh cut fruits
	Norovirus	Human feces, vomitus	

2.1.2. Quality

The quality of a food product is a general concept that involves maintenance or improvement of external and internal characteristics readily perceived by human senses (such as general appearance, texture, color, taste and aroma), and hidden attributes (such as nutritional value and microbial contamination) that may involve quantification and instrumental measurements [26, 27].

The product's characteristics should satisfy the final consumer and there is a range of attributes that can be analyzed for quality assessment. The relative importance of each of the quality components depends on the commodity and on the individual's interest [28]. The most relevant attributes and detection methods that can be used for quality evaluation of fresh fruits and vegetables are included in Table II.

Table II. Quality attributes and some detection methods that can be used for quality assessment of fresh fruits and vegetables [29, 30]

Quality attributes	*Based on*	*Examples*	*Detection methods or equipment's*
External	Appearance (sight)	Size and shape	Grade standards and eye differentiation between items
		Colour	Visual or mechanical methods (i.e., colourimetry and spectrophotometry)
	Feel (touch)	Firmness (softening of the product)	Mechanical means (i.e., texture analyzers)
	Defects	Defects due to production, handling, environment, diseases and other factors	Visual and mechanical methods (i.e., ultrasounds and machine vision)
Internal	Odor/Aroma	Aromatic compounds (sum of compounds perceived by nose)	Gas chromatography, mass spectrometry or similar mechanisms
	Taste	Sweet, sour, bitter and astringent (perception of chemical compounds on the tongue and other mouth's nerve endings)	Spectrophotometric and gravimetric methods
	Texture (mouthfeel)	Tenderness, crispness, crunchiness, chewiness and fibrousness	Mechanical means (i.e., texture analyzers)
Hidden	Wholesomeness (freshness)	May be brought about by the food itself or external factors such as environment or handling	Microscopic, microbiological and X-ray technology methods
	Nutritive value	Vitamins, minerals, fibers, antioxidants and other phytochemicals (presence and levels of components that support life)	Wet chemistry, chromatographic methods and other chemical and physical methods
	Safety	Pathogenic microorganisms	Microbiological examination

Consumers can scrutinize external quality attributes. Appearance and defects can be easily evaluated by vision and touch (evaluation of size, shape, state of maturation, general coloration, softening and defects of the product). Internal quality attributes can be more difficult to evaluate, although fresh fruits and vegetables have characteristic aroma, taste and texture that can only be analyzed if the consumer tastes the product before acquiring it. Hidden quality parameters like nutritive value and wholesomeness are impossible to measure at purchase [29].

Quality control (QC) is the process of keeping an acceptable quality level of what is offered to the consumer. Quality assurance (QA) is the system that assures that the overall QC is being done effectively [31]. QA and QC are often used interchangeably to cover the planning, development, and implementation of inspection and testing techniques; they are time consuming and demand a great deal of training. A successful QA/QC system cannot be flexible, but it must be subjected to constant review and improvement as conditions change [31]. An effective QA/QC system including all the handling steps between harvest and retail display is essential to provide a consistently good-quality supply of fresh horticultural crops to the consumers and to protect the reputation of a given marketing label. [28, 32-34].

To keep a high quality standard, it is important to control microbial spoilage, enzymatic degradation, and chemical degradation. However, law enforcement can provide consumer protection in what concerns quality parameters that are often difficult to evaluate during the buying process.

2.1.3. Legislative Aspects

Increased attention has been given to the quality and safety of foods, not only from a consumer's perspective but also from governmental and non-governmental organizations. To ensure the consumers satisfaction and health protection, some food standards were devised to give information about the product and how to preserve quality. Food standards were also developed to prevent economic fraud and to establish market value of products [29].

Existing specific legislative and legal standards for frozen vegetables are mainly regulatory advisory establishments. These specific documents are guidelines for proper practice of handling and processing. There are several agencies that establish regulatory standards for frozen fruits and vegetables based on import-export regulations of countries around the world.

Some of the general regulations for consideration in the freezing of fruits and vegetables are summarized by the Canadian Food Inspection Agency Liaison (Preparedness and Policy Coordination). In terms of hygiene, environmental issues, working conditions, labeling, HACCP, packaging, storage and distribution, the existing EU legislation for food production also applies to frozen food [35]. The Council Directive 89/108/EEC defined the approximation of the member states relating to quick-frozen foodstuffs for human consumption, which was then implemented through the Commission Directive 92/1/EEC and 92/2/EEC, respectively on the monitoring of temperatures in transport, warehousing and storage, and on laying down the sampling procedures and method of analyses for the official control of temperatures.

Several Codex Standards (FAO/WHO Food Standards) were created for a good number of quick-frozen products, such as strawberries, peaches, and other berries, spinach, broccoli,

Brussels sprouts, peas, leek, cauliflower, green beans and potatoes. Only sound and wholesome raw materials should be used and should be in prime condition just before processing. After preparation, the product should be quick frozen without delay, using appropriate equipment to minimize physical, biochemical and microbial changes. Furthermore, the range of temperatures for maximum crystallization should be passed quickly and the process should be achieved only when the product temperature reached -18 ºC after thermal stabilization. Thereafter, the temperature has to be controlled to be kept at -18 °C and repackaging can be done under controlled conditions. Sørensen [35] presented a critical analysis on the available EU legislation, mentioning also the US and Australian laws, and concluded that it is limited due to the enormous amount of different products to give few guidelines to producers, and that more detailed information can only be found in some countries.

Concerning legislation applied to innovative pre-treatments to sanitize fresh fruits and vegetables (such as ozone, ultrasounds or UV radiation), there is still a long work to be developed. There is very little information available on the application of UV-C radiation. However, the U.S. Food and Drug Administration (FDA) recognized the use of such technology as a sanitizer method, especially to process liquid food or water, but no limits for the intensities/frequencies or exposure times have been established.

2.2. Freezing Processes Applied to Vegetables

Fruits and vegetables are perishable foods with extremely rapid deterioration. Therefore, lowering the products' temperature is a key procedure, once under such conditions the quality degradation reaction rates and microbial growth rate is prevented after harvesting and during storage [36, 37]. Freezing is one of the oldest and most widely used methods of food preservation over long storage periods, which allows the preservation of agricultural products characteristics better than any other method [37, 38].

However, the physical characteristics and nutritive value of the raw material highly affects the final product's quality (frozen vegetables). Vegetables appropriateness to be frozen depends on their specific features in raw, such as: species and varieties involved, crop production, ripening degree, kind of harvesting practice, crop storage, transportation, factory reception and time delay from harvesting till processing starts [39, 40, 2].

2.2.1. Enzymes Related to Frozen Vegetables Quality

The main food quality-related enzymes have been discussed by [41]. According to this author, the enzymes can be separated into four groups, related to changes in flavor, color, texture/consistency and nutritive value (Table III).

Although little is known about the key enzymes that are responsible for quality deterioration in frozen vegetables, it is well known that the inactivation of enzymes extends storage life. Unfortunately, there is no single key enzyme and the information is not clear. It was soon shown that enhanced storage resulted from the inactivation of enzymes. Olson and

Dietrich [42] were among the first to note that blanching could prolong vegetables high quality storage life.

The enzymes used as indicators of effectiveness of the blanching treatment are peroxidase, catalase, and more recently lipoxygenase. Peroxidase (POD, donor: hydrogen peroxide oxireductase, E.C.1.11.1.7) inactivation is commonly used in vegetable processing, since peroxidase is easily detected, occurring in a considerable number of vegetables, and is the most heat stable of these enzymes [1, 43, 44]. Nevertheless, the use of peroxidase is not unrestricted, and can regain activity under certain conditions. Besides, there is no evidence that this enzyme is associated to quality deterioration in most plant materials [45].

Table III. Food quality enzymes [41]

Quality Parameter	*Enzyme*	*Catalyzed reaction*	*Quality Effect*
Flavor	Lipolytic acyl hydrolase (lipase, esterase, etc) Lipoxygenase	Hydrolysis of lipids Oxidation of polyunsaturated fatty acids	Hydrolitic rancidity (soapy flavor) Oxidative rancidity (green flavour)
Color	Peroxidase/catalase Protease Polyphenol oxidase	Hydrolysis of protein Oxidation of phenols	Off-flavour Bitterness Dark color
Texture	Amylase Pectin methylesterase Polygalacturonase	Hydrolysis of pectin to pectic acid and methanol Hydrolysis of L-1,4 glycosidic linkages in pectic acid	Softness/loss in viscosity
Nutritional Values	Ascorbate oxidase Thiaminase	Oxidation of L-ascorbic acid Hydrolysis of thiamine	Loss in vitamin C content Loss in vitamin B1 content

Inactivation of all enzymes through heat treatment is easily achieved. However, there are drastic losses in color, texture, aroma and nutritive value. To reach the maximum quality retention, the intensity of heat treatment to stabilize the product against quality deterioration should be properly designed, but, at the same time, quality losses as a direct result of heating should be minimized [46]. Consequently the U.S. Department of Agriculture in Technical Procedures for Enzyme Inactivation Test (1975) recommended an endogenous enzyme, such as peroxidase, as indicator of treatment adequacy.

Despite the importance of enzymes on the quality of frozen vegetables, it seems that much information is still missing. It is necessary to identify, for each plant material, the decisive quality factor specific to retain during the freezing process and respective enzyme. It is also important to define the optimal blanching conditions for these specific enzymes/products and related quality factors. A database would be profitable.

2.2.2. Effects on Microbial Growth

The selection for freezing only clean and sound raw material is a good practice but not the only one to keep a product sound and hygienic. There are also many field practices and factors during crop growth that can significantly improve quality. The agro technical factors, including soil type, climatic conditions, irrigation, methods of cultivation, agricultural chemicals to control pests or diseases and fertilizer composition and application, can and have a significant impact on the nutrient content and texture of the frozen products [47].

On the produce surface one can find several bacteria, yeast and molds that may affect the quality of raw materials and of the products derived from them. Some vegetables should be prepared prior to the freezing process in terms of peeling, slicing or cutting [48]. All these operations must be carried out with utmost care, under the most stringent hygienic conditions, in order to prevent product contamination and chemical damage. Some of these operations destroy the protective barrier provided by cells membranes allowing oxygen access that causes color and texture changes [49, 50].

In general, the freezing process does not significantly destroy microorganisms that may be present in fruits and vegetables. The blanching process applied prior to freezing, reduces the levels of microorganisms present on product surfaces, contributing to safety in subsequent operations and there is a gradual decline in the number of microorganisms during freezing storage. [51 – 53]. A blanching treatment of 3 minutes reduced the total bacteria counts to < 14 000/g, coliforms to < 10/g and eradicated *Salmonella* in boiling water. However, the number of thermoresistant organisms may not be reduced, such as *Bacillus stearothermophilus* which has been isolated from blanching water, used at 90 °C for 5 minutes [54]. Tijskens et al. [55] analyzed the effect of blanching on Brussels sprouts and concluded that usually 5 logarithmic reductions were obtained on total viable counts and coliforms at 30 °C when the initial contamination present in the fresh product were nearly to 10^6 CFU/g. Li et al. [56] reported that immersion for 90 seconds in water at 50 °C reduced the mesophilic microflora on lettuce leaves by 1.73 to 1.96 (expressed as $\log_{10}$). However, a sufficient number of survivors are still present that can multiply and cause spoilage of the product during thawing. Experimental data on microbial reduction for different vegetables, different blanching techniques and conditions are still lacking.

2.2.3. Classical Freezing Systems

The rate of heat transfer from the freezing medium to the product is important in defining the freezing time of the product. Therefore, the equipment selected for freezing process characterizes the rate of freezing. There is a variety of industrial freezing systems which can be classified in many ways, namely as equipment used for batch or in-line operation, heat transfer systems (air, contact, cryogenic), and product stability. For most products, more than one type of freezer can be used on food industry, according to their necessities. Hence, in selecting a freezing system initially, a cost-benefit analysis should be conducted based on three important factors (economy, functionality and feasibility) but their relative importance may change depending on the process [57].

2.2.4. Innovative Freezing Methods

Advances in freezing technology have been focusing mainly in speeding up the freezing operation process in order to improve the products' quality and to reduce costs [58].

Ice crystal nucleation and growth are one of the main causes of degradation of quality in fruits and vegetables during freezing. Therefore, a series of strategies for controlling ice crystals in food have been attempted.

Examples of some new techniques are: high pressure freezing, which leads to instantaneous and homogeneous formation of ice throughout the whole volume of product, improving its quality [16, 59]; impingement jet freezing, characterized by its high turbulence, results in a very rapid freezing process without the high costs associated to cryogenic methods [60]; addition of anti-freeze proteins, which lower the freezing temperature and slow down recrystallization rate during frozen storage [16, 61 – 63]; addition of bacterial ice nucleators, by strains of *Pseudomonas*, *Erwinia* and *Xantonomas* (which have been detected and investigated since the 70's and recognized as one of the major causes of frost injury in plants), increases the nucleation temperature, may reduce freezing time and improves the products' quality (e.g. the flavor and textural properties) [64]; and dehydrofreezing, which diminish the tissues' damages by reduction of the product's water content, especially on fruits and vegetables [15].

2.3. Freezing Pre-Treatments

Preparatory, or pre-freezing treatments, are usually necessary to obtain a product ready for consumption, giving optimum conditions for withstanding storage [40, 2]. Grading, sorting, washing, removal of defective produce and inspecting, peeling and, in some cases, shelling, trimming, chopping and slicing are the main common operations for vegetables [65]. These operations reduce the microbial load, remove foreign material and minimize product variation. However, some damage may be caused by these preparatory treatments.

Furthermore, many microbial contaminants are part of the environment and fruits and vegetables may be inadvertently contaminated. The usual sources of pathogens contamination are from irrigation or wash water, chemical or animal fertilizers and municipal biosolids, infected operators, and operation of facilities with poor sanitization [66]. Unless decontamination procedures are adopted to decontaminate them, their safety may not be assured. Food-borne outbreaks associated with these contaminated foods are documented [67, 68].

2.3.1. Chemical Sanitizers

Washing cleans the product of dirt and impurities and removes pesticides residues. When the risk of infection or intoxication associated with raw fruits and vegetables cannot be excluded, the application of the most effective decontamination processes should be considered. A number of sanitizing agents may be used to reduce the risk of contamination that can achieve 10 to 100-fold reductions. Viruses and protozoan cysts generally exhibit higher resistance to disinfectants than do bacteria or fungi. Traditionally, fresh fruits and vegetables are rinsed in a commercial scale with tap water or chlorinated water, hydrogen peroxide and acid solutions. However, the impact of those treatments varies greatly with the type and pH of disinfectant, contact time, temperature, chemical and physical properties of the fruit or vegetable surface, and diversity of initial microbial loads. For high contamination and especially in the case of pathogens, these washing solutions may not be efficient. Little is known about the efficiency of disinfectants in relation to the roughness of fruit and vegetable surfaces, although higher amounts of cuticle material may protect against embedding of cells, thereby increasing the need for exposure to chemical treatments [69, 70]. The legal use of various treatments differs from country to country.

2.3.2. Thermal Processes

Thermal processes are commonly used in the food industry with the main purpose of inactivating pathogenic and spoilage organisms, and enzymes responsible for products deterioration. Additionally, anti-nutritional components of foods are destroyed (e.g. trypsin inhibitors in legumes) thus improving the digestibility of proteins, gelatinization of starches and release of niacin. When high temperatures are applied for short time periods, the lethal effect and shelf life extension is similar to the ones obtained if products are submitted to lower temperatures for longer periods. However, the nutritional and sensorial quality is better retained [71].

Table IV. Vegetable freezing guide [73]

Vegetable	*Preparation*	*Blanching Time (min)*	
		Water	Steam
Asparagus	Wash and sort by size. Snap off tough ends. Cut stalks into 5-cm lengths.	2	3
Beans	Wash and trim the ends. Cut if desired.	Whole: 3 Cut: 2	Whole: 4 Cut: 3
Broccoli	Wash and cut into pieces.	3	3
Cabbage	Wash and cut into wedges.	3	4
Carrots	Wash, peel and trim. Cut if desired.	5	---
Cauliflower	Discard leaves; steam and wash. Break into flowerets.	5	7
Corn	Remove husks and silks. Trim ends and wash.	Whole: 5	Whole: 7
Greens	Select young tender greens. Wash and trim the leaves.	2	3
Herbs	Wash.	No heat treatment is needed.	
Mushrooms	Wipe and damp with paper towel. Trim hard tip of stems. Sort and cut large mushrooms.	May be frozen without heat treatment.	
Peas	Shell garden peas.	1-1/2	1-1/2
Peppers	Wash, remove stems and seeds.	Freeze whole or cut as desired. No heat treatment is needed.	
Potatoes	Peel, cut or grate as desired.	Whole: 5 Cut: 2-3	---

Heat-based methods commonly applied to fruits and vegetables include hot water (usually named blanching), heat steam and hot dry air. However, other processes such as infrared, radio, and microwave radiation heating can also be used. The choice of the most convenient method should consider the heat tolerance of the commodity, its thermophysical properties, the size and shape of the product, the characteristics of the target to be inactivated (that define lethal temperature and time) and the position on the product where the organism is located [72].

For example, blanching times recommended for various vegetables are given in Table IV, which indicates that the operation time can vary depending on the intended product use.

2.3.3. Alternative Non-Thermal Treatments

Technologies such as ozonation, ultrasonication and ultraviolet radiation are examples of non-thermal technologies that can be applied in the food industry. The combination of these processes with less intensive heat treatments or with traditional sanitizers may be promising preservation methods, more economic and environmental friendly [7].

2.4. Ultraviolet Method

2.4.1. Disinfection Mechanism

The electromagnetic spectrum consists of different types of radiations that are classified according to their wavelengths. In order to classify the effect on cellular constituents is possible to distinguish non-ionizing radiation (e.g. UV-253.7 nm) from ionizing radiation (e.g. gamma rays, electron beams) which are based on entirely different mechanisms of action.

Ultraviolet (UV) radiation occupies a wide band in the non-ionizing region of the electromagnetic spectrum, with a wavelength between 100 and 400 nm. UV light is categorized on the basis of wavelength as UV-A (320–400 nm), UV-B (280–320 nm), UV-C (200–280 nm) and UV-vacuum (100–200 nm) [74]. UV-A radiation causes indirect damage to DNA, proteins and lipids through reactive oxygen intermediates, which are responsible for the changes in human skin that lead to tanning; though it can benefit humans by synthesizing vitamin D. UV-B radiation is capable of causing harm at the molecular level, and the cumulative exposure can cause skin burning and eventually lead to skin cancer. Short wavelength UV-C radiation is considered to be germicidal, promoting direct and indirect damage in the DNA of bacteria, viruses, protozoa, yeasts, molds and algae [75-80].

The germicidal effect of radiation is mainly due to the photochemical reactions that are induced inside microorganisms. The major lesion is related to DNA absorption of the UV light (253.7 nm is the most lethal), taking place by crosslinking pyrimidine nucleoside bases (thymine and cytosine) in the same DNA strand. The formation of thymine dimmers in DNA and RNA will compromise the cellular functions (impeaching the transcription and replication of the nucleic acids) leading to cell death [81-86].

2.4.2. Continuous and Pulses Application

Ultraviolet treatment as a non-thermal food processing technology can be carried out by exposure to continuous or pulsed light flux. The application of UV pulses, a promising alternative to the traditional sanitizers and thermal processing, uses short triggered peaks of electricity (seconds). This technology is referred frequently in literature reviews as "having potential for future use" or "imminent commercial applications". It is generally claimed that the application of a high intensity electric field for a specific time can produce microbial and enzyme inactivation on food [87, 88]. Explanations for its mechanism of action have been given in terms of structural changes of microbial DNA, comparable to the effect caused by continuous ultraviolet sources, but additional mechanisms seem to be involved, such as membrane disruption and vacuole extension observed in yeast cells [89, 90]. Since the intense light pulses decontamination effect seems to depend on the light absorption by microorganisms, certain food can also absorb the effective wavelengths and decrease the efficacy of this treatment which depends also on other parameters such as pulse frequency, pulse width and polarity [91-94]. In fact, McDonald (95) treated *Bacillus subtilis* spores in aqueous solutions and surfaces with pulsed UV-light and concluded that the inactivation was significantly higher than continuous wave UV light.

2.4.3. Effects on Quality and Safety

Ultraviolet treatment does not produce by-products or generate chemical residues that could change the sensory characteristics in the final product [74]. Another advantage is that it does not deliver residual radioactivity as ionizing radiation. Also, it is a cold and dry process requiring very low maintenance [96].

Due to its strong germicidal effect, exposure to low doses of UV-C have been reported to be used in drinking water, wastewater and air disinfection, as well as in the disinfection of food containers and post-harvest decay of fruits and vegetables [83, 85, 97-101]. The germicidal effect of UV-C radiation light has been demonstrated to be effective against some common foodborne pathogenes such as *Campylobacter*, *Salmonella*, *Listeria* and *Escherichia coli* [102-105]. However, low dosages may not be sufficient to inactivate some viruses and spores.

No legislation about the application of UV-C radiation to sanitize fresh fruits and vegetables was found. However, U.S. Food and Drug Administration recognize the use of such technology as a sanitizer method, especially to process liquid food or water, but no limits for the intensities/frequencies or exposure times have been established.

However, several microorganisms possess repair mechanisms in response to the damages due to ultraviolet light exposure. Microorganisms can repair their DNA by light-independent (dark-repair) mechanisms as well as light dependent (photoreactivation) mechanisms. Dark repair mechanisms are attained through the nucleotide excision repair, post replication recombinational repair or by error-prone repair. The induction of those mechanisms is dependent on DNA damage. The photoreactivation can be developed by inactivating microorganisms that recover their activity by repairing pyrimidine dimers in the DNA. This phenomenon is light dependent because it uses a photolyase enzyme that can be activated in the presence of different wave lengths as UV-A and photosynthetic active radiation (400 to 700 nm) [78, 79, 102, 106].

2.4.4. UV Systems

Industrially, two types of UV lamps are of main interest: (i) the usual low-pressure mercury arc lamps with a monochromatic output at 254 nm (UV-C range), and (ii) the high intensity and medium pressure mercury arc lamps with a polychromatic output in all UV wavelengths (however, concentrated at certain peaks inside of the UV-C region). Lamps that emit polychromatic light continuously have a medium light intensity. For a high light intensity, UV light pulses are often used [81, 88, 107].

2.4.5. Disadvantages

UV-C light has limited applications, once:

1) it has a poor penetrative capacity, leading to food surface applications only;
2) its ability to catalyze oxidative changes that can lead to rancidity and discoloration, leaves only room to its application to low fat products. Due to the wide variety of organisms present on food surfaces, the most critical factor is to determine the required radiation dose. A low dose may not be effective, whereas a high dose, although greatly reducing the number of microorganisms, may produce discoloration and accelerate the senescence of the product. The flux intensity of UV light is usually measured in J/m^2 and depends on the distance from the source of UV to the product and the exposure time. In some cases, several minutes are needed to achieve a sufficient decrease in pathogen populations, which could make the technique unfeasible for certain applications [98, 108].

2.4.6. UV-C Radiation Combined with Heat and Other Non-Thermal Pre-Treatments

In agreement with the industrial needs regarding frozen vegetables, the application of UV-C radiation as a pre-treatment can be done individually with good results in terms of safety and quality. However, there are some limitations and factors which influence the effectiveness of this methodology, as mentioned above. All technologies have advantages and negative aspects, thus, the combination of a mild thermal pre-treatment, with processes considered as non-thermal such as ultrasounds and ozone, may be promising. The intensity of each technology can decrease by combining two or more treatments, attaining target microbial parameters without compromising the quality standards.

Several researchers have studied the synergetic effects of processes. Pan et al. [109] and Marqueni et al. [98] investigated the effect of UV-C and heat treatment on strawberries and sweet cherries throughout storage. Neves et al. [10] analyzed the same combinations of technologies on zucchini inactivation of peroxidase. Allende et al. [110] combined UV-C light, gaseous ozone, superatmosferic oxygen and high carbon dioxide on heath promoting compounds and shelf life of strawberries.

CONCLUSION

Increasing the consumption of vegetables and fruits as part of a healthy diet, and a transition toward convenience foods, such as ready-to-eat, requires continuing efforts to conserve the nutritious value and extend shelf life without compromising food safety and quality. The few recent outbreaks of produce-associated illness reinforces that the agro-food industry generally produces high quality and safe foods all year round. Although no one can guarantee that food safety will not be breached at some point during the production chain, producers and the food industry should make every effort to minimize such events.

In order to prevent contaminations, producers and the food industry should develop and apply more efficacious decontamination treatments focused on current and newly introduced chemicals and techniques that will preserve nutritional values, as well as organoleptic and esthetic attributes. There are several new or revisited techniques to reduce the enzyme activity and microbial load on products. Non-thermal methods have emerged as attractive alternatives to conventional thermal processing.

These new treatments might include ultraviolet radiation, which may have potential applications instead of the traditional sanitizers and thermal treatments. Due to its strong germicidal effect, UV-C radiation has a great potential in food processing. Moreover, it requires minor energy consumption, and appears to be a much more economic and environmentally friendly technology.

It seems that no single method or treatment would be sufficient to prevent the decay of fruits and vegetables and attain safety targets. In general, the combined use of several disinfectant agents has been widely reported in the last few years to have a synergistic effect and results in a better sensory and microbial quality of the product.

However, the study and application of ultraviolet radiation is still insufficiently explored. Due to its potential, it is revealed as an area of great academic and economic interest, which should be developed in the future.

REFERENCES

[1] Arthey, D. (1993). Freezing of vegetables and fruits. In: Mallett, C.P. ed., Frozen Food Technology Chapman and Hall, London, UK.

[2] Canet, W. (1989). Quality and stability of frozen vegetables. In Developments in food preservation (Ed. S. Thorne), vol. 5, *Elsevier Science Publishing Inc.*, New York.

[3] Mallett, C. P. (1993). Frozen Food Technology. Chapman and Hall, London, UK.

[4] Fu, B., and Labuza, T. P. (1997). Shelf life testing: procedures and prediction methods for frozen foods. *Fscn.che.umn.edu.* [2009].

[5] Cruz, R. M., Vieira, M. M. C., and Silva, C. L. M. (2009c). Effect of cold chain temperature abuses on the quality of frozen watercress (Nasturtium officinale R. Br.). *Journal of Food Engineering*, 94 (1), 90 - 97. doi: 10.1016/j.jfoodeng.2009.03.006.

[6] Oliveira, S. R., Cruz, R. M. S., Vieira, M. M. C., Gaspar, M. N., and Silva, C. L. M. (2009). Enterococcus faecalis and Pseudomonas aeruginosa behaviour in frozen watercress (Nasturtium officinale) temperature abuses. *International Journal of Refrigeration*, 32 (3), 472-477. doi:10.1016/j.ijrefrig.2008.08.002.

[7] Piyasena, P., Dussault, C., Koutchma, T., Ramaswamy, H. S., and Awuah, G. B. (2003). Radio frequency heating of foods: principles, applications and related properties – a review. *Critical Reviews in Food Science and Nutrition*, 43, 587-606.

[8] Knorr, D., Zenker, M., Heinz, V., and Lee, D. U. (2004). Applications and potential of ultrasonic in food processing. *Trends in Food Science and Technology*, 15, 261-266.

[9] Cruz, R. M. S., Vieira, M. M. C., and Silva, C. L. M. (2006). Effect of heat and thermosonication treatments on peroxidase inactivation kinetics in watercress (Nasturtium officinale*). Journal of Food Engineering*, 72 (1): 8-15.

[10] Neves, F. I. G., Vieira, M. M. C., and Silva, C. L. M. (2012). Inactivation kinetics of peroxidase in zucchini (Cucurbita pepo L.) by heat and UV-C radiation. *Innovative Food Science and Emerging Technologies*, 13, 2012: 158 - 162. doi: 10.1016/j.ifset. 2011.10.013.

[11] Lee, C. Y., Bourne, M. C., and Van Burnen, J. P. (1979). Effect of blanching treatment on the firmness of carrots. *Journal of Food Science*, 44(2), 615-616.

[12] Kuldiloke, J. (2002). Effect of ultrasound, temperature and pressure treatments on enzyme activity and quality indicators of fruit and vegetables juices. Thesis, Technischen Universität, Berlin, Germany.

[13] Zhao, Y. (2005). Pathogens in fruit. In: Improving the safety of fresh fruit and vegetables. W. Jongen (ed), Woodhead Publishing Limited, Cambridge: 44-43.

[14] Gorny, J. (2006). Microbial contamination of fresh fruits and vegetables. In: Microbiology of fruits and vegetables. G. M. Sapers, J. R. Gorny and A. E. Yousef (ed), CRC Press, Taylor & Francis, Boca Raton: 3-32.

[15] Maestrelli, A., Lo Scalzo, R., Lupi, D., Bertolo, G., and Torreggiani, D. (2001). Partial removal of water before freezing: cultivar and pre-treatments as quality factors of frozen muskmelon (Cucumis melo, cv reticulatus Naud.). *Journal of Food Engineering*, 49 (2-3), 255-260.

[16] Li, B, and Sun, D. W. (2002). Novel methods for freezing and thawing of foods – a review. *Journal of Food Engineering*, 54(3), 175-182.

[17] Andrews, J. H., and Harris, R. F. (2000). The ecology and biogeography of microorganisms on plant surfaces. *Annual Review Phytopathology*, 38, 145–180.

[18] Janisiewicz,W. J., and Korsten, L. (2002). Biological control of postharvest diseases of fruits. *Annual Review of Phytopathology*, 40, 411–441.

[19] Oomah, B. D., and Mazza, G. (2000). Functional foods. In: Francis, F.J. (Ed.), *The Wiley Encyclopedia of Science and Technology*, Vol. 2, 2nd ed.Wiley, New York, NY, pp. 1176–1182.

[20] Kim J. G., Youself, A. E., and Dave, S. (1999). Application of ozone for enhancing the microbial safety and quality of foods: a review. *Journal of Food Protection*, 62: 1071-1087.

[21] Alloway, B. J. (2004). Contamination of soils in domestic gardens and allotments: a brief overview. *Land Contamination & Reclamation* 12(3): 179-187.

[22] Torres, C. M., Picó, Y. and Mañes, J. (1996). Determination of pesticide residues in fruit and vegetables. *Journal of Chromatography* A, 754(1-2): 301-331.

[23] Knezevic, Z. and Serdar, M. (2009). Screening of fresh fruit and vegetables for pesticides residues on Croatian marked. *Food Control* 20(4): 419-422.
[24] Anonymous (2001). Chapter IV. Outbreaks associated with fresh produce: incidence, growth and survival of pathogens in fresh and fresh-cut produce. In: Analyses and evaluation of preventive control measures for the control and reduction/elimination of microbial hazard on fresh and fresh-cut produce C.F.F.S.A.A. Nutrition, U.S. Food and Drug Administration. Available from: URL:http:www.//fda.gov/Food/ScienceReserch/ReserchAreas/SafepracticesforFoodProcesses/ucm091265.htm.
[25] Warriner, K. (2005). Pathogens in vegetables. In: Improving the safety of fresh fruit and vegetables. W. Jongen (ed), Woodhead Publishing Limited, Cambridge: 3-43.
[26] Shewfelt. R. L., Erickson, M.C., Hung, Y. C. and Malundo, T. M. M. (1997). Applying quality concepts in frozen food development. *Food Technology* 51(2): 56-59.
[27] Abbot, J. A. (1999). Quality measurement of fruits and vegetables. *Postharvest Biology and Technology* 15(3): 207-225.
[28] Kader, A. A. (1988). Influence of preharvest and postharvest environment on nutritional composition of fruits and vegetables, In: B. Quebedeaux and F.A. Bliss (eds.).*Horticulture and human health, contributions of fruits and vegetables*. Prentice-Hall, Englewood Cliffs, NJ, pp. 18-32.
[29] Anonymous (2002). Food safety and quality assurance issues. In: Improving the safety and quality of fresh fruits and vegetables: A training manual for trainers. University of Maryland, Food and Drug Administration: V1-V27. Available from: _URL:http:www.//groups.ucanr.org/UC_GAPs/documents/Other_Training_Resources2659.pdf.
[30] Rico, D., Martín-Diana, A. B., Barat, J. M. and Barry-Ryan, C. (2007). Extending and measuring the quality of fresh-cut fruit and vegetables: a review. *Trends in Food Science & Technology* 18(7): 373-386
[31] Hubbard, M. R. (1999). Choosing a quality control system. *Technomic Publ. Co.*, Lancaster, PA, pp. 207.
[32] Cavalieri, R.P. (ed.) (1999). Fresh fruits and vegetable: quality and food safety. *Postharvest Biol. Technol.* 15:195-340. (Proceedings of Beltsville Symposium XXIII, May 1998).
[33] Kader, A. A. (ed.) (1992). Postharvest Technology of Horticultural Crops. Second edition, *Univ. Calif., Div. of Agr. and Nat. Resources*, Publ. 3311, pp. 296.
[34] Shewfelt, R. L., and Prussia, S.E. (eds.) (1993). Postharvest handling: *A systems approach*. Academic Press, San Diego, pp. 358.
[35] Sørensen, L. B. (2002). Frozen Food Legislation. *Bulletin of the International Institute of Refrigeration*, 2002-2004.
[36] Delgado, A. E., and Sun, D. W. (2000). Heat and mass transfer for predicting freezing processes, a review. *Journal of Food Engineering*, 47, 157-174.
[37] Zhang, M., Duan, Z. H., Zang, J. F., and Peng, J. (2004). Effects of freezing conditions on quality of areca fruits. *Journal of Food Engineering*, 61(3): 393-367.
[38] Silva, C. L. M., Gonçalves, E. M., and Brandão, T. R. S. (2008). Freezing of fruits and vegetables. In: *Frozen Food Science and Technology*. J. A. Evans (ed). Blackwell, United Kingdom: 165-183.
[39] Fennema, O. (1996). An over-all view of low temperature food preservation. *Cryobiology*, 3(3), 197-213.

[40] Munoz-Delgado, J. A. (1977). Effect of freezing, storage and distribution on quality and nutrition attributes of foods, in particular of fruits and vegetables. *In food quality and nutrition: Research priorities* (Ed. W. K. Downey), Applied Sciences Publishers Ltd., London, p.353-384.

[41] Svensson, S. (1977). Inactivation of enzymes during thermal processing. In: Physical, chemical and biological changes in food caused by thermal processing. Ed. T. Hoyem and O. Kvale, *Applied Science Publishers LTD*, London, p. 202-217.

[42] Olson, R. L., and Dietrich, W. C. (1969). Quality and stability of frozen vegetables. In: Quality and Stability of Frozen Vegetables, Ed. By W. B. Van Arsdel, M. J. Copley and R. L. Olson, *Wiley Interscience*, New York, p: 117-141.

[43] Barrett, D. M., and Theerakulkait, C. (1995). Quality indicator in blanched, frozen, stored vegetables. *Food Technology*, 1(62), 64-64.

[44] Bahçeci, K. S., Serpen, A., Gökmen, V., and Acar, J. (2005) Study of lipoxygenase and peroxidase as indicator enzymes in green beans: change of enzyme activity, ascorbic acid and chlorophylls during frozen storage. *Journal of Food Engineering*, 66(2), 187-192.

[45] Lu, A. T., and Whitaker, J. R. (1974). Some factors affecting rates of heat inactivation of horseradish peroxidase. *Journal of Food Science*, 39(6), 1173-1178.

[46] Velasco, P. J., Lim, M. H., Pangborn, R. M., and Witaker, J. R. (1989). Enzyme responsible for off-flavour and off-aroma in blanched and frozen-stored vegetables. *Biotechnology and Applied Biochemistry*, 11, 118-127.

[47] Hui, Y. H., Ghanzala, S., Graham, D. E., Murrell, K. D. and Nip, W. K. (2004). *Handbook of Vegetable Preservation and Processing*. Marcel Dekker, Inc.

[48] Spiess, W. E. L. (1984). *Changes in ingredients during production and storage of deep frozen food-a review of the pertinent literature*. ZFL 8:625.

[49] Moreira, M. R., Roura, S. I., and del Valle, C. E. (2003). Quality of Swiss chard produced by conventional and organic methods, Lebensmittel Wissenschaft und-Technologie, 36(1), 135-141.

[50] Reid, D. S. (1993).Basic physical phenomena in the freezing and thawing of plant and animal tissue. In: *Frozen food technology* (Ed. C. P. Mallet), Blackie Academic & Professional, Glasgow, p. 1-19.

[51] Archer, G. P., Kennedy, C. J., and Wilson, A. J. (1995). Towards predictive microbiology in frozen food systems-a framework for understanding microbial population dynamics in frozen structures and in freeze-thaw cycles. International *Journal of Food Science and Technology*, 30, p. 711.

[52] Préstamo, G., Fuster, C., and Risueno, M. C. (1998). Effects of blanching and freezing on the structure of carrots cells and their implications for food processing. *Journal of the Science Food and Agriculture*, 77(2), 223-229.

[53] Selman, J. D. (1987). The blanching process. In: Developments in Food Preservation (Ed. Stuart Thorme), 4 Vol., *Elsevier Applied Science Publishers*, London and New York, p. 2005-250.

[54] Marcy, G., and Adam, W. (1977). Soy sprouts salad as a health hazard. *Archiv. Fur Lebernsmittelhygiene*, 28(5), 197-203.

[55] Tijskens, L. M. M., Koek, P. C., Van Der Meer, M. A., Schijvens, E. P. H. M., and De Witte, Y. (1979). Quality changes in frozen Brussels sprouts during storage. II

Objective quality parameters: texture, colour, ascorbic acid content and microbiological growth. *Journal of Food Tec*hnology, 14, 3001-313.

[56] Li, Y., Brackett, R. E., Shewfelt, R. L., and Beuchat, L. R. (2001). Changes in appearance and natural microflora on iceberg lettuce treated in warm, chlorinated water and then stored at refrigeration temperature. *Food Microbiology*, 18(3), 299-308.

[57] Bevilacqua, M., D'Amore, A., and Polonara, F. (2004). A multi-criteria decision approach to choosing the optimal blanching-freezing system. *Journal of Food Engineering*, 63 (3), 253-263.

[58] Barbosa-Cánovas, G. V., Altunakar, B., and Mejía-Lorío, D. J. (2005). Freezing fruits and vegetables. Food and Agriculture Organization – *Agricultural Services Bulletin* 158, Rome.

[59] Van Buggenhout, S., Lille, M., Messagie, I., Van Loey, A., Autio, K., and Hendrickx, M. (2006). Impact of pretreatment and freeing conditions on the microstructure of frozen carrots: qualification and relation to texture loss. *European Food Research Technology*, 222, 543-553.

[60] Soto, V., and Bórquez, R. (2001). Impingement jet freezing of biomaterials. *Food Control*, 12, 515-522.

[61] Griffith, M., and Ewart, K. V. (1995). Antifreeze proteins and their potential use in frozen foods. *Biotechnology Advances*, 13(3), 375-402.

[62] Cruz, R. M. S., Vieira, M. M. C., and Silva, C. L. M. (2009a). Effect of heat and thermosonication treatments on watercress (Nasturtium officinale) vitamin C degradation kinetics. *Innovative Food Science and Emerging Technologies* 9 (4), 483-488. doi:10.1016/j.ifset.2007.10.005.

[63] Cruz, R. M. S., Vieira, M. M. C., and Silva, C. L. M. (2009b). The response of watercress (Nasturtium officinale) to vacuum impregnation: effect of an antifreeze protein type I. *Journal of Food Engineering*, 95, 339–345. doi:10.1016/jjfoodeng. 2009.05.013.

[64] Li, J., and Lee, T. C. (1995). Bacterial ice nucleation and its potential application in the food industry. *Trends in Food Science and Technology*, August (6), 259-265.

[65] Cioubanu, A., and Niculescu, L. (1976). Cooling technology in the food industry (Ed. A. Cioubanu, G. Lascu, V. Bercescu and L. Niculescu), Abacus Press, Kent.

[66] Food and Drug Administration (FDA), US Department of Agriculture (USDA), and Centers for Disease Control and Prevention (CDC) (1998). *Guidance of industry-Guide to minimize microbial food safety hazards for fresh fruits and vegetables*, Oct. 26, 1998, http://vm.cfsan.fda.gov/dms/prodguid.html.

[67] Odemuru. J. A., Mitchell, S. J., Alves, D. M., Lynch, J. A., Yee, A. J., Wang, S. L., Styliadis, S., and Farber, J. M. (1997). Assessment of the microbiological quality of ready-to-use vegetables for health-care food services. *Journal of Food Protection*, 60, 954-960.

[68] Gill, N. (1998). Cyclosporiasis outbreaks yet again in North America. *Eurosurveillance* Weekly, 2, 610-618.

[69] Beuchat, L. R. (1998). Surface decontamination of fruits and vegetables eaten raw: a review. Food safety issues, Food Safety Unit, World Health Organization: 1-42.

[70] Kennedy, C. (2003). Developments in freezing. Food preservation techniques. P. Zeuthen and L. Bogh-Sorensen (ed). Cambridge, England, Woodhead Publishing Limited and CRC Press LLC: 228-240.

[71] Barbosa-Cánovas, G., Fernández-Molina, J. J., Alzamora, S. M., Tapia, M. S., López-Malo, A., and Chanes, J. W. (2003). Handling and preservation of fruits and vegetables by combined methods for rural areas – Technical manual. *Food and Agriculture Organization of the United Nations*, 52-53.

[72] Geysen, S., Verlinden, B. E., and Nicolai, B. M. (2005). Thermal treatments of fresh fruit and vegetables. In: Improving the safety of fresh fruit and vegetables. W. Jongen (ed). Woodhead Publishing Limited, Cambridge: 429-453.

[73] Archuleta, M. (2003). Freezing Vegetables Guide E-320. New Mexico State University, college of Agricultural Home Economics Extension Service.

[74] Guerrero-Beltrán, J. A., and Barbosa-Cánovas, G. V., (2004). Advantages and limitations on processing foods by UV light. *Food Science and Technology International*, 10, 137–147.

[75] Britt, A. B. (1996). DNA damage and repair in plants. Annu. Rev. Plant Physiol. Plant Mol. Biol, 47: 75–100.

[76] Tuteja, N., Singh, M. B., Misra, M. K., Bhalla, P. L., and Tuteja, R. (2001). Molecular mechanisms of DNA damage and repair: progress in plants. *Crit. Rev. Biochem. Mol. Biol.*, 36: 337–397.

[77] Hollósy, F. (2002). Effects of ultraviolet radiation on plant cells. *Micron* 33(2): 179-197.

[78] Matak, K. E. (2004). Effect of UV irradiation on the reduction of the bacterial pathogens and chemical indicators of milk. Thesis, Polytechnic Institute and State University of Virginia.

[79] Zenoff, V. F., Sineriz, F., and Farías, M. E. (2006). Diverse responses to UV-B radiation and repair mechanisms of bacteria isolated from high-altitude aquatic environments. *Applied and Environmental Microbiology*, 72(12): 7857-7863.

[80] Keyser, M., Muller, I. A., Cilliers, F. P., Nel, W., and Gouws, P. A. (2008). Ultraviolet radiation as a non-thermal treatment for the inactivation of microorganisms in fruit juice. *Innovative Food Science and Emerging Technologies*, 9(3): 348-354.

[81] Giese, N., and Darby, J. (2000). Sensitivity of microorganisms to different wavelengths of UV light: implications on modeling of medium pressure UV systems. *Water Research*, 34(16): 4007-4013.

[82] Ohlsson, T. (2002). Minimal processing of foods with non-thermal methods. In: *Minimal processing technologies in the food industry*. T. Ohlsson and N. Bengtsson (ed). Woodhead Publishing Limited, Cambridge: 34-60.

[83] Wang, T., MacGregor, S. J., Anderson, J. G., and Woolsey, G. A. (2005). Pulsed ultraviolet inactivation spectrum of Escherichia coli. *Water Research*, 39(13): 2921-2925.

[84] Sharifi-Yazdi, M. K., and Darghahi H. (2006). Inactivation of pathogenic bacteria using pulsed *UV-light and its application in water disinfection and quality control*. Acta Medica Iranica, 44(5): 305-308.

[85] Altic, L. C., Rowe, M. T., and Grant, I. R. (2007). UV light inactivation of Mycobacterium avium subs. paratuberculosis in milk as assessed by FASTPlaque TB Phage Assay and Culture. *Applied and Environmental Microbiology*, 73(11): 3728-3733.

[86] Unluturk, S., Atilgan, M. R., Handan, B. A., and Tari, C. (2008). Use of UV-C radiation as a non-thermal process for liquid egg products (LEP). *Journal of Food Engineering*, 85(4): 561-568.

[87] Dunn, J. (2001). Pulsed electric field processing: an overview. In G. V. Barbosa-Canovas, & Q. H. Zhang (Eds.), Pulsed electric fields in food processing: *Fundamental aspects and applications* (pp. 1–30). Lancaster, PA: Technomic Publishing Co.

[88] Morris, C., Brody, A. L., and Wicker, L. (2007). Non-thermal food processing/preservation technologies: a review with packaging implications. *Packaging Technology and Science*, 20(4), 275–286.

[89] Takeshita, K., Shibato, J., Sameshima, T., Fukunaga, S., Isobe, S., Arihara, K., and Itoh, M. (2003). Damage of yests cells induced by pulsed light irradiation. *Int. J. Food Microbiology*, 85, 151-158.

[90] Wuytack, E. Y., Thi Phuong, L. D., Aertsen, A., Reyns, K.M.F., Marquenie, D., De Ketelaere, B., Masschalk, B., Van Opstal, I., Diels, A. M. J., and Michiels, C. W. (2003). Comparison of sublethal injury induced in Salmonella enteric serovar typhimurium by heat and by different nonthermal treatments. *J. Food Prot.,* 66, 31-37.

[91] Mosqueda-Melgar, J., Raybaudi-Massilia, R. M., and Martín-Belloso, O., (2008a). Combination of high-intensity pulsed electric fields with natural antimicrobials to inactivate pathogenic microorganisms and extend the shelflife of melon and watermelon juice. *Food Microbiology*, 25 (3), 479-491.

[92] Mosqueda-Melgar, J., Raybaudi-Massilia, R. M., and Martín-Belloso, O. (2008b). Inactivation of *Salmonella enterica* Ser. *Enteritidis* in tomato juice by combining of high-intensity pulsed electric fields with natural antimicrobials. *Journal of Food Science*, 73 (2), M47-M53.

[93] Mosqueda-Melgar, J., Raybaudi-Massilia, R. M., and Martín-Belloso, O. (2008c). Nonthermal pasteurization of fruit juices by combining high-intensity pulsed electric fields with natural antimicrobials. *Innovative Food Science and Emerging Technologies*. doi: 10.1016/j.ifset.2007.09.003.

[94] Aguiló-Aguayo, I., Odriozola-Serrano, I., Quintão-Teixeira, L.J., and Martín-Belloso, O. (2008). Inactivation of tomato juice peroxidase by high-intensity pulsed electric fields as affected by process conditions. *Food Chemistry*, 107, 949-955.

[95] McDonald, K. F., Curry, R. D., and Clevenger, T. E. (2000). A comparison of pulsed and continuous ultraviolet light sources for the decontamination of surfaces. *IEEE Trans Plasma Science*, 28, 1581–1587.

[96] Morgan, R. (1990). UV "green" light disinfection. *Dairy Industries International*, 54, 33-35.

[97] Lu, J. Y., Stevens, C., Yakabu, P., Loretan, P. A., and Eakin, D. (1987). Gamma, electron beam and ultraviolet radiation on control of storage rots and quality of Walla Walla onions. *Journal of Food Process Press*, 12, 53-62.

[98] Erkán, M., Wang, C. Y., and Krizek, D. T. (2001). UV-C irradiation reduces microbial populations and deterioration in Cucurbita pepo fruit tissue. *Environmental and Experimental Botany,* 45, 1-9.

[99] Marquenie, D., Lammertyn, J., Geeraerd, A. H., Soontjens, C., Van Impe, J. F., Nicolaï, B. M., and Michiels, C. W. (2002). Inactivation of conidia of Botrytis cinerea and Monilinia fructigena using UV-C and heat treatment. *International Journal of Food Microbiology*, 74 (1-2): 27-35.

[100] Allende A., and Artés, F. (2003). UV-C radiation as a novel technique for keeping quality of fresh processed Lollo Rosso lettuce. *Food Research International*, 36(8), 739-756.

[101] FDA (2009). Kinetics of microbial inactivation for alternative food processing technologies – Executive Summary. Food and Drug Administration. Available from: http://www.fda.gov/Food/ScienceResearch/ResearchAreas/SafePracticesforFoodProcesses/ucm100204.htm.

[102] Yaun, B. R. (2002). *Efficacy of ultraviolet treatments for the inhibition of pathogens on the surface of fresh fruits and vegetables*. Thesis, Polytechnic Institute and State University, Virginia.

[103] Allende, A., McWvoy, J. L., Luo, Y., Artes, F., and Wang, C. Y. (2006). Effectiveness of two-sided UV-C treatments in inhibiting natural microflora and extending the shelf life of minimally processed 'Red Oak Leaf' lettuce. *Food Microbiology*, 23(3): 241-249.

[104] Gómez, P. L., Alzamora, S. M., Castro, M. A., and Salvatori, D. M. (2010). Effect of UV-C light dose on quality of cut-apple: microorganisms, color and compression behavior. *Journal of Food Engineering*, 98(1): 60-70.

[105] Manzocco, L., Pieve, S. D. and Maifreni, M. (2011). Impact of UV-C light on safety and quality of fresh-cut melon. *Innovative Food Science and Emerging Technologies*, 12(1):13-17.

[106] Sanz, E. N., Dávila, I. S., Balao, J. A. A., and Alonso, J. M. Q. (2007). Modeling of reactivation after UV disinfection: Effect of UV-C dose on subsequent photoreactivation and dark repair. *Water Research*, 41(14): 3141-3151.

[107] Sassi J., Viitasalo, S., Rytkonrn, J., and Leppakoski, E. (2005). Experiments with ultraviolet light, ultrasound and ozone technologies for on board ballast water treatment. VTT TIEDOTTEITA – research notes 2313, Abo Akademi University.

[108] Maharaj, R. (1995). The effect of ultraviolet radiation (UV-C) on the post-harvest storage behavior of tomato (Lycopersicon esculentum Mill cv. Capello). Ph.D. Thesis. Laval University, Quebec, Canada.

[109] Pan, J., Vicente, A. R., Martinez, G. A., Chaves, A. R., and Civello, P. M. (2004). Combined use of UV-C irradiation and heat treatment to improve postharvest life of strawberry fruit. *Journal of the Science of Food and Agriculture*, 84, 1831–1838.

[110] Allende, A., Tomas-Barberan, F. A., and Gil, M. I. (2007). Minimal processing for healthy traditional foods. *Trends in Food Science and Technology*, 17, 513–519.

In: UV Radiation: Properties, Effects, and Applications
Editor: James A. Radosevich
ISBN: 978-1-63321-090-5

Chapter 3

IMPACTS OF UV RADIATION ON CARIBBEAN COASTAL MARINE ECOSYSTEMS

Juan L. Torres-Pérez[1,*], Roy A. Armstrong[2,†] and Yasmín Detrés[2,‡]
[1]Bay Area Environmental Research Institute,
NASA Ames Research Center, Moffett Field, CA, US
[2]University of Puerto Rico at Mayagüez,
Department of Marine Sciences, Mayagüez, Puerto Rico

ABSTRACT

Global assessments of ultraviolet radiation (UVR) by the end of the 20th century indicated worldwide increases in UVR. While with the implement of the Montreal Protocol ozone-depleting substances are starting to decrease, studies in coral reef platforms have shown a 1% increase in UVR during the past two decades. Despite the exponential light attenuation in the water column, reef areas not exposed to a high sediment influx and other particulate and dissolved matter are continually exposed to high levels of Photosynthetically Active Radiation (PAR) and UVR. In the tropics, UVR is also enhanced due to the low solar zenith angle, shorter atmospheric path length and natural thinning of the ozone layer. Nonetheless, coral reefs and other associated marine ecosystems (i.e., mangroves and seagrasses) have thrived for millennia due to the evolution and development of UV-protecting mechanisms. In the case of reef corals, such mechanisms exist in the form of mycosporine-like amino acids (MAAs) thanks to their symbiotic relationship with unicellular photoautotrophic dinoflagellates known as zooxanthellae. Mangroves and seagrasses have a wide range of flavonoids which absorb in the UV region of the electromagnetic spectrum. Nonetheless, these organisms are still susceptible to physiological damages caused by UVR including DNA and protein damage, production of reactive oxygen species, and photo-oxidation of chlorophyll and other photosynthetic pigments. This chapter presents a summary of a series of experiments performed by researchers of the Bio-optical Oceanography Laboratory of the Department of Marine Sciences of the University of Puerto Rico aimed to study the

* E-mail: juan.l.torresperez@nasa.gov.
† E-mail: roy.armstrong@upr.edu.
‡ E-mail: yasmin.detres@upr.edu.

effects of actual, lowered and enhanced levels of UVR on the shallow-water reef branching corals *Acropora cervicornis* and *Porites furcata*, the turtle grass *Thalassia testudinum* and the red mangrove *Rhizophora mangle* in La Parguera, Puerto Rico between 1997-2006.

Our results show that slight increases in UVR can terminate the coral's reproductive process, reduce linear extension rates, skeletal density and photosynthetic pigments production, and increase MAAs concentrations while not necessarily affecting zooxanthellae densities within the coral tissue. Both species of macrophytes responded to ambient and increased levels of UV by changes in their photosynthetic and photoprotective pigments.

Also, mangroves exposed to full solar radiation showed changes in leaf optical properties. Some of the responses observed in tropical coastal marine organisms indicate that under actual or enhanced UV scenarios most of the available energy is diverted into adaptive and survival strategies.

Keywords: UV radiation, MAAs, flavonoids, corals, macrophytes

Abbreviations

CDOM	Colored Dissolved Organic Matter
CPDs	cyclobutane pyrimidine dimers
DOM	Dissolved Organic Matter
HPLC	High Pressure Liquid Chromatography
NASA	National Aeronautics and Space Administration
NOAA	National Oceanic and Atmospheric Administration
POC	particulate organic carbon
PAR	Photosynthetically Active Radiation
MAAs	mycosporine-like amino acids
R	reflectance
T	transmittance
UV	Ultraviolet
UVR	Ultraviolet radiation

3.0. Introduction

Tropical tropospheric ozone trends for 1979-2005 have shown great uncertainties depending on the methodology used to analyze these trends (see 1-3 for discrepancies). These discrepancies are related to the difficulty of obtaining good measurements due to the low ozone density and strong vertical gradients [4]. For instance, stratospheric ozone depletion assessments from almost two decades ago showed increases of UVR, particularly unweighted UV-B radiation (280-320 nm) at the poles and at temperate latitudes, with less variation in the tropics [5].

Changes in atmospheric circulation patterns, mediated by global climate change, may lead to increases in the UV-B irradiance in the order or 2-3% in the lower latitudes although these projections must be treated with caution as they strongly depend on changes in cloud cover, air pollutants, and aerosols, influenced by climate change [6]. Comparisons in UV-B doses observed in Brazil, Puerto Rico and Patagonia showed the highest summer daily dosage in Puerto Rico [7].

Because the recovery of the ozone column can take decades, the amount of ultraviolet radiation reaching the Earth's surface is likely to remain elevated for as long as the ozone levels remain below historically normal values. While there is still uncertainty regarding the actual stratospheric ozone loss in the tropics, coral reefs continue to receive probably the higher irradiances of UVR anywhere on Earth at sea level. For instance, a 10-year climatology study for the Great Barrier Reef in Australia revealed a small but steady increase of near 1% in solar radiation in 1995-2005 corresponding closely with coral bleaching events occurring in the area during those years [8]. This chapter presents a summary of physical and biochemical factors that influence the penetration of UVR in seawater with a particular emphasis in tropical coastal waters. The deleterious effects of UVR on marine organisms is exemplified here with a series of experiments with shallow-water branching corals, seagrasses and mangroves performed during the past decade in the La Parguera Natural Reserve in southwest Puerto Rico.

3.1. Attenuation of UV Radiation in Ocean Waters

3.1.1. Oceanic vs. Coastal Waters

Up until the early work of Jerlov [9] it was assumed that UVR was rapidly attenuated by water and had no significant effect on aquatic organisms. Jerlov found that in clear oceanic waters the biological important UVR could penetrate up to 20 m depth. More recently, Smith and Baker [10] found that UVR is of considerable ecological significance to aquatic biology since it can penetrate to tens of meters in natural waters. Biologically damaging levels of UVR (particularly UV-A) have been measured in the Atlantic Ocean up to 37 m depth [11] with UV-B penetrating up to 15 m in the clear waters of the Bahamas [12].

The amount of incident solar UVR and the depth of penetration are key factors responsible for potential damages to aquatic organisms. The penetration of UVR in the water column differs greatly among aquatic ecosystems. For instance, attenuation (as a function of absorption and scattering) reduces the water transparency to UVR particularly in eutrophic freshwater ecosystems and coastal areas [13]. Typically, UV-B penetrates into the upper 10% of the coastal marine euphotic zone before it is reduced to 1% of its surface irradiance [9]

We have measured total UVR at the La Parguera insular shelf edge reaching at least 20 m [14]. Further inshore, at a shallow back reef lagoon, the attenuation coefficients (Kd) of UVR were found to be variable, depending for the most part, on wind speed and the resulting re-suspension of sediment particles by wave action. Temporal changes in the attenuation coefficients of UVR, which decrease with increasing wavelength, can vary by up to two-fold (Figure 1). Although relatively high values of colored dissolved organic matter (CDOM) are typically found along the coastal mangrove channels in this area, they are very low at this reef site (Figure 2), further supporting the potential role of sediment re-suspension in modulating UVR attenuation in these coral reef environments [15].

3.1.2. Colored Dissolved Organic Matter (CDOM) and Suspended Sediments

Maximum penetration of UV-B occurs in clear ocean waters having minimum concentration of chlorophyll and dissolved organic matter (DOM) [16]. In coastal waters, UVR penetration diminishes dramatically mostly due to the presence of dissolved and particulate organic carbon (DOC and POC, respectively) and particularly CDOM [17] influenced by riverine influx. Studies in the Caribbean have shown that DOC and CDOM originated in the Orinoco River plume in the northeastern South America can travel as far as Puerto Rico with its photo-oxidation directly related to the penetration of UVR in the water column [18, 19].

As DOC degrades in the water column by photo-oxidation [19, 20] and it becomes fragmented by solar UVR to smaller subunits, is later consumed by bacterioplankton [21, 22] resulting in an increase in the UVR transparency of the water column. The results of Kuwahara et al. [23] support other studies that show that chromophorically active concentrations of DOM and POC can significantly alter the amount of UV penetrating coral reef waters and may contribute to the regulation of intricate biogeochemical cycles surrounding benthic coral communities.

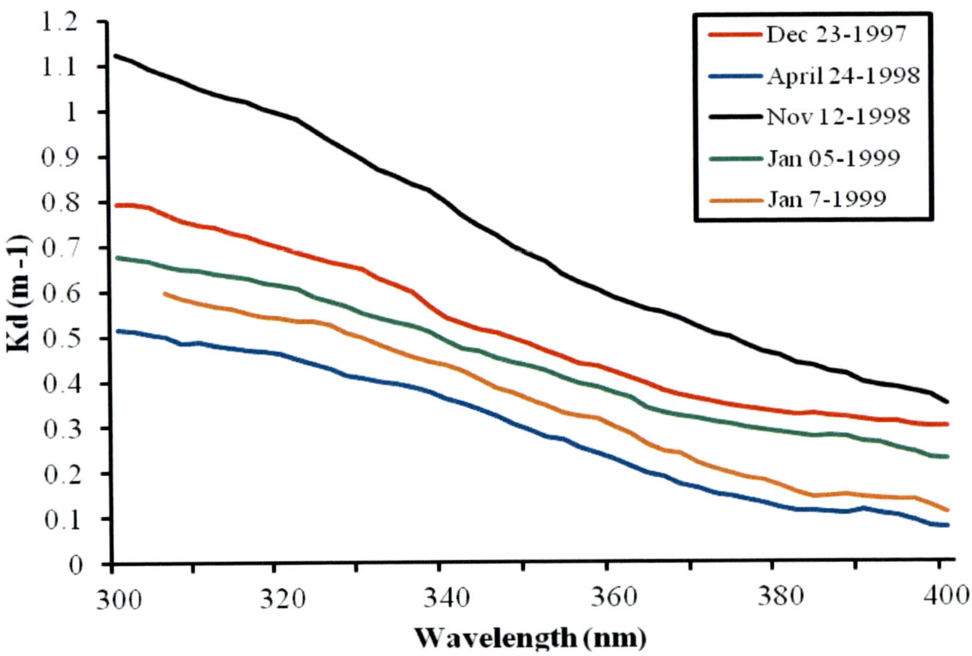

Figure 1. Temporal variability of spectral attenuation coefficients of UVR measured at 2 m depth in the back reef lagoon of Cayo Laurel, southwestern Puerto Rico.

Besides the attenuation caused by the water itself, CDOM is the major attenuator of UVR in oceanic and coastal waters with POC having a significant contribution also in some coastal areas [24, 25]. CDOM has indeed been proposed as a mediator in climate-UVR interactions [26] as its filtering capacities are analogous to that of the ozone layer; this is CDOM absorbs more efficiently the shorter wavelengths UV-B radiation than UV-A [27]. However, it has been proposed that with global warming and other deleterious climate change-related consequences, a decrease in oceanic CDOM is possible leading to greater penetration of UVR in the water column [28] affecting both planktonic and benthic organisms [29].

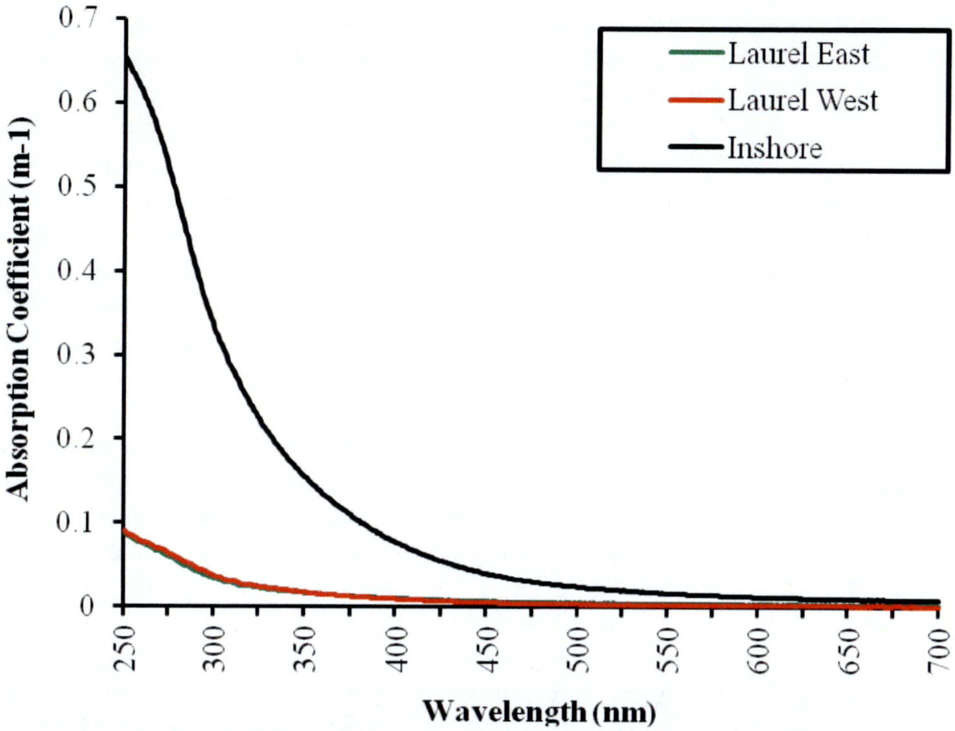

Figure 2. CDOM spectral absorption coefficients for typical coral reef waters in La Parguera as compared to inshore values along mangrove channels.

Waters surrounding coral reefs in some coastal areas around the world typically show a low penetration of UVR, due to the influence of CDOM, as in the case of Florida and the Gulf of Mexico [30] where the 10% irradiance depth of UV-A (defined as Z10% or the depth where 10% of the surface UV-A remains) reaches only up to 5 m [31]. It was recently proposed that reef areas associated with intact shorelines tend to be exposed to lower intensities of UVR than areas associated with developed shorelines where natural sources of CDOM are reduced [32]. In fact, over the years, several working groups have shown that areas not exposed to a high input of terrigenous sediments and organic matter are notably transparent not only to the longer wavelength Photosynthetically Active Radiation (PAR; 400-700 nm), but also to solar UVR (280-400 nm) [9, 14, 16, 33, 34]. In these areas, shallow-water benthic organisms, such as reef corals, might continually be exposed to high doses of both PAR and UVR. High rates of CDOM breakdown have been proposed as a possible important factor influencing the susceptibility of reef corals to bleaching in clear waters [32]. Zepp et al. [35] showed that absorption coefficients of CDOM are 6-25 times larger than particulate absorption coefficients in the UV region indicating that CDOM controls the UV penetration in inshore coastal waters in the Florida Keys. Higher UV penetration was measured during low wind doldrum conditions contributing to the higher bleaching incidence at summertime.

3.2. UVR and Tropical Marine Organisms

UVR affects an array of aquatic organisms from different kingdoms. UVR can affect diversity and biomass with species succession in hard-bottom shallow marine communities worldwide [36]. The apparent impact of increased UVR at the ecosystems level can be relatively low, but the effect on the individual species is considerably different [28]. Solar UVR can reduce productivity [37], impaired or reduce reproduction and development, and increase mutation rates in phytoplankton [38], macroalgae [39], zooplankton and fishes [40]. Although overfishing and pollution are some of the main causes of reductions of the main fisheries populations worldwide, UV-B also plays an important role as the larvae and juveniles of many fish species are particularly vulnerable [40, 41].

The negative effects of UVR for diverse coral reef organisms include reduced growth and skeletal density, impaired reproduction and larvae settlement, bleaching and photoinhibition, DNA damage, oxidative stress through the production of reactive oxygen species (ROS), and mortality just to mention a few [42]. The production of ROS is particularly toxic as it can damage DNA, enzymes, membrane proteins and lipids, and the photosynthetic apparatus components [43]. The formation and further accumulation of cyclobutane pyrimidine dimers or CPDs in bacterioplankton under high solar radiation constitutes by far the most frequent DNA damage induced by UVB followed by single- and double-strand breaks [44]. The formation of CPDs, larval damage and even cell apoptosis has also been shown in higher organisms such as sea urchins [45, 46].

The effects of UVR on reef-building Scleractinian corals have received widespread attention in the Pacific during the last few decades (see 47-60 for some examples). Nevertheless, the literature on the response of Caribbean corals to UVR is scarcer [12, 61-66].

Likewise, despite considerable work has been done in general on the effects of UVR in the physiology and morphology (i.e., photosynthesis, photoinhibition, leaf area, growth, primary production) of terrestrial plants [67-73], relatively little literature exists particularly on tropical marine and/or coastal macrophytes. For instance, it was found [74] that there are various degrees of sensitivity to UVR in five seagrass species from Australia with thicker-leaf species (such as those of the genus *Zostera* and *Syringodium* in their case) providing greater morphological protection for UV-sensitive organelles than those with thin leaves (i.e., *Halophila* sp. and *Halodule* sp.). UV-B also seems to impair or reduce the growth of salt marsh vegetation as was demonstrated with the genus *Salicornia* in high and low latitude regions [7]. A recent study performed with different types of Caribbean macrophytes (phaeophytes, chlorophytes and seagrasses) found that photoinhibition can be induced under full solar, full UV and UV-B (UV-A depleted) radiation conditions [75].

3.3. Defensive Mechanisms of Marine Organisms against UV Radiation

3.3.1. Sessile vs. Mobile Organisms

Organisms living millions of years ago possessed different strategies and mechanisms to prevent or repair UV-induced damage [76]. The development of precursors of MAAs like scytonemin probably allowed cyanobacteria to colonize shallow-water, terrestrial and high-latitude habitats during the Precambrian period [77, 78]. Bacteria and other very small organisms, which are major producers and recyclers of energy and nutrients in aquatic ecosystems, can overcome the stress of UVR by fast cell division and growth. However, if growth is impaired and enzyme activities are stopped, DNA damage can occur through the formation of pyrimidine dimers causing cell death if cell repair mechanisms cannot keep up with the damage [79,80].

Other strategies which provide UVR protection in planktonic organisms, such as cyanobacteria, include the migration into reduced-light habitats, formation of crusts, self-shading [81], the production of quenching agents such as carotenes and superoxide dismutase [82-83], photoreactivation and DNA repair, and the production of antioxidant enzymes [84]. Studies have found that while melanin and carotenoids protect non-sessile groups such as zooplankton they use vertical migration as their main way to avoid high UVR levels [85, 86]. Protective and mitigating mechanisms in cyanobacteria include the formation of crusts, self-shading and vertical migration [81]. Decreased leaf area, epiphytic shading and increases in leaf thickness, leaf epicuticular wax and leaf pigments play a role in plants protection from UV-B radiation [67, 68, 87-89].

3.3.2. UV-Absorbing Compounds

It is generally accepted that the development of phenolic polymer metabolism, ultimately resulting in the evolution of UV-absorbing and other defensive compounds such as flavonoids, lignin, tannins and MAAs, among others, was partly induced by the stimulation of UV-B in the early Archean atmosphere [90]. The complexity of polyphenolics apparently increased as photosynthetic organisms evolved from early cyanobacteria and algae to the higher plants.

Cyanobacteria, which originally moved from the aquatic to the terrestrial environment, still protect themselves with MAAs (see below) and scytonemins [91]. Independently of their complexity, it seems that the more simple UV-absorbing compounds such as MAAs are as efficient as the more complex flavonoids of terrestrial plants [90].

3.3.3. Mycosporine-Like Amino Acids (MAAs)

UV-absorbing compounds presumably evolved for other physiological roles and later adapted to UV-screening functioning under selection pressure. MAAs, for instance, function as osmotic regulators in some cyanobacteria [92] and such role may have given rise to the first UV-screening MAAs [93] as UVR increased in the atmosphere [91]. MAAs have a putative role in UV photoprotection.

MAAs were originally identified in fungi as having a role in UV-induced sporulation [94]. MAAs are small (<400 Da), colorless, water-soluble compounds composed of a cyclohexenone or cyclohexenimine chromophore conjugated with the nitrogen substituent of an amino acid or its imino alcohol. Differences in the absorption spectra of MAAs are due to variations in the attached side groups and nitrogen substituents [95]. MAAs typically have wide absorption bands ranging between 40-50 nm wide with peak maxima between 309-360 nm [96].

So far, over 30 different MAAs have been identified [97]. Table 1 shows a list based on Carreto et al. [98].

Mycosporine-glycine is structurally the simplest MAA and the one postulated to give rise to the other MAAs by amino acid substitutions [99, 100]. This is based on its simple structure and similarity to the probable precursor 4-deoxygadusol [101], its wide occurrence among phototrophs [102], and its rapid appearance in response to UVR [99].

The presence of MAAs in phytoplankton has been established in a number of ecologically important groups including several types of microalgae [103], free-living dinoflagellates [104] and diatoms [105]. As such, MAAs have been reported in phytoplanktic Dinophyceae, Bacillariophyceae, Haptophyceae, and in benthic Rhodophyceae, Phaeophyceae, and Chlorophyceae [91, 92, 106-108]. MAAs have also been identified in CDOM of coastal waters, where they might contribute up to 10% of the CDOM absorption at 330nm [109].

Table 1. Some of the most common MAA's found in marine organisms

Mycosporine-like Amino Acid (MAA)	Wavelength of maximum absorption (λ_{max}) (nm)
Mycosporine-taurine	309
Mycosporine-glycine	310
Mycosporine sulfate ester	318
Palythine	320
Palythine-serine	321
Palythine-serine sulfate	321
Mycosporine-methylamine-serine	327
Mycosporine-methylamine-threonine	329
Asterina-330	330
Palythinol	330
Shinorine methyl ester	332
Mycosporine-2-glycine	332
Shinorine	333
Porphyra-334	334
Z-palythenic acid	335
E-palythenic acid	337
Usujirene	357
Palythene	360

In macroalgae while they can produce MAAs [110], photosynthetic processes can also be severely damage by UVR [111-113]. As in the case of reef corals [14, 114], macroalgae have a depth-dependent susceptibility to UVR with deep species being more susceptible than shallow-water species [115, 116]. MAAs have also been found in zooplankton of high-altitude lakes [117], presumably obtained from their diets. Coral reef fishes can adapt to high UVR by incorporating UV-absorbing substances, through their diet, into their eyes and the epidermal slimy cover [118].

Often protective mechanisms pose conflicting selection pressures on species making UV radiation an additional stressor on the organism [29]. While MAAs are photoprotective, their production entails a high energetic cost leading to reduced growth [14] while recovering from stress [119].

High solar radiation synergistically with thermal stress may depress MAA production [65] leading to bleaching and damage to the photosynthetic apparatus not only in Scleractinian corals [65], but also in Gorgonaceans [120]. Approximately 19% of the total cellular energy budget is required to synthesize 100 mol m^{-3} of palythine [121]. Despite this cost, the importance of MAAs as photoprotectants in shallow-water corals and their symbionts is implied as all the coral species analyzed contained at least one MAA [96].

Evidence shows that MAAs may also function as: 1) antioxidant scavengers of toxic oxygen radicals [122, 123], 2) be compatible solutes following salt stress [92], 3) a mechanism to counteract desiccation [124] and thermal stress [99], 4) be an intracellular nitrogen reservoir [125] and, 5) even have some involvement in fungal reproduction [94, 126]. As such, MAAs are considered secondary metabolites due to their apparent multiple functions [127, 128].

MAAs can also protect dinoflagellate cells by accumulation and packaging around vital cellular organelles such as the nucleus and chloroplasts [129]. MAAs are highly photostable compounds [13] even when exposed to high temperatures [130]. As such several are being tested for potential use in human skin-care and cosmetic products [131, 132].

3.3.4. Flavonoids

Higher plants, including marine, rely on the production of UV absorbing pigments, usually flavonoids and other phenolic compounds, for protection against harmful UV radiation.

Flavonoids are water-soluble phenolic derivatives, synthesized by the phenylpropanoid pathway [68]. Each molecule has a C_{15} skeleton formed from condensation of three malonate units with a phenylalanine-derived C_6-C_3 precursor [133]. They have a wide distribution in higher plants and are located in vacuoles in the epidermis and stems, modifying the optical properties of the plant tissue and protecting UV sensitive molecules from damage [74, 134, 135]. The presence of sulphated flavones and sulphated phenolic acids has been reported in seagrasses [136-138]. Studies with five seagrass species from Australia [74] indicated that their responses to high light intensity are comparable to those exhibited by terrestrial plants.

Studies with mangroves from Australia [139] established the protective role of flavonols in the epidermis of leaves. Their study indicated that these phenolic compounds and increments in leaf succulence could protect the mangroves leaves from the dangerous effects of UV radiation.

Flavonoids, particularly those that absorb within the UV-blue light region (320-500 nm), have been found to provide protection from photoinhibtion and photooxidation in *Arabidopsis* [140]. Other functions associated with polyphenolics such as the flavonoids include allelopathy, signal transduction, plant hormones, chemical defense and cell structural rigidity [141].

3.4. Case Study: UVR Impacts on the Coral Reefs and Associated Ecosystems of La Parguera, Puerto Rico

During 1997-2006 a series of studies were funded by NASA and NOAA to investigate the effects of UVR on Caribbean coastal and marine organisms, mostly shallow-water reef corals, seagrasses, and mangroves in Puerto Rico. The following is a description of the study sites and a summary of the methodology used and the results.

Study sites were located in La Parguera, southwestern Puerto Rico (Figure 3). This area has minimal input of suspended sediments or dissolved organic matter from terrigenous sources due to low precipitation and the absence of riverine input [142]. Annual precipitation ranges from 500 to 1200 mm with a dry period from December through April. The wind field has its strongest component from the east (10-15 km/h) and southeast (15-25 km/h) with a maximum from May through July [143]. This region has numerous fringing reefs that protect the shore from intense wave action resulting in extensive seagrass meadows and a coastline dominated by mangroves.

3.4.1. Study Sites

A. San Cristóbal Reef (17°56.683'N; 67°04.633'W) is located in La Parguera Natural Reserve, approximately 3.2 km southwest of the Magueyes Island Field Station (University of Puerto Rico, Department of Marine Sciences). Here, *Acropora cervicornis* and *Porites furcata* dominate the coral cover at the shallow waters of the back-reef zone in less than 5 m of water. Other coral species present in lower abundances are: *Acropora prolifera, Montastraea annularis, Porites astreoides, Favia fragum, Diploria clivosa, D. strigosa* and *D. labyrinthyformis*. Other benthic components include calcareous algae and octocorals. Relatively calm and clear water is found year-long in this area.

B. The Old Buoy site is located at La Parguera shelf edge (17°53.183'N; 66°59.850'W) approximately 9.6 km south of Magueyes Island. The site is characterized by a spur and groove topography with medium-sized colonies of *M. faveolata, M. franksi, M. cavernosa, Colpophyllia natans, Porites astreoides, Agaricia* spp. and *Mycetophyllia* spp. dominating the scleractinian fauna.
Colonies of *A. cervicornis* can be found scattered in the area, but the coverage is less than that of the above-mentioned species. Depth in the shelf varies between 19 m on the platform and 24 m at the shelf edge. See Figure 3 for an image showing the location of all study sites.

C. UV reduction and exclusion field experiments with the turtle seagrass *Thalassia testudinum* were carried out in the back-reef lagoon of Laurel Reef (17°56.527'N; 67°03.450'W) located 3.6 km south of Magueyes Island. The backreef lagoon at this site is dominated by an extensive seagrass meadow dominated by extensive homogeneous patches of *T. testudinum.*

D. Due to the logistics of working in mangrove swamps and the height of trees, field experiments with the red mangrove *Rhizophora mangle* were performed at a disturbed site along the coast of La Parguera was selected (17°58.392'N;

67.03.000'W). The mangrove stand at this site is a relatively even-aged canopy of *R. mangle*, of approximately 2 m in height that have been subjected to cropping for at least 20 years.

For several years we studied the effects of UVR on various aspects of the shallow-water corals physiology.

These physiological aspects included: a) growth (in terms of linear extension rates and skeletal density); b) production of photosynthetic and UV-absorbing compounds (e.g., MAAs); and c) fecundity.

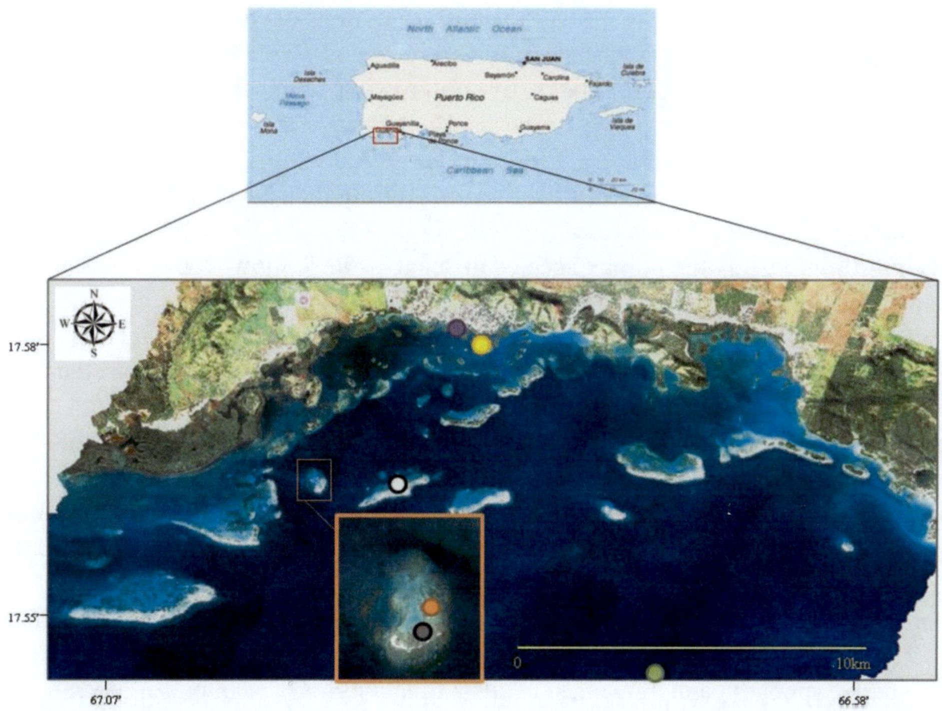

Figure 3. Study sites in La Parguera, Puerto Rico. San Cristóbal Reef is enlarged. The circles indicate specific study areas as follows: Grey – *P. furcata* collection site, Orange – *A. cervicornis* collection site and underwater domes location for the exclusion experiment, Green – Old Buoy site at the shelf edge, White – *T. testudinum* collection site, Purple – *R. mangle* collection site and exclusion experiment location, Yellow – Magueyes Island Field Station (University of Puerto Rico – Department of Marine Sciences). Image provided by Liane Guild (NASA Ames Research Center).

Our studied species were the branched finger coral *Porites furcata*, and the threatened staghorn coral *Acropora cervicornis* [14, 34, 66]. The studies comprised a series of experiments including: a) relocation of selected colonies of both species under a series of underwater treatments with different degrees of exposure to UVR; b) relocation and transplantation of selected colonies of *A. cervicornis* between shallow (1 m depth) and deep waters (20 m); and c) relocation of *P. furcata* colonies to outdoor aquaria with minimal, normal and enhanced exposure to UVR.

At the same time our group was also investigating the effects of UVR on the tropical Caribbean macrophytes, the turtle seagrass (*Thalassia testudinum*) [15] and the Red mangrove (*Rhizophora mangle*) [144] under enhanced and reduced UVR scenarios. Changes in the leaf optical properties, photosynthetic pigments and production of UV-absorbing compounds (e.g., flavonoids) were the focus of these investigations [15, 144, 145]. A summary of methodology and main results is presented below.

3.4.2. Species Description

A. cervicornis and *P. furcata* (Figure 4 A, B) are both perforate ramose corals [146] with relatively rapid growth rates [147, 148], typical of shallow-water zones of Caribbean coral reefs. Both species usually show polyp expansion during daytime. In general, highly phototrophic genera with small polyps, such as *Acropora* and *Porites* [148, 149], seem to be more successful in competing for light by outgrowing competitors, while the more heterotrophic genera with larger polyps, such as *Montastraea*, can persist in the understory layer [150].

Extensive populations decreases of *A. cervicornis* during the past decades led to its listing as a threatened species in the Endangered Species Act in 2006 (50CFR223).

Thalassia testudinum is the climax species of seagrass in shallow tropical waters (Figure 4C). This is a perennial seagrass with up to 30 cm long and 2 cm wide leaf blades typical of shallow-water sandy mounds in the Caribbean [151]. Along with other seagrasses such as *Syringodium filiforme*, *T. testudinum* forms extensive meadows that serve as essential habitats for the development and establishment of dozens of invertebrates and fish species.

Rhizophora mangle is the dominant species of mangrove along the coastline and on offshore keys in southwestern Puerto Rico (Figure 4D). It is a halophytic woody spermatophyte that inhabits the land-sea interface of Caribbean coastal zones. The plants are distinguishable by their prominent prop roots, viviparous "seeds" called propagules, and waxy leaves.

Under ideal conditions of light and nutrients, some trees can grow more than 20 m in height [152]. Its root system provides an attachment site for many sessile invertebrates including corals, sponges, and tunicates. It also functions as a shelter for juveniles of a number of fish species [153].

3.4.3. Summary of the Methodology Used for the Study of UVR Effects on Marine Organisms in Puerto Rico

3.4.3.1. Exclusion Experiments on Corals (UV-E Hereafter)

Colonies of *A. cervicornis* and *P. furcata* were subjected to depleted UVR and reduced PAR levels (Treatment), and compared to colonies living at normal UVR and PAR levels (Controls) at 1 m depth in the back-reef zone of San Cristóbal Reef. All the colonies were collected on-site at a minimum of 5 m distance to their nearest intraspecific neighbor to reduce pseudoreplication due to clone mates. All other physical factors remained unaltered (salinity, water motion, temperature, etc.). To block UVR and some of the incident PAR, hemispherical open-ended steel rod frames (1 m width X 1 m length X 1 m height) covered by a 4 mm thick Hyzod® SR-Polycarbonate Sheet (DSM Sheffield Plastics) were constructed.

Figure 4. Photos of all the studied species in La Parguera, Puerto Rico: A) staghorn coral *A. cervicornis*; B) finger coral *P. furcata*; C) turtle grass *Thalassia testudinum*; and D) red mangrove *Rhizophora mangle*.

All frames were located at the same depth as the control colonies (1 m). All frames were aligned to the mean sun azimuth angle calculated for the experimental period to ensure that direct solar radiation reached the colonies only after passing through the filter.

3.4.3.2. Exclusion Experiments on Marine Macrophytes

Reduction and total exclusion of UVR experiments were conducted in the field to determine the effects of ambient levels of UV-B radiation on *T. testudinum* and *R. mangle* without disturbing their natural environment. In the *T. testudinum* experiments, semispherical domes (1.2 m x 1.2 m) were covered with UV filters and aligned to the mean sun azimuth angle calculated for the experimental period. Total UV exclusion was achieved using 3.12 mm clear polycarbonate material (Lexan MR5, GE Plastics) with less than 3% transmittance of wavelengths below 400 nm and an average of 82% transmittance in the PAR region. An OP4 acrylic filter, which absorbs 14% in the UV region and 10% of PAR was used for the UV reduction treatment. In the *R. mangle* experiments the UV reduction treatment used a shading fabric (Saran) that filtered approximately 25% PAR and UV radiation. The control treatment in both experiments was exposed to unfiltered ambiance irradiance fluxes.

3.4.3.3. Supplementation Experiments on Corals (UV-S Hereafter)

Due to past failed attempts of maintaining *A. cervicornis* colonies healthy in outdoor aquariums we decided to use a different approach for enhancing the levels of UVR reaching the corals. In the case of *A. cervicornis*, colonies were exposed to an increase in UVR and PAR in the field by transplanting colonies from the La Parguera shelf edge at 20 m depth to 1m depth at the back-reef area of San Cristóbal Reef. Colonies of *A. cervicornis* living at San Cristóbal Reef at 1 m were also transplanted to the shelf edge at 20 m for exposure to reduced UVR and PAR levels. Additionally, other colonies were used as controls at San Cristóbal (1 m depth) and at the shelf edge (20 m depth). All colonies studied (controls and transplants) were tied to plastic-coated cyclone fence wires to hold them in place. The cyclone fence wires were cleaned of epiphytes on a weekly basis to avoid interactions with the coral colonies.

In the case of *P. furcata*, early attempts to maintain the colonies in the outdoor aquariums were very successful. Hence, the supplementation experiment consisted in enhancing the UVR with an artificial light system placed over an open-system running seawater outdoor aquariums setup built at the Magueyes Island Field Station. Several 75-liter aquariums were set under three different treatments. Colonies were exposed to mimicked, as much as possible, the levels of UVR and PAR received at collection depth by placing a neutral density mesh over the aquariums and served as controls. Other colonies were exposed to enhanced UV radiation by supplementing the solar incident levels of UVR with UV-B 313 (Q-Panel Co.) fluorescent lamps located 0.5 m over the top of the aquariums. To contrast, other colonies were exposed to depleted UVR in aquariums located under similar Hyzod® sheets as those used in the field for the UV-E experiment. In this sense, both scenarios (enhanced UVR and depleted UVR) could be examined and compared to the controls at the same time.

3.4.3.4. Supplementation Experiments on Seagrasses

Thalassia testudinum transplants were gently collected from shallow areas (depth 0.5-1 m) to minimize damage to roots or rhizomes. Healthy shoots were transplanted to outdoor aquariums with open-systems in which filtered sea water was continuously renewed. After an acclimation period plants were exposed to different lamp supplementation regimes in an unmodulated on/off outdoor UV supplementation facility constructed at the Magueyes Island Field Station. The UV-B 313 fluorescent lamp arrays were suspended at a distance of 25 cm above the aquariums and were turn on at a full output for a specific period centered on solar noon.

3.4.3.5. UVR and PAR Measurements

Irradiance measures between 280-700 nm at 2 nm intervals for all the experiments (corals, seagrass and mangroves) were obtained during the UV-E and UV-S experiments with an Optronic OL-754 spectroradiometer (Optronic Laboratories) connected to a submersible integrating sphere through a quartz fiber optics cable. All measures were obtained between 11:00-13:00 hrs on randomly selected days. Total UVR was measured at the beginning, middle and end of the UV-S experiment at San Cristóbal Reef and the Old Buoy site with a Solar Light radiometer connected to an underwater 2π cosine collector. This radiometer measures unweighted UVR in a single band (bandwidth: 280-400 nm). Measurements were recorded at 1 m intervals between 0-10 m and every 2 m between 10-20 m depth at the Old Buoy site and at 0.5 m intervals from the surface to the bottom at San Cristóbal Reef (Figure 3). PAR measurements were obtained at the same depths intervals with a LiCor 182s cosine collector connected to a LiCor 1400 data logger. Additional 2π cosine collectors were located on the research boat to normalize irradiance underwater measures and correct for atmospheric changes during the sampling. Sampling was conducted at similar times on back-to-back days at both sites during all sampling dates for consistency. Both measurements were later used to obtain diffuse vertical attenuation coefficient for PAR and UVR (Kd_{PAR} and Kd_{UVR}, respectively) values for both sites using:

$$Kd = 1/(Z_2 - Z_1)[\ln(Ed_2/Ed_1)] \tag{1}$$

where, Ed_1 and Ed_2 are the downwelling irradiances measured at depths Z_1 and Z_2, respectively [154].

3.4.3.6. Other Physical Parameters

Temperature was recorded hourly with a HOBO® Temperature Data Logger (Onset Computer Corporation) at all experimental sites (corals, seagrass and mangroves). The data were recovered at approximately monthly intervals and analyzed statistically.

3.4.4. Biological Parameters Studied in Corals and Macrophytes

3.4.4.1. Corals

3.4.4.1.1. Skeletal Growth Analysis

Coral growth was measured as a function of skeletal linear extension rates and skeletal density. Before locating the colonies haphazardly at any particular treatment, they were stained with Alizarin Red S and let to rest for five days. To measure linear extension rates the tissue was digested with a 5% NaOCl solution for 24 hrs, washed with fresh water and the growth above the stain line was measured with a Vernier caliper to the nearest 0.1 mm. Linear extension rates were estimated by dividing the skeletal extension (above the stain line) by the days of exposure (mm day^{-1}). Skeletal density was measured using a liquid displacement technique. The tissue-free piece of each colony used for the pigment analysis (see below) was left to dry for a week at room temperature, weighed to the nearest 0.0001 g in an analytical balance and covered with a previously weighed aluminum foil piece. Then, it was quickly submerged in a test tube containing deionized water and the volume recorded to the nearest 0.1 ml. The skeletal density was obtained by dividing the weight of the piece by the volume of displaced water.

3.4.4.1.2 MAAs Analysis

One branch from each colony (both species) was removed during each of the three sampling dates (beginning, one month after, and end of the experiment). All samples were transported in sterilized Whirl-Pak plastic bags in a dark container. Each branch was broken into three pieces. The proximal piece (closer to the base of the branch) was used for histological analysis of the gonads [34], the middle piece was used for zooxanthellae density (data not shown) and the distal piece was used for the MAAs analysis. The white tip of each branch was discarded as it usually either lacks zooxanthellae or pigments. To avoid differences in the concentration of pigments with increasing distance from the branch tip, this procedure was kept constant for each branch. Samples were extracted following established procedures [61, 63, 64, 155] using 20 ml of menthanol:tetrahydrofuran (80:20, v/v) solution at 4°C in the dark for 24 h followed by a second 20 min extraction to remove any remaining pigments. Photosynthetic pigments were separated from MAAs following the procedures of [64] using Sep-Pak C_{18} 900 mg cartridges, and later removed from the cartridge by injecting 4 ml of the extraction solvent. MAA extracts were stored at -70°C until High Performance Liquid Chromatography (HPLC) analysis was performed.

Individual MAAs were quantified using reversed-phase HPLC. MAAs were separated following published extraction protocols [63, 64]. Aliquots of 30-80 μl were injected into a guarded Phenosphere® 250 x 4.6 mm C_8 column connected to a Shimadzu LC-10AT Liquid Chromatographer coupled to a Shimadzu SPD-M10AV diode-array detector, and passed through an isocratic solution consisting of 55% MeOH, 0.1% Acetic Acid and 44.9% DIW ran at 0.8 ml/min for 20 min. Individual MAAs were identified from representative samples by Dr. Daniel Gleason at Georgia Southern University. Quantification of MAAs was performed using an extension of Beer's Law and their respective published molar extinction coefficients. The MAAs concentrations were normalized to soluble protein and expressed as nmol mg $prot^{-1}$. For protein analysis we followed a similar procedure to that described by [156]. Protein analysis is also described in detail in [14].

3.4.4.1.3. Fecundity

Corals branched were fixed, decalcified, cleaned and soaked in 75% ethanol, and histologically analyzed following the procedures of [157]. The maturation stage of eggs and spermaries present, number of each by mesenteries, number of mesenteries containing gonads per polyp, and average number of gonads per polyp was noted. The percentage of polyps containing gonads (either eggs or spermaries) was obtained by analyzing ten haphazardly chosen polyps within the tissue sample. Data of all the colonies within a treatment were pooled for later statistical analysis.

3.4.4.2. Macrophytes

3.4.4.2.1. Leaf Optical Properties

Mature leaves of *R. mangle* and *T. testudinum* were collected, cleaned and transported to the laboratory in an ice chest on a monthly basis for optical properties and pigments processing.

Leaves were processed immediately upon arrival to the laboratory. Leaf reflectance (R) and transmittance (T) within the 400-800 nm range were obtained with a GER-1500 spectroradiometer (Geophysical and Environmental Research Corp.) attached to a LI-COR 1800-12 integrating sphere with a fiber optics cable. Derivative analysis of the reflectance spectra was used to determine differences in the red-edge inflection point.

3.4.4.2.2. Leaf Photosynthetic Pigments and UV-Absorbing Compounds

Rizophora mangle pigments were extracted in 100% acetone using a tissue grinder following the procedures of [158]. *Thalassia testudinum* leaf discs were extracted in 100% methanol and microhomogenized to a fine powder. Extracts were filtered using Gelman PTFE Accrodiscs CR 0.2 μm and analyzed using a diode array spectrophotometer. Absorbance was measured in the 400-700 nm spectral range. Absorbance of chlorophyll *a*, chlorophyll *b* and carotenoids were calculated according to equations given by Lichtenhaler and Wellburn [159].

UV-absorbing compounds were extracted in 70% methanol following Lovelock's et al. [160] procedures and their absorbance spectra were obtained with spectrophotometric analysis within the 250-400 nm region.

3.4.5. Summary of Results and Discussion

3.4.5.1. Physical Parameters

With respect to the UV-E experiment, the polycarbonate material (Hyzod®) absorbed 99% of the incident UVR at 1 m as well as 12% of PAR (Figure 5). Downwelling irradiance for total UVR (Ed_{UVR}) at the shelf edge and at San Cristóbal Reef (Figure 6) showed an exponential decay with depth similar to that of Ed_{PAR}. Nonetheless, Ed_{UVR} showed a 2.5-fold increase in UVR but only a 9% increase in Ed_{PAR} experienced by the colonies transplanted from 20 to 1 m depth.

Depending on the characteristics of the water column being studied, typical values of the diffuse attenuation coefficient for UVR (Kd_{UVR}) range from 0.08 to 0.50 m^{-1} [14, 35, 161]. Kd_{UVR} values measured during the UV-S experiment fluctuated between 0.19-0.27 m^{-1} for San Cristóbal and 0.13-0.16 m^{-1} for the shelf edge, whereas Kd_{PAR} values were between 0.09-0.10 m^{-1} for San Cristóbal and 0.06-0.08 m^{-1} for the shelf edge showing the lower transparency to both UVR and PAR at the nearshore site.

Variations in Kd_{UVR} during the experimental period were probably influenced by sediment re-suspension at this shallow water site or the influx of CDOM from the nearby coastal mangroves.

Our results compare to earlier studies performed in oligotrophic waters off La Parguera [162] and other studies performed after the completion of ours [163].

Many areas, particularly those located far from river mouths such as La Parguera, may receive minimal amounts of CDOM [33] and solar irradiances in the PAR region can reach more than 100 m with UV radiation reaching at least 20 m in the clearest waters (see Figure 6 above). In the shallow areas within the back-reef zone (collection site), high levels of UVR typically reach the substrate [14].

Doldrum conditions are typical on this reef platform during summertime influenced by low tides and windless conditions with shallow-water corals being exposed to high UVR and PAR levels sometimes for weeks. In the outdoor aquariums, measurements were taken just before and after the experiment. The neutral density mesh located above the control aquariums reduced the UVR and PAR levels to 10% higher than those measured at the collection depth (Figure 7).

Total UVR in the aquariums held under the enhanced UVR was 22.2% higher than at the controls at the beginning of the experiment and 9.3% higher at the end due to the lamps decay during the experiment. Individually, UV-B was enhanced between 13.7% at the beginning and 4.8% at the end, and UV-A between 8.5and 4.5% at the beginning and the end. There was no net change in the levels of UVR in the control aquariums (-0.20% change between July and March 2003).

Similar aquariums setups have been used in the past. For example, others have (60 and 164) exposed colonies of *Stylophora pistillata* to enhanced UVR levels, but using Q-Panel 340nm UV lamps for 30 days. In our case, we chose to use lamps with a peak emission in the UV-B range (Q-Panel 313nm) as it provides a more appropriate setup for studying effects of UVR in marine organisms, especially considering the more deleterious effects of short-wavelength energy (UV-B) in terrestrial and marine organisms [165] rather than UV-A.

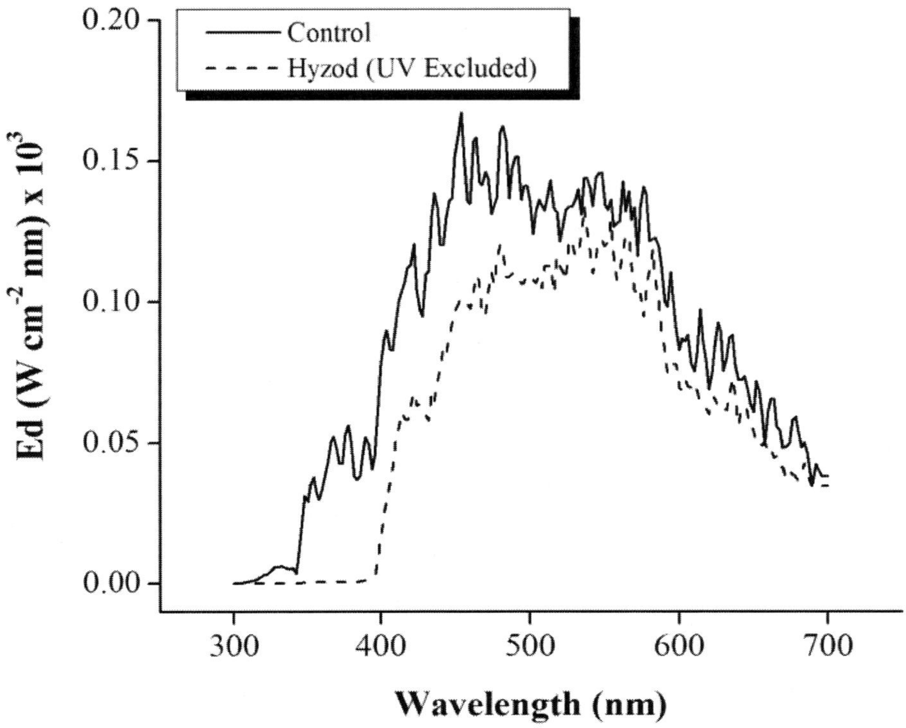

Figure 5. Unweighted UV irradiance measured at 1m depth at the back-reef lagoon of San Cristóbal Reef in La Parguera, Puerto Rico.

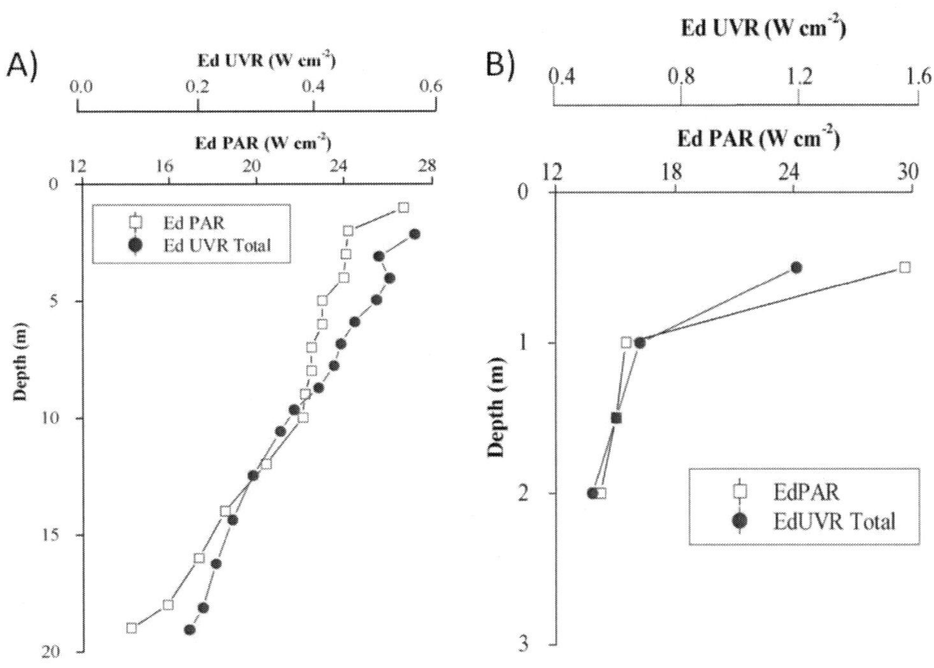

Figure 6. Average downwelling irradiance (Ed_{PAR} and Ed_{UVR}) measured at: A) the shelf edge site, and B) the back-reef lagoon of San Cristóbal Reef in La Parguera, Puerto Rico.

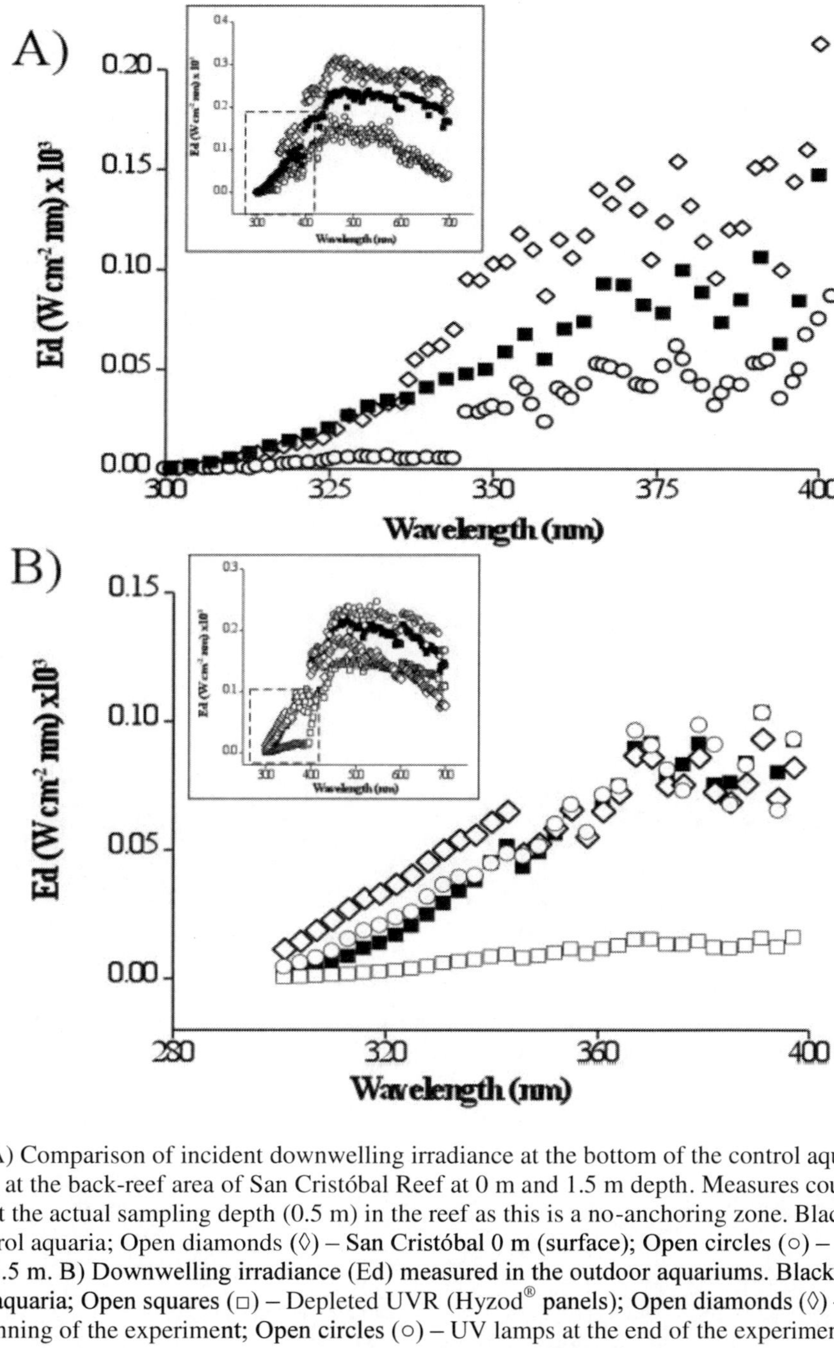

Figure 7. A) Comparison of incident downwelling irradiance at the bottom of the control aquariums (0.5 m) vs. at the back-reef area of San Cristóbal Reef at 0 m and 1.5 m depth. Measures could not be collected at the actual sampling depth (0.5 m) in the reef as this is a no-anchoring zone. Black squares (■) – Control aquaria; Open diamonds (◊) – San Cristóbal 0 m (surface); Open circles (○) – San Cristóbal 1.5 m. B) Downwelling irradiance (Ed) measured in the outdoor aquariums. Black squares (■) – Control aquaria; Open squares (□) – Depleted UVR (Hyzod® panels); Open diamonds (◊) – UV lamps at the beginning of the experiment; Open circles (○) – UV lamps at the end of the experiment.

The Kd values measured at sampling depth during the *T. testudinum* experiment in the back-reef zone of Laurel Reef were wavelength-dependent with values decreasing between 300-400 nm reaching a minima around 500-550 nm and then increasing in the red region showing the typical light absorption of seawater (Figure 8). Within the UV-exclusion treatments, UVR varied from 0-53% attenuation depending on the treatment.

Temperature was recorded at hourly intervals during all experiments. No statistically significant differences were found among temperature means, minima, or maxima recorded in both the UV-E and UV-S experiments.

Hence, at least for these experiments, temperature was discarded as a possible environmental factor influencing the response of the species. Similarly, in the outdoor aquariums experiment no differences in temperature were observed among aquariums under the same treatment, between aquariums under different treatments and between aquariums and the temperatures measured at the collection site. Similar results were found in the macrophytes experiments with no statistically significant differences found among treatments.

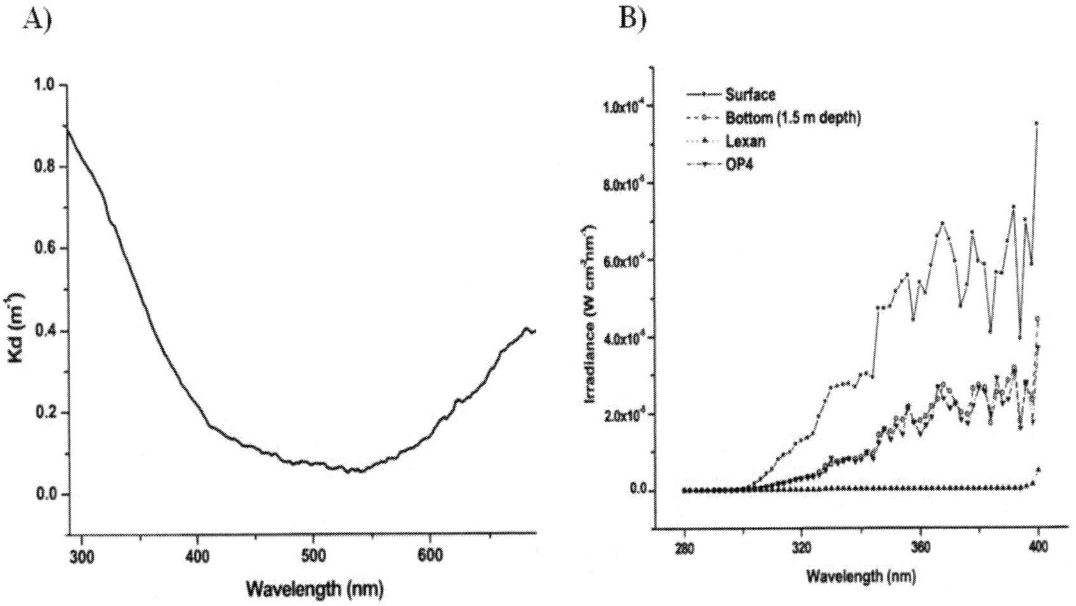

Figure 8. A) Vertical downwelling attenuation coefficient (Kd) measured within a spectral range of 290-700 nm at the back-reef area of Laurel Reef in La Parguera. B) UVR downwelling irradiance (280-400 nm; Ed_{UVR}) measured just above the ocean surface, at the seagrass sampling depth, and under the Lexan and OP4 filters.

3.4.5.2. Biological Parameters

3.4.5.2.1. Corals

3.4.5.2.1.1. Skeletal Growth

The linear extension rates (mm d^{-1}) of *A. cervicornis* were 22% higher in colonies growing under UV-depleted conditions compared to those growing at ambient levels of UVR (control colonies). The skeletal density of *A. cervicornis* showed no differences between colonies exposed to normal and those under 99% depleted UVR conditions during the UV-E experiment. Conversely, during the UV-S experiment the linear extension rates of those colonies transplanted from 20 m to 1 m were reduced by a 66%. This reduction was most likely the result of a decrease in the possible photosynthates of the zooxanthellae which suffered an almost immediate reduction in photosynthetic pigment content evidenced by the visually evident paling of these colonies and reduced concentration of photosynthetic pigments (data not shown) due to a photo-acclimation response by the corals caused by the increase in UVR and probably oxidative stress. Colonies transplanted to shallow-waters also produced a significantly less dense skeleton compared to the control colonies at 1 m (Figure 9A, B). The results compare to those of one group [114] which found that UVR and high PAR irradiances promoted a reduction in the photosynthetic capacity of the symbiotic dinoflagellates of the Caribbean coral *Montastraea faveolata*. Furthermore, histological evidence [34] suggested that these colonies may have survived by, among other ways, by re-absorbing their gametes. Alternatively, the phycobiliproteins, which serve as the light-harvesting antennae of the photosynthetic apparatus of many autotrophic organisms, are especially bleached with exposure to UVB [166]. This might have also affected the photosynthetic apparatus of the *A. cervicornis* symbionts ultimately leading to a reduction in the translocation of resources to their cnidarian host. In any case, the significant reduction in linear extension rates and the increased fragility of the skeletons evidenced by the reduction in skeletal density after transplantation during the UV-S experiment to shallower waters are evidence of the detrimental effects that increased levels of UVR can cause to shallow-water scleractinian corals.

In the case of *P. furcata*, linear extension rates and skeletal density were significantly lower in colonies under enhanced UVR (Figure 9C, D). Conversely, those colonies living under depleted UVR exhibited significantly greater linear extension rates and similar increases in skeletal density. Gleason [61] found similar results in the growth of transplanted colonies of the brown morphs of its cogenerate *Porites astreoides* in St Croix, USVI and attributed his results in part to the high metabolic cost of producing large quantities of MAAs to block UVR.

3.4.5.2.1.2. MAAs

During the UV-E experiment, the total concentration of MAAs was reduced to 16% of that of those colonies of *A. cervicornis* under excluded UVR environment (64.5 ± 6.6 compared to 400.9 ± 35.7 nmol mg protein^{-1} in the control colonies). During the UV-S experiment, colonies of *A. cervicornis* transplanted from 20 m to 1 m, despite an almost immediate paling, contained 40-88 times higher total MAAs than those living at 20 m at the shelf edge (Figure 10A). This is the largest increase in MAA concentrations documented in reef corals and mediated by increases in UVR levels either in the field or laboratory.

The significant increase in total MAAs in colonies transplanted from 20 m to 1 m shows an immediate response to higher UVR levels for host and symbionts protection contrary to earlier results with the massive species *M. annularis* in the Bahamas [12]. Our findings indicate that, at least in *A. cervicornis*, the increase in MAAs is directly related to UVR protection. Nevertheless, based on a similar study by Shick et al. [59], the effect of PAR while little, cannot be discarded. Similar increases have only been reported for the red alga *Chondrus crispus* [167]. Those colonies transplanted to deeper reef zones drastically reduced their respective total and individual MAA concentrations with similar values to those living naturally at those zones by the end of the experiment. This supports earlier findings that were reported [168] of an inverse relationship between MAA concentration in four Hawaiian scleractinian corals (*Montipora verrucosa, M. patula, Pocillopora meandrina,* and *Porites compressa*) with depth of occurrence and level of UVR. MAAs can also act as antioxidants [122] against ROS. This formation is typically mediated by UV-related oxidative stress in the hyperoxic media of coral tissues [169-171]. The extremely rapid and continuous increase in

MAAs in the *A. cervicornis* colonies transplanted from deep to shallow waters suggests that these compounds were formed as a result of a sudden increase in UVR and formation of ROS within the coral tissues. ROS play a central role in both, injury to the partners (coral host and zooxanthellae separately) and to the inter-partner (coral and zooxanthellae symbiosis) communication of a stress response. The expulsion of the zooxanthellae ultimately causing bleaching can occur through a variety of mechanisms including: exocytosis, host cell detachment, host cell apoptosis, *in situ* degradation and host cell necrosis [172].

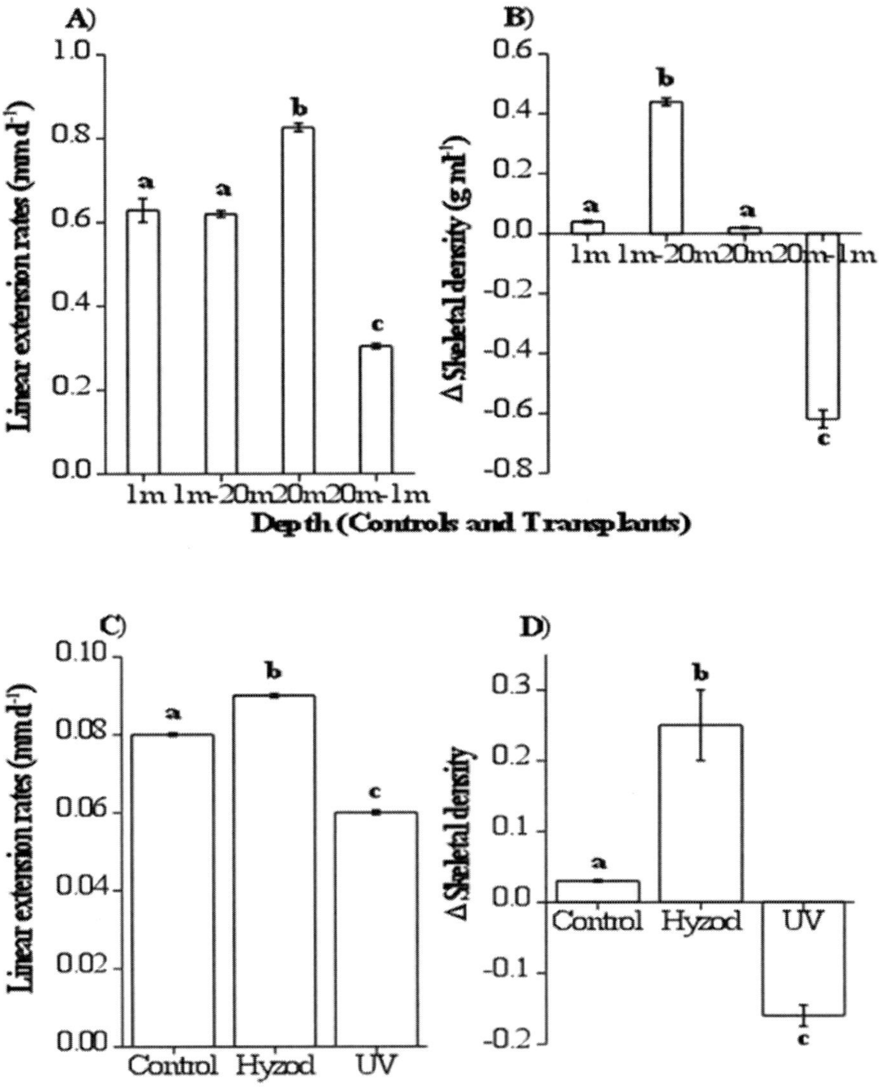

Figure 9. UVR effect on the growth of both species during the UV-S experiment. A) Average linear extension rates and, B) change in skeletal density of *A. cervicornis* 115 days after transplanting the colonies to the different sites. C) Average linear extension rates and, D) change in skeletal density of *P. furcata* after 128 days of exposure to the different treatments in outdoor aquaria. Vertical lines represent ± 1SD. Different letters represent significantly different data ($\alpha = 0.05$). Different letters indicate statistically significant values.

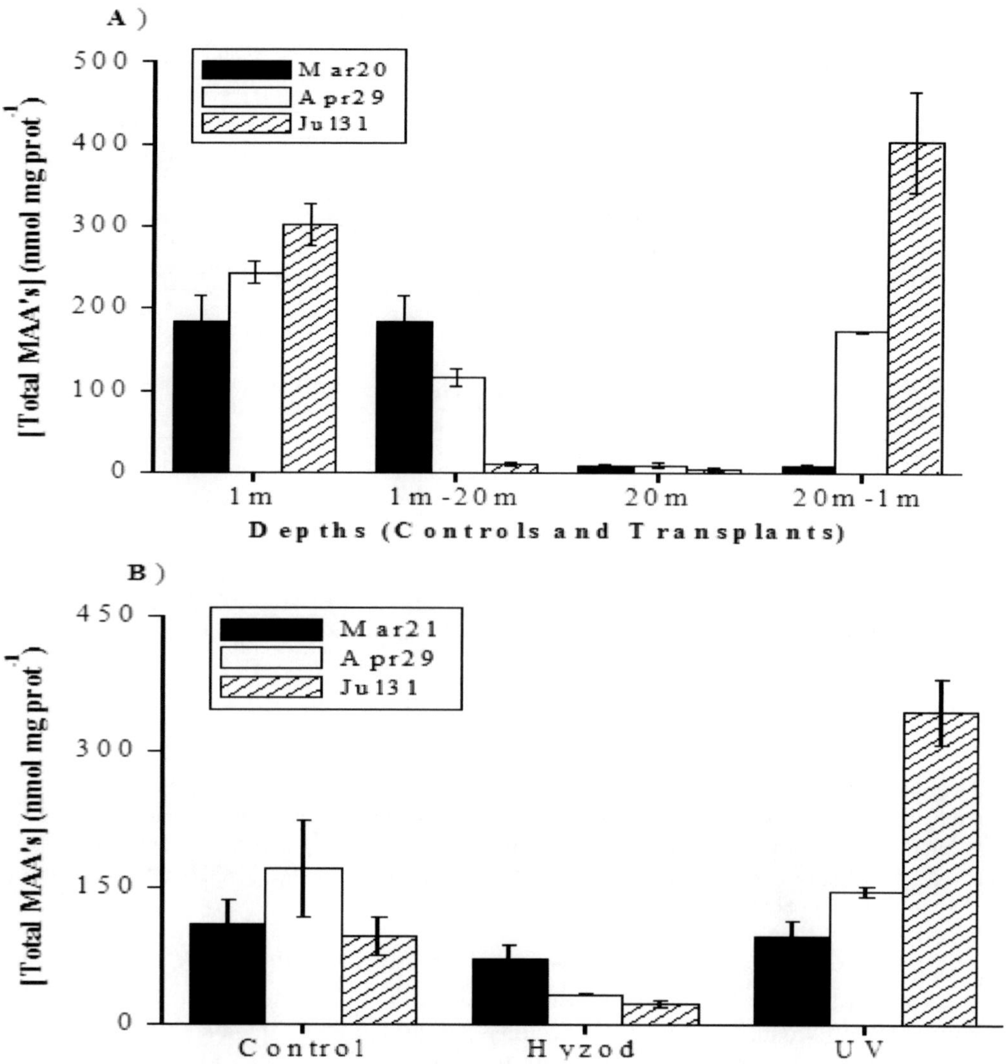

Figure 10. A) Total MAAs concentration in *A. cervicornis* measured at the beginning, one month later, and at the end of the experimental period during the UV-S experiment. B) Total MAAs concentration in *P. furcata* measured in the outdoor aquariums. Vertical lines represent ± 1SE. ($\alpha = 0.05$).

Colonies of *P. furcata* subjected to enhanced UVR levels significantly increased their total MAA concentration relative to their initial levels. By the end of the experiment, their total MAAs was 3.5x than those under "normal" UVR conditions (controls) and 15.7x than those under depleted UVR (Hyzod®) (Figure 10B above).

The MAA composition in *A. cervicornis* was consistent during the UV-E and UV-S experiments. A representative chromatogram showing the different MAAs found in *A. cervicornis* and *P. furcata* is shown in Figure 11. In order of dominance, the following MAAs were identified in *A. cervicornis*: palythine, asterina-330, mycosporine-glycine, palythene, shinorine, usujirene, and porphyra-334. All the MAAs showed a systematic decrease through the UV-E experiment in colonies exposed to reduced levels of UVR.

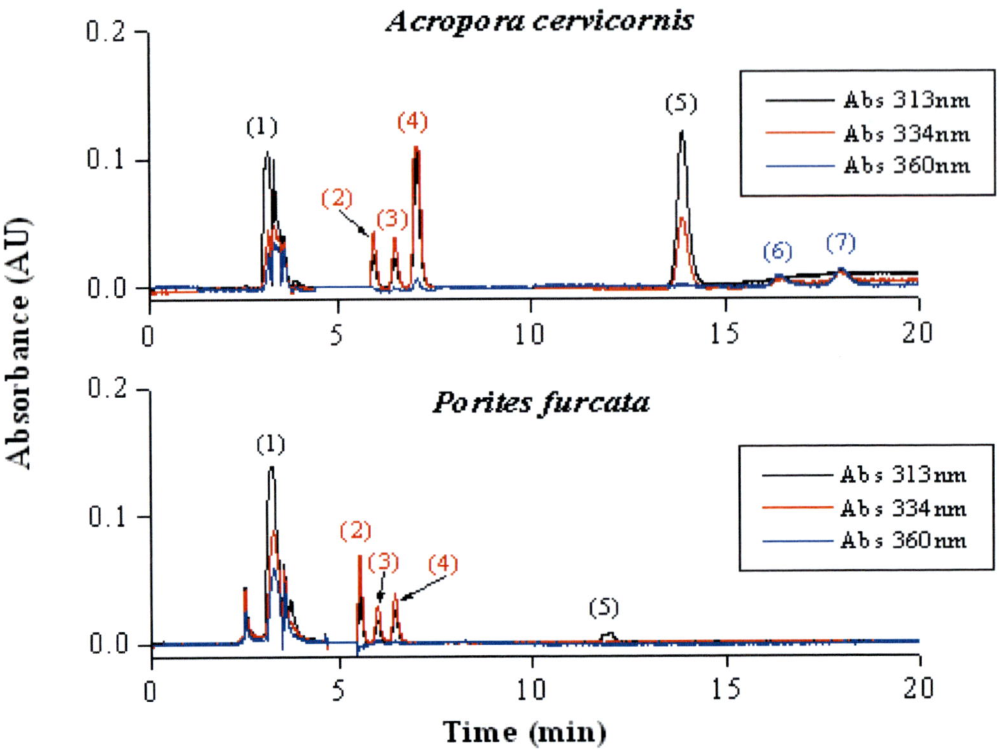

Figure 11. Representative chromatogram showing the different MAA's found in both species. The colors of the numbers indicate, according to the legend, at which wavelength absorbance was read for each pigment. (1) mycosporine-glycine (λ_{max} = 310 nm); (2) porphyra-334 (λ_{max} = 334 nm); (3) shinorine (λ_{max} = 333 nm); (4) asterina-330 (λ_{max} = 330 nm); (5) palythine (λ_{max} = 320 nm); (6) usujirene (λ_{max} = 357 nm); and (7) palythene (λ_{max} = 360 nm).

In contrast, the control colonies showed an increase in all the MAA concentrations, associated with an increase in both solar day and UVR doses as the experiment approached summertime. All individual MAAs concentrations increased under enhanced UVR levels with mycosporine-glycine as the dominant MAA followed by shinorine, porphyra-334, asterina-330 and palythine. MAAs, such as mycosporine-glycine, are known to have moderate antioxidant activity and can provide some relief against photooxidative stress in the hyperoxic tissues of algal-invertebrate symbiosis [122,123]. In *P. furcata*, mycosporine-glycine was the dominant MAA found and moreover it is one of the few MAAs with a peak absorption in the UV-B region (λ_{max} = 310nm); hence providing a direct defense against these harmful wavelengths. Another interesting finding was the fact that by the end of the experimental period, while still in low concentrations related to primary MAAs, secondary MAAs doubled or tripled their concentration probably at the expense of the conversion of primary MAAs. A regulation of the *de novo* synthesis of mycosporine-glycine could not have occurred in *P. furcata* as the colonies maintained the concentration of this primary MAA in high concentrations at all times contrary to the results of Shick [58] who found that decrease in four primary MAAs between 2-4 weeks of exposure was balanced by stoichiometric increases in six additional MAAs suggesting a precursor-product relationship. A possible explanation to this difference can be related to the use of different light sources, in our case using those emitting very near the λ_{max} of mycosporine-glycine. On the other hand, mycosporine-glycine has the smallest extinction coefficient of all the identified MAAs in *P. furcata* (ε_{307}=28100) which might also explain its high concentrations. Once accumulated, MAAs have a long residence time and are photochemically stable [173]. In plants, downstream enzymes involved in the biosynthesis of flavonoids are coordinately induced by UVR within hours and their mRNAs accumulate sequentially during 2-7 days of exposure [174]. A similar timing of the induction of early enzymes in the shikimate pathway and the branchpoint to MAAs [99] seems likely and consistent with the accumulation of primary MAAs. Similarly, it has been foumd that [175] on a daily basis, the individual and total MAAs concentration in several Scleractinian species from Okinawa varied, typically rising at midday and dropping at night with secondary MAAs showing the highest increase at midday whereas primary MAAs such as mycosporine-glycine had little variation implying the higher energetic cost of producing primary MAAs *de novo.*

The production of MAAs (in terms of the number of distinct MAAs produced) in both species was particularly different (see Figure 11 above). For example, *A. cervicornis* contained seven MAAs whereas *P. furcata* contained only five MAAs. None of the MAAs that absorb beyond 340 nm (i.e., palythene and usujirene) were found in *P. furcata. A. cervicornis* hosts clade A while *Porites furcata* hosts clade A4a in shallow areas. Clades B, C, and D symbionts are analogous to shade plants as they are usually found in deeper areas whereas clade A is found in shallow water corals probably because of a higher resistant to solar irradiance and enhanced light-harvesting capability. It has been reported that up to ten different MAAs were found in the Pacific species *Stylophora pistillata* [58]. The differences in MAAs are most probably a direct result of harboring different zooxanthellae clades [96, 176].

3.4.5.2.1.3. Fecundity

The fecundity of *A. cervicornis* was studied during the UV-S experiment. None of the colonies of *A. cervicornis* contained any visible eggs one month after being transplanted in March, 2003, and only empty holes were found in the mesenteries (Figure 12). None of the colonies exposed to enhanced UVR (i.e., those transplanted from 20 m to 1 m) released any egg/sperm bundles. These corals contained visible stage II oocytes just before being transplanted to 1 m; hence, this suggests a possible re-absorption of the gametes by the coral polyps as an energy source. Corals can survive periods of decreased photosynthetic productivity between 1-4 months through catabolism of lipid resources, such as gametes [149], although this might represent a reduction in fertilization rates [177]. This behavior was reported in the past in bleaching-stressed colonies of Caribbean [178] and Pacific [179-180] coral species. Nonetheless, another possibility is the early expulsion of eggs either due to stress caused by the manipulation, increases in PAR or UVR, or any other aspect. In any case, this suggests that fecundity can be used as an indicator of sub-lethal stress in reef coral [34].

a.

b.

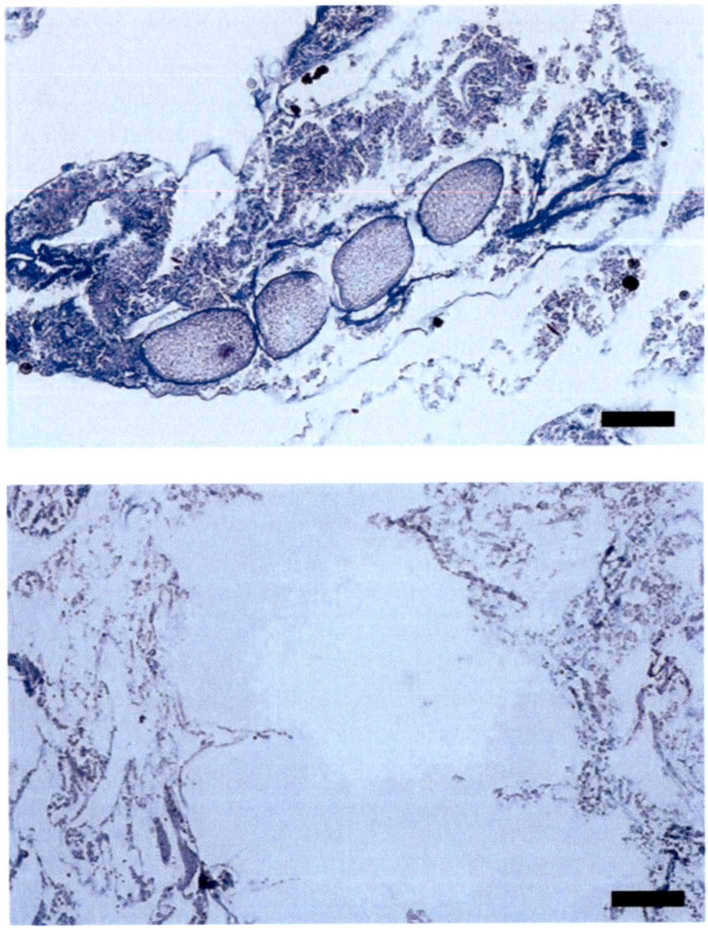

Figure 12. A) Histological sections of samples from colonies living at 20 m at La Parguera shelf edge the day just before being transplanted to 1 m depth at San Cristóbal Reef in La Parguera (March 20, 2003) showing four eggs developing within a mesentery. Scale bar = 300 μm. B) Photo taken one month after the transplantation (April 29, 2003) showing the empty spaces within the mesentery where the eggs were located before. Scale bar = 500 μm.

On the other hand, those colonies transplanted from 1 m to 20 m showed a significant increase in the number of eggs per polyp and these colonies did spawned as expected.

In *P. furcata*, none of the colonies exposed to enhanced UVR contained eggs and larvae by the end of the experiment. On the other hand, the colonies under reduced UVR levels contained eggs and sperm packages during all the experiment. *P. furcata* is a year-long brooding species [181] and it was possible that some colonies released larvae during part of the experiment.

Contrary to earlier suggestions regarding the apparent insufficiency of using one reproductive cycle to detect differences in the tolerance of corals to changes in UVR [182], the exposure to enhanced UVR for 4 months resulted in different gametic pools in both species than the ones presented by their colonies living under natural conditions. Nonetheless, contrary to *A. cervicornis*, the colonies of *P. furcata* did not completely lose all the gametes suggesting that brooding colonies such as *P. furcata*, are apparently more resistant to changes in their respective reproductive structures as a result of increased UVR than broadcasting species like *A. cervicornis*. This also suggests that individual species responses should be taken into considerations when addressing or extrapolating the effects of stressors such as UVR at the ecosystem level.

3.4.5.2.2. Macrophytes

3.4.5.2.2.1. Leaf Optical Properties of Macrophytes

The leaves of *R. mangle* showed a typical green plants reflectance and transmittance curves with reflectance decreasing from 400-530 nm and from 600-700 nm with exposure to UVR (Figure 13).

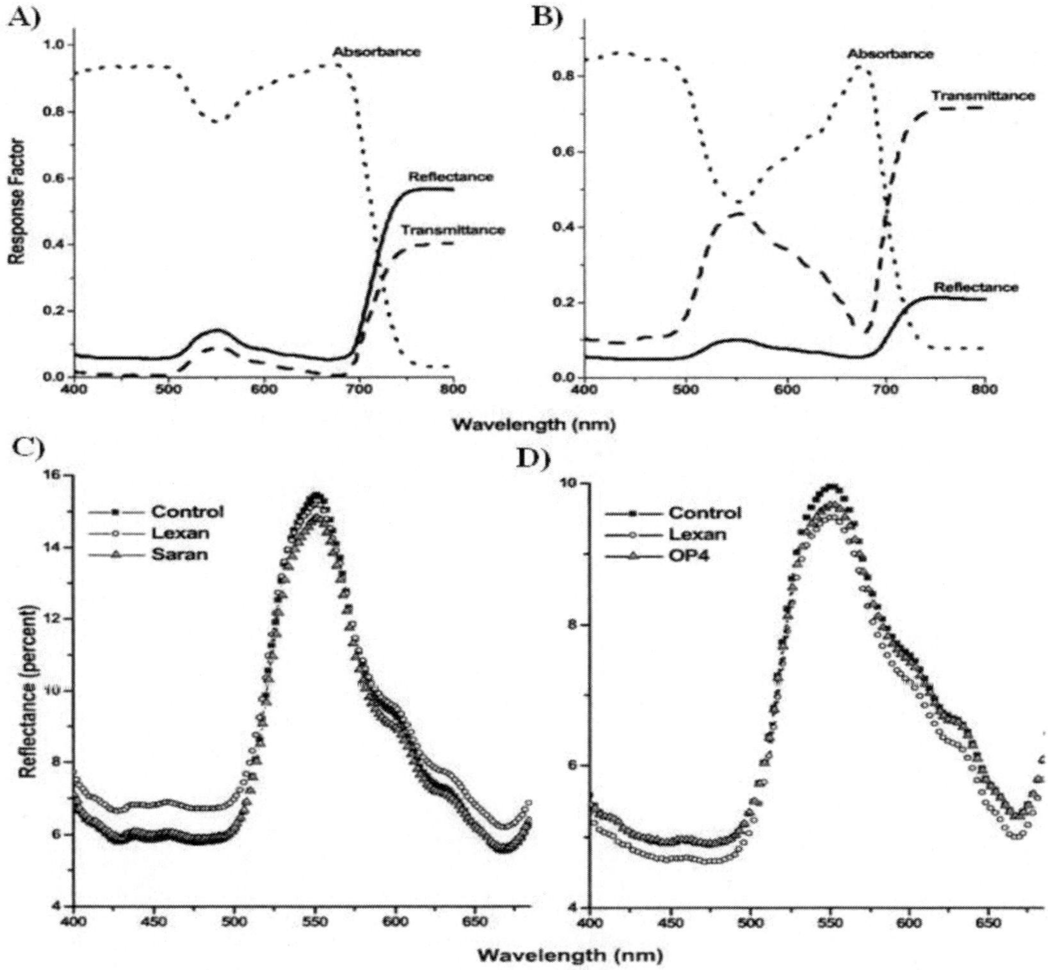

Figure 13. A and B) Optical properties (400-800 nm), and C and D) Leaf reflectance averages (400-700 nm) of *R. mangle* and *T. testudinum*, respectively.

Derivative analysis showed a change in the position of the red edge (708-713 nm) with UV radiation exposure. A blue shift of approximately 5 nm was observed in UV exposed leaves. While no statistically significant differences were found between treatments, leaf thickness seems to influence the reflectance of *T. testudinum*. In order to explain the varied results observed between the two species it is necessary to first consider that both species live under considerably different irradiance environments with *R. mangle* being exposed to a less filtered light regime than *T. testudinum* which inhabits shallow-waters. The attenuation of solar radiation in the water column affects the spectral distribution of downwelling irradiance and underwater organisms, such as seagrasses, may have developed adaptations to this distribution.

3.4.5.2.2.2. Macrophytes Pigments

No differences were found in the overall concentration of photosynthetic pigments of *R. mangle* between UV-excluded and natural light conditions treatments. Nonetheless, a trend of reductions in the concentration of chlorophylls was indeed found in *R. mangle*. Contrarily, in *T. testudinum* significant differences were found in the concentrations of chlorophylls and carotenoids with UV-excluded samples showing the highest concentrations. The results agree with those of others [74] who reported similar decreases in chlorophyll concentrations in five seagrass species in Australia.

UV-absorbing compounds showed an increase in both studied macrophyte species with exposure to UVR (Figure 14).

Absorption maxima were measured between 280-330 nm in *R. mangle* and 272-332 nm in *T. testudinum* with significantly higher absorbances in the natural light-exposed samples compared to the UV-excluded ones.

Individual flavonoids absorption was not measured during these experiments with these macrophytes. Nonetheless, the results provide insights on the effects of UVR on the metabolism of these plants and are in agreement with other studies [74].

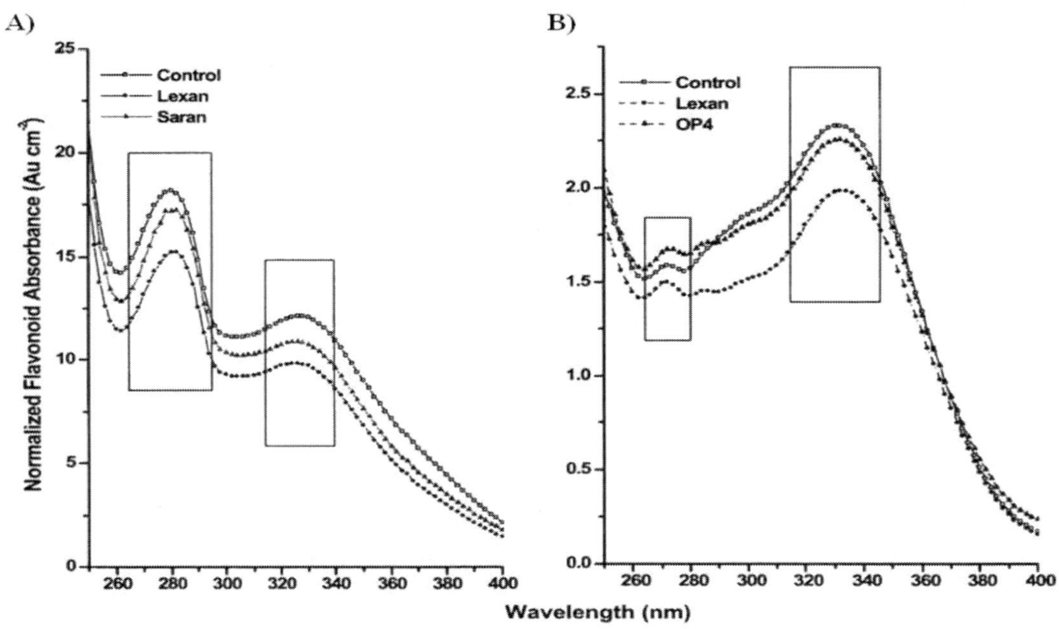

Figure 14. Normalized flavonoid absorbance for A) *R. mangle* and B) *T. testudinum* at the conclusion of both experiments.

Similarly, in a comparative high- and low-latitude study, the salt marsh genus *Salicornia* responded to increased levels of UV-B radiation by increasing the concentration of UV-absorbing flavonoids up to 330% with up to seven different flavonoids detected by HPLC [7]. This exposure also resulted in reduced growth rates, lower chlorophyll *a* concentrations, and higher absorbance within the 420-470 nm region (presumably as a result of an increase in the concentration of several photoprotective carotenoids). Interestingly, it has been reported that shallow-water high-UVR-adapted adapted species recovered less under depleted solar UV-B, suggesting that UV-B not only causes negative effects, but also may facilitate or induce recovery processes in aquatic macrophytes [75].

Conclusion

The proportion of corals threatened with extinction has increased dramatically in the past decades with the Caribbean having the largest proportion of corals in high extinction risk categories [183]. Their assessment covered all the Scleractinian species known plus octocorals and hydrocorals. Species in the family Acroporidae are at the top of the high risk list with the Poritidae being less threatened. Deeper water species appear to be in a lesser risk than those at shallow-waters. The proportion of threatened coral species exceeds that of most terrestrial animal groups, except amphibians, probably because of the corals' apparent susceptibility to climate change.

Global climate change has severely impacted coral reefs during the last decades [184]. Coral reef calcification is predicted to decrease 20-60% by 2100, relative to pre-industrial levels [185, 186], and that increased UVR may induce the formation of less dense skeletons with reduced linear extension rates, the scenario for reef corals, especially for threatened species, such as *A. cervicornis*, remains uncertain. A combination of acidification and climate change may lead to dramatic increases in underwater UV penetration [187]. Reduced growth rates, relatively rapid and intense bleaching episodes in shallow-water corals can occur with moderate increases in UVR. Most importantly, as ultimately the survival of the species depends on this, the fecundity of reef-building branching corals can be drastically reduced or impaired.

Our results indicate that not only reef Scleractinian corals are affected by changes in the UVR reaching coastal zones but also important ecosystem framework building macrophytes as well. Changes in pigment composition might lead to decreased photosynthetic rates and primary production affecting the growth and ultimately the resilience of these important tropical ecosystem builders. Unless strong mitigation measures on climate change are taken, coral reefs and their associated ecosystems established on tropical latitudes may continue to be affected by UVR for the next decades. Ultimately, it will depend on the continued severity of climate change, the synergistic effect of other environmental disturbances, and the ability of the organisms to adapt to future physical and environmental conditions.

ACKNOWLEDGMENTS

This research was funded by NASA grant NCC5-595 (Tropical Center for Earth and Space Studies) and NOAA grant NA17AE1623 (NOAA Center for Atmospheric Sciences).We greatly appreciate and acknowledge all the help provided by the employees at the Magueyes Island Field Station during all these years. We thank Xana Connelly for the analysis of the red mangrove data. The Department of Natural and Environmental Resources kindly lend their Histology Laboratory facilities to Dr. Torres-Pérez for processing the fecundity samples at no cost.

REFERENCES

[1] Lelieveld J., van Aardenne J., Fisher H. et al (2004) Increasing ozone over the Atlantic Ocean. *Science,* 304:1483-1487.

[2] Bortz S. E., Prather M. J., Cammas J.-P. et al (2006) Ozone, water vapor, and temperature in the upper tropical troposphere: Variations over a decade of MOZAIC measurements. *J. Geophys. Res.,* 111:D05305.

[3] Ziemke J. R., Chandra S., Bhartia P. K. (2005) A 25-year data record of atmospheric ozone in the Pacific from Total Ozone Mapping Spectrometer (TOMS) cloud slicing: Implications for ozone trends in the stratosphere and troposphere. *J. Geophys. Res.,* 110:D15105.

[4] Randell W. J., Wu F. (2007) A stratospheric ozone profile data set for 1979-2005: Variability, trends and comparisons with column ozone data. *J. Geophys. Res.*, 112:D06313.

[5] Madronich S., McKenzie R. L., Björn L. O. et al (1995) Changes in ultraviolet radiation reaching the Earth's surface. *Ambio.*, 24:143-152.

[6] McKenzie R. L., Aucamp P. J., Bais A. F. et al (2011) Ozone depletion and climate change: impacts on UV radiation. *Photochem. Photobiol. Sci.,* 10:182-198.

[7] Costa C. S. B., Armstrong R., Detrés Y. et al (2006) Effect of Ultraviolet-B radiation on salt marsh vegetation: Trends of the genus *Salicornia* along the Americas. *Photochem. Photobiol.*, 82:878-886.

[8] Masiri I., Nuñez M., Weller E. (2008) A 10-year climatology of solar radiation for the Great Barrier Reef: implications for recent mass coral bleaching events. *Int. J. Remote Sens.*, 29:4443-4462.

[9] Jerlov N. G. (1950) Ultra-violet radiation in the sea. *Nature,* 4211:111-112.

[10] Smith R. C., Baker K. S. (1981) Optical properties of the clearest natural waters (200-800). *Applied Optics*, 20(2):177-184.

[11] Weinbauer M. G., Wilhelm S. W., Suttle C. A. et al (1997) Photoreactivation compensates for UV damage and restores infectivity to natural marine virus communities. *Appl. Environ. Microbiol.*, 63:2200-2205.

[12] Gleason D. F., Wellington G. M. (1993) Ultraviolet radiation and coral bleaching. *Nature,* 365:836-838.

[13] Conde D., Aubriot L., Sommaruga R. (2000) Changes in UV penetration associated with marine intrusions and freshwater discharge in a shallow coastal lagoon of the Southern Atlantic Ocean. *Mar. Ecol. Prog. Ser.*, 207:19-31.

[14] Torres J. L., Amstrong R. A., Corredor J. E. et al (2007) Physiological responses of *Acropora cervicornis* to increased solar irradiance. *Photochem. Photobiol.*, 83:839-851.

[15] Detrés Y. (2001) Effects of ultraviolet radiation in the tropical marine macrophyte *Thalassia testudinum* (König). PhD Thesis. Department of Marine Sciences, University of Puerto Rico, Mayagüez Campus.

[16] Smith R. C., Baker K. S. (1979) Penetration of UV-B and biologically effective dose-rates in natural waters. *Photochem. Photobiol.*, 29:311-323.

[17] Ziegler S., Benner R. (2000) Effects of solar radiation on dissolved organic matter cycling in a subtropical seagrass meadow. *Limnol. Oceanogr.*, 45:257-266.

[18] Del Castillo C. E., Coble P. G., Morell J. M. et al (1999) Analysis of the optical properties of the Orinoco River plume by absorption and fluorescence spectroscopy. *Mar. Chem.*, 66:35-51.

[19] Morell J. M., Corredor J. E. (2001) Photomineralization of fluorescent dissolved organic matter in the Orinoco River plume: Estimation of ammonium release. *J. Geophys. Res.,* 106:16,807-16,813.

[20] Skoog A., Wedborg M., Fogelqvist E. (1996) Photobleaching of fluorescence and the organic carbon concentration in a coastal environment. *Mar. Chem.*, 55:333-345.

[21] Wetzel R. G., Hatcher P. G., Bianchi T. S. (1995) Natural photolysis by ultraviolet irradiance of recalcitrant dissolved organic matter to simple substrates for rapid bacterial metabolism. *Limnol. Oceanogr.*, 40:1369-1380.

[22] Zagarese H. E., Diaz M., Pedroso F. et al (2001) Photodegradation of natural organic matter exposed to fluctuating levels of solar radiation. *J. Photochem. Photobiol. B.,* 61:35-45.

[23] Kuwahara V. S., Nakajima R., Othman B. H. R. et al (2010) Spatial variability of UVR attenuation and bio-optical factors in shallow coral-reef waters of Malaysia. *Coral Reefs,* 29(3):693-704.

[24] Nelson J. R., Guarda S. (1995) Particulate and dissolved spectral absorption on the continental shelf of the south eastern United States. *J. Geophys. Res.*, 100:8715-8732.

[25] Siegel D. A., Michaels A. F. (1996) On non-chlorophyll light attenuation in the open ocean: Implications for biogeochemistry and remote sensing. *Deep-Sea Res. II,* 43:321-345.

[26] Leavitt P. R., Cumming B. F., Smol J. P. et al. (2003) Climatic control of ultraviolet radiation effects on lakes. *Limnol. Oceanogr.*, 48:2062-2069.

[27] Blough N. V., Zafiriou O. C., Bonilla J. (1993) Optical absorption spectra of waters from the Orinoco River outflow-terrestrial input of colored organic matter to the Caribbean. *J. Geophys. Res.*, 98:2271-2278.

[28] Häder D. P., Kumar H. D., Smith R. C. (2003) Aquatic ecosystems: effects of solar ultraviolet radiation and interactions with other climatic factors. *Photochem. Photobiol. Sci.*, 2:39-50.

[29] Häder D. P., Kumar H. D., Smith R. C. et al (2007) Effects of solar UV radiation on aquatic ecosystems and interactions with climate change. *Photochem. Photobiol. Sci.,* 6:267-285.

[30] Tedetti M., Sempéré R. (2006) Penetration of ultraviolet radiation in the marine environment: A review. *Photochem. Photobiol.*, 82:389-397.

[31] Wilhelm S. W., Jeffrey W. H., Suttle C. A. (2002) Estimation of biologically damaging UV levels in marine surface waters with DNA and viral dosimeters. *Photochem. Photobiol.,* 76:268-273.

[32] Ayoub L., Hallock P., Coble P. (2008) Colored dissolved organic material increases resiliency of coral reefs by controlling exposure to UVR. Proc. 11th Int. Coral Reef Symp. Ft. Lauderdale, FL. 572-576.

[33] Miller R. L., Belz M., Del Castillo C. et al (2002) Determining CDOM absorption spectra in diverse coastal environments using a multiple pathlength, liquid core waveguide system. *Continental Shelf Res.*, 22:1301-1310.

[34] Torres J. L., Armstrong R. A., Weil E. (2008) Enhanced ultraviolet radiation can terminate sexual reproduction in a Caribbean broadcasting species. *J. Exp. Mar. Biol. Ecol.*, 358:39-45.

[35] Zepp R. G., Shank S. G., Stabenau E. et al (2008) Spatial and temporal variability of solar ultraviolet exposure of coral assemblages in the Florida Keys: importance of colored dissolved organic matter. *Limnol. Oceanogr.*, 53:1909-1922.

[36] Wahl M., Molis M., Davis A. et al (2004) UV effects that come and go: a global comparison of marine benthic community level impacts. *Global Change Biol.*, 10:1962-1972.

[37] Worrest R. C. (1982) Review of literature concerning the impact of UV-B radiation upon marine organisms. In: Calkins J (ed) The role of solar ultraviolet radiation in marine ecosystems. Plenum Press. New York. Pp. 429-458.

[38] Bracher A. U., Wiencke C. (2000) Simulation of the effects of naturally enhanced UV radiation on photosynthesis of Antarctic phytoplankton. *Mar. Ecol. Prog. Ser.*, 196: 127-141.

[39] Huovinen P. S., Oikari A. O. J., Soimasuo M. R. et al. (2000) Impact of UV radiation on the early development of the giant kelp (*Macrocystis pyrifera*) gametophytes. *Photochem. Photobiol.*, 72:308-313.

[40] Kouwenberg J. H. M., Browman H. I., Cullen J. J. et al. (1999) Biological weighting of ultraviolet (280-400 nm) induced mortality in marine zooplankton and fish. I. Atlantic cod (*Gadus morhua*) eggs. *Mar. Biol.*, 134:269-284.

[41] Lesser M. P., Farrell J. H., Walker C. W. (2001) Oxidative stress, DNA damage and p53 expression in the larvae of Atlantic cod (*Gadus morhua*) exposed to ultraviolet (290-400 nm) radiation. *J. Exp. Biol.*, 204:157-164.

[42] Banaszak A. T., Lesser M. P. (2009) Effects of solar ultraviolet radiation on coral reef organisms. *Photochem. Photobiol. Sci.*, 8:1276-1294.

[43] Lesser M. P. (2006) Oxidative stress in marine environments: biochemistry and physiological ecology. *Annu. Rev. Physiol.*, 68:253-278.

[44] Cadet J., Sage E., Douki T. (2005) Ultraviolet radiation-mediated damage to cellular DNA. *Mutat. Res.*, 571:3-17.

[45] Lesser M. P., Kruse V. A., Barry T. M. (2003) Exposure to ultraviolet radiation causes apoptosis in developing sea urchin embryos. *J. Exp. Biol.*, 206:4097-4103.

[46] Lesser M. P., Barry T. M. (2003) Survivorship, development, and DNA damage in echinoderm embryos and larvae exposed to ultraviolet radiation (290-400 nm). *J. Exp. Mar. Biol. Ecol.*, 292:75-91.

[47] Siebeck O. (1981) Photoreactivation and depth-dependent U.V. tolerance in reef coral in the Great Barrier Reef/Australia. *Naturwissenschaften,* 67:426-427.

[48] Jokiel P. L., York Jr. R. H. (1982) Solar ultraviolet photobiology of the reef coral *Pocillopora damicornis* and symbiotic zooxanthellae. *Bull. Mar. Sci.*, 32:301-315.

[49] Dunlap W. C., Chalker B. E., Bandaranayake W. M. (1988) Ultraviolet light absorbing agents derived from tropical marine organisms of the Great Barrier Reef, Australia. Proc 6th Int Coral Reef Symp. 3:89-93.

[50] Kinzie III R. A. (1993) Effects of ambient levels of solar ultraviolet radiation on zooxanthellae and photosynthesis of the reef coral *Montipora verrucosa. Mar. Biol.,* 116:319-327.

[51] Masuda K., Goto M., Maruyama T. et al (1993) Adaptation of solitary corals and their zooxanthellae to low light and UV radiation. *Mar. Biol.*, 117:685-691.

[52] Krupp D. A., Blanck J. (1995) Preliminary report on the occurrence of mycosporine-like amino acids in the eggs of the Hawaiian scleractinian corals *Montipora verrucosa* and *Fungia scutaria.* In: Gulko D, Jokiel PL (eds) Ultraviolet Radiation and Coral Reefs. HIMB Tech. Report #41. UNIHI-Sea Grant-CR-95-03. Pp.129-134.

[53] Lewis S. (1995) Response of a Pacific stony coral to short-term exposure of ultraviolet and visible light. In: Gulko D, Jokiel PL (eds) Ultraviolet Radiation and Coral Reefs. HIMB Tech. Report #41. UNIHI-Sea Grant-CR-95-03. Pp.89-106.

[54] Lesser M. P., Lewis S. (1996) Action spectrum for the effects of UV radiation on photosynthesis in the hermatypic coral *Pocillopora damicornis. Mar. Ecol. Prog. Ser.,* 134:171-177.

[55] Drollet J. H., Teai T., Faucon M. et al (1997) Field study of compensatory changes in UV-absorbing compounds in the mucus of the solitary coral *Fungia repanda* (Scleractinia: Fungiidae) in relation to solar UV radiation, sea-water temperature, and other coincident physico-chemical parameters. *Mar. Freshwater Res.*, 48:329-333.

[56] Kuffner I. B. (2001a) Effects of ultraviolet radiation and water motion on the reef coral *Porites compressa* Dana: a flume experiment. *Mar. Biol.*, 138:467-476.

[57] Kuffner I. B. (2001b) Effects of ultraviolet (UV) radiation on larval settlement of the reef coral *Pocillopora damicornis. Mar. Ecol. Prog. Ser.,* 217:251-261.

[58] Shick J. M. (2004) The continuity and intensity of ultraviolet irradiation affect the kinetics of biosynthesis, accumulation, and conversion of mycosporine-like amino acids (MAAs) in the coral *Stylophora pistillata. Limnol. Oceanogr.* 49:442-458.

[59] Shick J. M., Ferrier-Pagès C., Grover R. et al (2005) Effects of starvation, ammonium concentration, and photosynthesis on the UV-dependent accumulation of mycosporine-like amino acids (MAAs) in the coral *Stylophora pistillata. Mar. Ecol. Prog. Ser.,* 295:135-156.

[60] Ferrier-Pagès C., Richard C., Forcioli D. et al (2007) Effects of temperature and UV radiation increases on the photosynthetic efficiency in four Scleractinian coral species. *Biol. Bull.*, 213:76-87.

[61] Gleason D. F. (1993) Differential effects of ultraviolet radiation on green and brown morphs of the Caribbean coral *Porites astreoides. Limnol. Oceanogr.*, 38:1452-1463.

[62] Gleason D. F., Wellington G. M. (1995) Variation in UVB sensitivity of planula larvae of the coral *Agaricia agaricites* along a depth gradient. *Mar. Biol.*, 123:693-703.

[63] Muszynski, F. Z., Bruckner A., Armstrong R. A. et al (1998) Within-colony variations of UV absorption in a reef building coral. *Bull. Mar. Sci.*, 63:589-594.

[64] Corredor J. E., Bruckner A. W., Muszynski F. Z. et al (2000) UV-absorbing compounds in three species of Caribbean zooxanthellate corals: depth distribution and spectral response. *Bull. Mar. Sci.,* 67:821-830.

[65] Lesser M. P., Farrell J. H. (2004) Exposure to solar radiation increases damage to both host tissues and algal symbionts of corals during thermal stress. *Coral Reefs*, 23:367-377.

[66] Torres-Pérez J. L., Armstrong R. A. (2012) Effects of UV radiation on the growth, photosynthetic and photoprotective components, and reproduction of the Caribbean shallow-water coral *Porites furcata*. *Coral Reefs*, 31:1077-1091.

[67] Mirecki R. M., Teramura A. H. (1984) Effects of ultraviolet-B irradiance on soybean. V. The dependence of plant sensitivity on the photosynthetic photon flux density during and after leaf expansion. *Plant Physiol.*, 74:475-480.

[68] Bornman J. F., Teramura A. J. (1993) Effects of ultraviolet-B radiation on terrestrial plants. In: Young A. R., Björn L. O., Moan J. et al. (Eds) Environmental UV Photobiology. Plenum Press, New York. Pp.427-462.

[69] Day T. A. (1993) Relating UV-B radiation screening effectiveness of foliage to absorbing-compound concentration and anatomical characteristics in a diverse group of plants. *Oecologia*, 95:542-550.

[70] Middleton E. M., Teramura A. (1993) The role of flavonols glycosides and carotenoids in protecting soybean from ultraviolet-B damage. *Plant Physiol.*, 103:741-752.

[71] Ziska L. H., Teramura A. H., Sullivan J. H. et al (1993) Influence of ultraviolet-B (UV-B) radiation on photosynthetic and growth characteristics in field grown cassava (*Manihot esculetum* Crantz). *Plant Cell Environ.*, 16:73-79.

[72] Searles P. S., Caldwell M. M., Winter K. (1995) The response of five tropical dicotyledon species to solar ultraviolet-B radiation. *Am. J. Bot.*, 82:445-453.

[73] Visser A. J., Tosserams M., Groen M. W. et al. (1997) The combined effects of CO_2 concentration and enhanced UV-B radiation on faba bean. Leaf optical properties, pigments, stomatal index and epidermal cell density. *Plant Ecol.,* 128:208-222.

[74] Dawson S. P., Dennison W. C. (1996) Effects of ultraviolet and photosynthetically active radiation on five seagrass species. *Mar. Biol.*, 125:629-638.

[75] Hanelt D., Roleda M. Y. (2009) UVB radiation may ameliorate photoinhibition in specific shallow-water tropical marine macrophytes. *Aquat. Bot.*, 91:6-12.

[76] Rothschild L. J., Manicelli R. L. (2001) Life in extreme environments. *Nature,* 409:1092-1101.

[77] Dillon J. G., Castenholz R. W. (1999) Scytonemin, a cyanobacterial sheath pigment protects against UVC radiation: implications for early photosynthetic life. *J. Phycol.,* 35:673-681.

[78] Sommaruga R., Garcia-Pichel F. (1999) UV-absorbing mycosporine-like compounds in planktonic and benthic organisms from a high-mountain lake. *Arch. Hydrobiol.*, 144:255-269.

[79] Douki T., Court M., Sauvaigo S. et al. (2000) Formation of the main UV-induced thymine dimeric lesions within isolated and cellular DNA as measured by high performance liquid chromatography-tandem mass spectrometry. *J. Biol. Chem.,* 275:11678-11685.

[80] Huot Y., Jeffrey W. H., Davis R. F. et al (2000) Damage to DNA in bacterioplankton: a model of damage by ultraviolet radiation and its repair as influenced by vertical mixing. *Photochem. Photobiol.*, 72:62-74.

[81] Wu H., Gao K., Villafañe V. E. et al (2005) Effects of solar UV radiation on morphology and photosynthesis of filamentous cyanobacterium *Arthrospira platensis*. *Appl. Environ. Microbiol.*, 71:5004-5013.

[82] Steiger S., Schäfer L., Sandmann G. (1999) High-light-dependent upregulation of carotenoids and their antioxidative properties in the cyanobacterium *Synechocystis* PCC 6803. *J. Photochem. Photobiol. B.*, 52:14-18.

[83] Canini A., Leonardi D., Caiola M. G. (2001) Superoxide dismutase activity in the cyanobacterium *Microcystis aeruginosa* after bloom formation. *New Phytol.*, 152:107-116.

[84] Araoz R., Häder D.-P. (1999) Enzymatic antioxidant activity in two cyanobacteria species exposed to solar radiation. *Recent Res. Devel. Photochem. Photobiol.*, 3:123-132.

[85] Vega M. P., Pizarro R. A. (2000) Oxidative stress and defense mechanisms of the freshwater cladoceran *Daphnia longispina* exposed to UV radiation. *J. Photochem. Photobiol. B.,* 54:121-125.

[86] Leech D. M., Williamson C. E. (2001) In situ exposure to ultraviolet radiation alters the depth distribution of *Daphnia*. *Limnol. Oceanogr.*, 46:416-420.

[87] Caldwell M. M., Robberecht R., Flint S. D. (1983) Internal filters: prospects for UV-acclimation in higher plants. *Physiol. Plant*, 58:445-450.

[88] Latimer J. G., Mitchell C. A., Mitchell G. A. (1987) UV-B radiation and photosynthetic irradiance acclimate eggplant for outdoor exposure. *Hort. Science*, 22(3):426-429.

[89] Murthy S. D. S., Rajagopal S. (1995) UV-B radiation induced alterations in the bioenergetic process of photosynthesis. *Photosynthetica,* 31:481-487.

[90] Rozema J., Björn L. O., Bornman J. F. et al (2002) The role of UV-B radiation in aquatic and terrestrial ecosystems – an experimental and functional analysis of the evolution of UV-absorbing compounds. *J. Photochem. Photobiol. B. Biology*, 66:2-12.

[91] Garcia-Pichel F. (1998) Solar ultraviolet and evolutionary history of cyanobacteria. *Orig. Life Evolut. Biosph.*, 28:321-347.

[92] Oren A. (1997) Mycosporine-like amino acids as osmotic solutes in a community of halophilic cyanobacteria. *Geomicrobiol. J.*, 14:231-240.

[93] Cockell C. S. (1998) The biological effects of UV radiation on early earth- a theoretical evolution. *J. Theoret. Biol.*, 193:717-729.

[94] Leach C. M. (1965) Ultraviolet-absorbing substances associated with light-induced sporulation in fungi. *Can. J. Bot.*, 43:185-200.

[95] Singh S. P., Kumari S., Rastogi R. P. et al. (2008) Mycosporine-like amino acids (MAAs): Chemical structure, biosynthesis and significance as UV-absorbing/screening compounds. *Indian J. Exp. Biol.*, 46:7-17.

[96] Banaszak A. T., Barba Santos M. G., LaJeunesse T. C. et al (2006) The distribution of mycosporine-like amino acids (MAAs) and the phylogenetic identity of symbiotic dinoflagellates in cnidarians hosts from the Mexican Caribbean. *J. Exp. Mar. Ecol. Biol.*, 337:131-146.

[97] Bandaranayake W. M. (1998) Mycosporines: are they nature's sunscreens? *Natural Product Reports,* 1998:159-171.

[98] Carreto J. I., Carignan M. O., Montoya N. G. (2005) A high-resolution reversed-phase liquid chromatography method for the analysis of mycosporine-like amino acids (MAAs) in marine organisms. *Mar. Biol.*, 146:237-252.

[99] Shick J. M., Dunlap W. C. (2002) Mycosporine-like amino acids and related gadusols: Biosynthesis, accumulation, and UV-protective functions in aquatic organisms. *Annu. Rev. Physiol.*, 64:233-262.

[100] Portwich A., F. Garcia-Pichel F. (2003) Biosynthetic pathway of mycosporines (mycosporine-like amino acids) in the cyanobacterium *Chlorogloeopsis* sp. Strain PCC 6912. *Phycologia*, 42:384-392.

[101] Favre-Bonvin J., Bernillon J., Salin N. et al. (1987) Biosynthesis of mycosporines: Mycosporine glutaminol in *Trichothecium roseum*. *Phytochemistry,* 29:2509-2514.

[102] Banaszak A. T., LaJeunesse T. C., Trench R. K. (2000) The synthesis of mycosporine-like amino acids (MAAs) by cultured, symbiotic dinoflagellates. *J. Exp. Mar. Ecol. Biol.*, 249:219-233.

[103] Jeffrey S. W., MacTavish H. S., Dunlap W. C. et al (1999) Occurrence of UVA- and UVB-absorbing compounds in 152 species (206 strains) of marine microalgae. *Mar. Ecol. Prog. Ser.,* 189:35-51.

[104] Klisch M., Sinha R. P., Richter P. R. et al. (2001) Mycosporine-like amino acids (MAAs) protect against UV-B-induced damage in *Gyrodinium dorsum* Kofoid. *J. Plant Physiol.*, 158:1449-1454.

[105] Moisan T. A., Mitchell B. G. (2001) UV absorption by mycosporine-like amino acids in *Phaeocystis antarctica* Karsten induced by photosynthetically available radiation. *Mar. Biol.,* 138:217-227.

[106] Apprill A. M., Lesser M. P. (2003) Effects of ultraviolet radiation on *Laminaria saccharina* in relation to depth and tidal height in the Gulf of Maine. *Mar. Ecol. Prog. Ser.*, 256:75-85.

[107] Gröniger A., Häder D.-P. (2002) Induction of the synthesis of an UV-absorbing substance in the green alga *Prasiola stipitata*. *J. Photochem. Photobiol. B. Biol.*, 66: 54-59.

[108] Korbee N., Figueroa F. L., Aguilera J. (2005) Effect of light quality on the accumulation of photosynthetic pigments, proteins and mycosporine-like amino acids in the red alga *Porphyra leucosticta* (Bangiales, Rhodophyta). *J. Photochem. Photobiol. B. Biol.*, 80:71-78.

[109] Whitehead K., Vernet M. (2000) Influence of mycosporine-like amino acids (MAAs) on UV-absorption by particle and dissolved organic matter in La Jolla Bay. *Limnol. Oceanogr.*, 45:1788-1796.

[110] Gröniger A., Sinha R. P., Kilsch M. (2000) Photoprotective compounds in cyanobacteria, phytoplankton and macroalgae – a database. *J. Photochem. Photobiol. B.,* 58:115-122.

[111] Grobe C. W., Murphy T. M. (1998) Solar ultraviolet-B radiation effects on growth and pigment composition of the intertidal alga *Ulva expansa* (Setch.) S. & G. (Chlorophyta). *J. Exp. Mar. Biol. Ecol.*, 225:39-51.

[112] Gröniger A., Hallier C., Häder D.-P. (1999) Influence of UV radiation and visible light on *Porphyra umbilicalis*: photoinhibition and MAA concentration. *J. Appl. Phycol.,* 11:437-445.

[113] Dring M. J., Wagner A., Lüning K. (2001) Contribution of the UV component of natural sunlight to photoinhibition of photosynthesis in six species of subtidal brown and red seaweeds. *Plant Cell Environ.*, 24:1153-1164.

[114] Lesser M. P. (2000) Depth-dependent photoacclimatization to solar ultraviolet radiation in the Caribbean coral *Montastraea faveolata*. *Mar. Ecol. Prog. Ser.,* 192:137-151.

[115] Häder D.-P., Lebert M., Helbling E. W. (2001) Effects of solar radiation on the Patagonian macroalga *Enteromorpha linza* (L.) J. Agardh-Chlorophyceae. *J. Photochem. Photobiol. B.,* 62:43-54.

[116] van de Poll W. H., Eggert A., Buma A. G. J. et al. (2001) Effects of UV-B-induced DNA damage and photoinhibition on growth of temperate marine red macrophytes: habitat-related differences in UV-B tolerance. *J. Phycol.,* 37:30-37.

[117] Tartarotti B., Laurion I., Sommaruga R. (2001) Large variability in the concentration of mycosporine-like amino acids among zooplankton from lakes located across an altitude gradient. *Limnol. Oceanogr.*, 46:1546-1552.

[118] Chatzifotis S., Pavlidis M., Jimeno C. D. et al. (2005) The effect of different carotenoid sources on skin coloration of cultured red porgy (*Pagrus pagrus*). *Aquacult. Res.,* 36: 1517-1525.

[119] Hoogenboom M. O., Anthony K. R. N., Connelly S. R. (2006) Energetic cost of photoinhibition in corals. *Mar. Ecol. Prog. Ser.*, 313:1-12.

[120] Drohan A. F., Thoney D. A., Baker A. C. (2005) Synergistic effect of high temperature and ultraviolet-B radiation on the gorgonian *Eunicea tuorneforti* (Octocorallia: Alcyonacea: Plexauridae). *Bull. Mar. Sci.*, 77:257-266.

[121] Raven J. A. (1991) Responses of aquatic photosynthetic organisms to increased solar UVB. *J. Photochem. Photobiol. B. Biol.,* 9:239-244.

[122] Dunlap W. C., Yamamoto Y. (1995) Small-molecule antioxidants in marine organisms: antioxidant activity of mycosporine-glycine. *Comp. Biochem. Physiol. B.,* 112: 105-114.

[123] Yakovleva I., Bhagooli R., Takemura A. et al. (2004) Differential susceptibility of oxidative stress of two scleractinian corals: antioxidant functioning of mycosporine-glycine. *Comp. Biochem. Physiol. B.,* 139:721-730.

[124] Tirkey J., Adhikary S. P. (2005) Cyanobacteria in biological soil crusts of India. *Curr. Sci.*, 89:515-521.

[125] Korbee Peinado N., Abdala Díaz R. T., Figueroa F. L. (2004) Ammonium and UV radiation stimulate the accumulation of mycosporine-like amino acids in *Porphyra columbina* (Rhodophyta) from Patagonia, Argentina. *J. Phycol.*, 40:248-259.

[126] Gorbushina A. A., Whitehead K., Dornieden T. et al. (2003) Black fungal colonies as units of survival: hyphal mycosporines synthesized by rock-dwelling microcolonial fungi. *Can. J. Bot.*, 81:131-138.

[127] Oren A., Gunde-Cimerman N. (2007) Mycosporines and mycosporine-like amino acids: UV protectants or multipurpose secondary metabolites? *FEMS Microbiol. Lett.*, 269: 1-10.

[128] Carreto J. L., Carignan M. O. (2011) Mycosporine-like amino acids: Relevant secondary metabolites, chemical and ecological aspects. *Mar. Drugs*, 9:387-446.

[129] Laurion I., Blouin F., Roy S. (2004) Packaging of mycosporine-like amino acids in dinoflagellates. *Mar. Ecol. Prog. Ser.*, 279:297-303.

[130] Sinha R. P., Klisch M., Gröniger A. et al. (2000) Mycosporine-like amino acids in the marine red alga *Gracilaria cornea* – Effects of UV and heat. *Environ. Exp. Bot.*, 43: 33-43.

[131] Dunlap W. C., Chalker B. E., Bandaranayake W. M. et al. (1998) Nature's sunscreen from the Great Barrier Reef, Australia. *Int. J. Cosmet. Sci.*, 20:41-51.

[132] Oyamada C., Kaneniwa M., Ebitani K. et al. (2008) Mycosporine-like amino acids extracted from scallop (*Patinopecten yessoensis*) ovaries: UV protection and growth stimulation activities on human cells. *Mar. Biotechnol.*, 10:141-150.

[133] Dakora F. D. (1995) Plant flavonoids: biological molecules for useful exploitation. *Aust. J. Plant Physiol.*, 22:87-99.

[134] Brandt K., Giannini A., Lercari B. (1995) Photomorphogenic responses to UV radiation III: a comparative study of UVB effects on anthocyanin and flavonoid accumulation in wild type and *aurea* mutant of tomato (*Lycopersicon eculentum* MILL). *Photochem. Photobiol.*, 62:1081-1087.

[135] Landry G. L., Chapple C. C. S., Last R. L. (1995) Arabidopsis mutants lacking phenolic sunscreens exhibit enhanced ultraviolet-B injury and oxidative damage. *Plant Physiol.*, 109:1159-1166.

[136] McMillan C. (1983) Sulphated flavonoids and leaf morphology of the *Halophila ovalis* – *H. minor* complex (Hydrocharitaceae) in the Pacific Islands and Australia. *Aquatic Bot.*, 16:337-347.

[137] McMillan C. (1986) Sulphated flavonoids and leaf morphology in the *Halophila ovalis* – *H. minor* complex (Hydrocharitaceae) of the Indo-Pacific ocean. *Aquatic Bot.*, 25: 63–72.

[138] McMillan C., Zapata O., Escobar L. (1980) Sulphated phenolic compounds in seagrasses. *Aquat. Bot.*, 8:267-278.

[139] Lovelock C. E., Clough B. F., Woodrow I. E. (1992) Distribution and accumulation of ultraviolet-radiation-absorbing compounds in leaves of tropical mangroves. *Planta.*, 188:143-154.

[140] Havaux M., Kloppstech K. (2001) The protective function of carotenoid and flavonoid pigments against excess visible radiation at chilling temperature investigated in *Arabidopsis* npq and tt mutants. *Planta.*, 213:953-966.

[141] Strack D. (1997) Phenolic metabolism. In: Plant Biochemistry. Academic Press. NY. Pp.387-416.

[142] Torres J. L., Morelock J. (2002) Effects of terrigenous sediments influx on the coral cover and linear extension rates of three Caribbean massive coral species. *Carib. J. Sci.*, 38(3-4):222-229.

[143] Glynn P. (1973) Ecology of a Caribbean coral reef. The *Porites* reef flat biotope: Part I. Meteorology and hydrography. *Mar. Biol.*, 20:297-318.

[144] Connelly-Pagán X. M. (1999) Responses of red mangrove *Rhizophora mangle* L. to solar ultraviolet radiation. MS Thesis. University of PR, Dept. of Marine Sciences.

[145] Detrés Y., Armstrong R. A., Connelly X. (2001) Ultraviolet-induced responses in two species of climax tropical marine macrophytes. *Photochem. Photobiol.*, 62.(1-2):.55-66.

[146] Barnes D. J. (1973) Growth in colonial scleractinians. *Bull. Mar. Sci.*, 23(2):280-298.

[147] Gladfelter E. H. (1984) Skeletal development in *Acropora cervicornis*: III. A comparison of monthly rates of linear extension and calcium carbonate accretion measured over a year. *Coral Reefs*, 3:51-57.

[148] Edmunds P. J., Davies P. S. 1989. An energy budget for *Porites porites* (Scleractinia), growing in a stressed environment. *Coral Reefs,* 8:37-43.

[149] Davies P. S. (1991) Effect of daylight variations on the energy budgets of shallow-water corals. *Mar. Biol.*, 108:137-144.

[150] Falkowski P. G., Jokiel P. L., Kinzie III R. A. (1990) Irradiance and corals. In: Dubinsky Z. (ed.) Ecosystems of the World: Coral Reefs. Elsevier. Amsterdam, The Netherlands. Pp. 89-107.

[151] van Tussenbroek B. I., Vonk J. A., Stapel J. et al. (2006) The biology of *Thalassia*: Paradigms and recent advances in research. In: Larkum A. W. D., Orth R. J. and Duarte C. M. (eds.) Seagrasses: Biology, Ecology and Conservation. Springer. The Netherlands. Pp. 409-439.

[152] Farnsworth E. J., Ellison A. M. (1996) Sun-shade adaptability of the red mangrove, *Rhizophora mangle* (Rhizophoraceae): changes through ontogeny at several levels of biological organization. *Am. J. Bot.,* 83(9):1131-1143.

[153] Sutherland J. P. (1980) Dynamics of the epibenthic community on roots of the mangrove *Rhizophora mangle*, at Bahía de Buche, Venezuela. *Mar. Biol.*, 58:75-84.

[154] Kirk J. T. O. (1994) Light and photosynthesis in aquatic ecosystems. Cambridge Univ. Press. New York.

[155] Dunlap W. C., Chalker B. E. (1986) Identification and quantification of near-UV absorbing compounds (S-320) in a hermatypic scleractinian. *Coral Reefs*, 5:155-159.

[156] Bradford M. M. (1976) A rapid and sensitive method for the quantitation of microgram quantities of protein utilizing the Principle of Protein-Dye Binding. *Anal. Biochem.*, 72:248-254.

[157] Szmant-Froelich A. M., Reutter M., Riggs L. (1985) Sexual reproduction of *Favia fragum* (Esper): Lunar patterns of gametogenesis, embryogenesis and planulation in Puerto Rico. *Bull. Mar. Sci.*, 37: 880-892.

[158] Corredor J. E., Morell J. M., Klekowski E. J. et al. (1995) Mangrove genetics. III. Pigment fingerprints for chlorophyll-deficient mutants. *Int. J. Plant Sci.*, 156:55-60.

[159] Lichtenhaler H. K., Wellburn A. R. (1983) Determination of total carotenoids and chlorophylls a and b of leaf extracts in different solvents. *Biochem. Soc. T.,* 603:591-592.

[160] Lovelock C. E., Clough B. F., Woodrow I. E. (1992) Distribution and accumulation of ultraviolet-radiation-absorbing compounds in leaves of tropical mangroves. *Planta.,* 188:143-154.

[161] Banaszak A. T., Lesser M. P., Kuffner I. B. (1998) Relationship between ultraviolet (UV) radiation and mycosporine-like amino acids (MAAs) in marine organisms. *Bull. Mar. Sci.,* 63:617-628.

[162] Dieppa-Ayala A. (1996) Bio-optical variability in tropical oceanic waters. MS Thesis. University of PR. 90pp.

[163] García-Urueña R. (2004) Dinámica de crecimiento de tres especies de coral en relación a las propiedades ópticas del agua. PhD Thesis. University of PR. 151pp.

[164] Shick J. M., Romaine-Lioud S., Ferrier-Pagès C. et al. (1999) Ultraviolet-B radiation stimulates shikimate pathway-dependent accumulation of mycosporine-like amino acids in the coral *Stylophora pistillata* despite decreases in its population of symbiotic dinoflagellates. *Limnol. Oceanogr.*, 44:1667-1682.

[165] Worrest R. C., Van Dyke H., Thomson B. E. (1978) Impact of enhanced simulated solar ultraviolet radiation upon a marine community. *Photochem. Photobiol.*, 27:471-478.

[166] Sinha R. P., Kumar A., Tyagi M. B. et al. (2005) Ultraviolet-B-induced destruction of phycobiliproteins in cyanobacteria. *Mol. Biol. Plants*, 11:313-319.

[167] Karsten U., Sawall T., Hanelt D. et al. (1998) An inventory of UV-absorbing mycosporine-like amino acids in macroalgae from polar to warm-temperate regions. *Botanica Marina*, 41:443-453.

[168] Kuffner I. B., Ondrusek M. E., Lesser M. P. (1995) Distribution of mycosporine-like amino acids in the tissues of Hawaiian Scleractinia: a depth profile. In: Gulko D and Jokiel PL (eds) Ultraviolet Radiation and Coral Reefs. HIMB Tech. Report #41. UNIHI-Sea Grant-CR-95-03. Pp.77-85.

[169] Shick J. M., Lesser M. P., Dunlap W. C. et al. (1995) Depth-dependent responses to solar ultraviolet radiation and oxidative stress in the zooxanthellate coral *Acropora microphthalma*. *Mar. Biol.*, 122:41-51.

[170] Lesser M. P. (1996) Exposure of symbiotic dinoflagellates to elevated temperatures and ultraviolet radiation causes oxidative stress and inhibits photosynthesis. *Limnol. Oceanogr.*, 41:271-283.

[171] Zhang P.-Y., Yu J., Tang X.-X. (2005) UV-B radiation suppresses the growth and antioxidant systems of two marine microalgae, *Platymonas subcordiformis* (Wille) Hazen and *Nitzschia closterium* (Ehrenb.) *W. Sm. J. Integrat. Plant Biol.*, 47:683-691.

[172] Weis V. M. (2008) Cellular mechanisms of Cnidarian bleaching: stress causes the collapse of symbiosis. *J. Exp. Biol.,* 211:3059-3066.

[173] Adams N. L., Shick J. M. (1996) Mycosporine-like amino acids provide protection against ultraviolet radiation in eggs of the green sea urchin, *Strongylocentrotus droebachiensis*. *Photochem. Photobiol.*, 64:149-158.

[174] Kubasek W. L., Shirley B. W., McKillop A. et al. (1992) Regulation of flavonoid biosynthetic genes in germinating *Arabidopsis* seedlings. *Plant Cell*, 4:1229-1236.

[175] Yakovleva I., Hidaka M. (2004) Diel fluctuations of mycosporine-like amino acids in shallow-water scleractinian corals. *Mar. Biol.*, 145:863-873.

[176] Rowan R. (1998) Diversity and ecology of zooxanthellae on coral reefs. *J. Phycol.,* 34:407-417.

[177] Omori M., Fukami H., Kobinata H. et al. (2001) Significant drop of fertilization of *Acropora* corals in 1999: an after-effect of heavy coral bleaching? *Limnol. Oceanogr.,* 46:704-706.

[178] Szmant A. M., Gassman N. J. (1990) The effects of prolonged "bleaching" on the tissue biomass and reproduction of the reef coral *Montastraea annularis*. *Coral Reefs*, 8:217-224.

[179] Rinchevich B., Loya Y. (1979) The reproduction of the Red Sea coral *Stylophora pistillata*. II. Synchronization in breeding and seasonality of planulae shedding. *Mar. Ecol. Prog. Ser.,* 1:145-152.

[180] Ward S. (1995) Two patterns of energy allocation for growth, reproduction and lipid storage in the scleractinian coral *Pocillopora damicornis*. *Coral Reefs*, 14:87-90.

[181] Soong K. (1991) Sexual reproductive patterns of shallow-water reef corals in Panama. *Bull. Mar. Sci.*, 49:832-846.

[182] Baker A. C. (1995) Solar UV-A inhibition of planula larvae in the reef-building coral *Pocillopora damicornis*. In: Gulko D, Jokiel PL (eds) Ultraviolet Radiation and Coral Reefs. HIMB Tech. Report #41. UNIHI-Sea Grant-CR-95-03. Pp. 149-163.

[183] Carpenter K. E., Abrar M., Aeby G. et al. (2008) One-third of reef-building corals face elevated extinction risk from climate change and local impacts. *Science,* 321:560-563.

[184] Hoegh-Guldberg O., Mumby P. J., Hooten A. J. et al. (2007) Coral reefs under rapid climate change and ocean acidification. *Science,* 318:1737-1742.

[185] Pelejero C., Calvo E., McCulloch M. T. et al. (2005) Preindustrial to modern interdecadal variability in coral reef pH. *Science,* 309:2204-2207.

[186] Orr J. C., Fabry V. J., Aumont O. et al. (2005) Anthropogenic ocean acidification over the twenty-first century and its impact on calcifying organisms. *Nature,* 437:681-686.

[187] Williamson C. E., Zagarese H. E. (2003) UVR effects on aquatic ecosystems: a changing climate perspective. In: Helbling E. W., Zagarese H. E. (eds) UV effects in aquatic organisms and ecosystems. Royal Society of Chemistry, Cambridge, UK. Pp. 547-567.

In: UV Radiation: Properties, Effects, and Applications
Editor: James A. Radosevich
ISBN: 978-1-63321-090-5

Chapter 4

MELATONIN: POSSIBLE ROLE AS LIGHT-PROTECTOR IN PLANTS

Marino B. Arnao* and Josefa Hernández-Ruiz
Department of Plant Physiology, University of Murcia, Murcia, Spain

ABSTRACT

Melatonin is a biogenic amine with hormonal nature found in animals. One of the most studied physiological functions is its role as a regulator of the sleep-wake cycles. In addition, melatonin is a potent antioxidant with demonstrated protective properties in various cellular dysfunctions and diseases. Its exceptional anti-oxidative qualities are mainly due to a cascade of reactions whereby generated-intermediates prolong and accentuate its anti-oxidant properties. In these reactions, light, especially UV radiation, seems to play a decisive role.

Melatonin was discovered in plants in 1995, and since then studies have focused on the identification and quantification of melatonin levels in many plant species and, also on the possible physiological roles that this indoleamine could have in plants. One of the aspects that has aroused much interest is their possible role as a regulator in light-dark cycles, as in animals, and as an indicator of light stress (at high PAR) and under UV radiation. In this work, the most significant data related to the effect that light, in natural and artificial conditions, can have on the regulation of endogenous levels of melatonin in plant tissues are presented. We discuss the physiological consequences that this natural anti-oxidant molecule has in stressed plants.

Keywords: Antioxidant; melatonin, phytomelatonin, plant growth, plant stress

* Corresponding author: Prof. Marino Bañón Arnao, Department of Plant Physiology, Faculty of Biology, University of Murcia, 30100-Murcia. Spain, Email: marino@um.es.

Abbreviations

AFMK	N^1-acetyl-N^2-formyl-5-methoxy-kynuramine
AMK	N^1-acetyl-5-methoxy-kynuramine
HIOMT	hydroxyindole O-methyltransferase
IAA	indolil-3-acetic acid
PAR	Photosynthetically Active Radiation
ROS	reactive oxygen
RNS	reactive nitrogen species
SNAT	serotonin N-acetyltransferase
T5H	tryptophan 5-hydroxylase decarboxylase
TDC	L-tryptophan decarboxylase
UV	Ultraviolet

4.0. Introduction

Melatonin (N-acetyl-5-methoxytryptamine) is an indolamine first purified from the bovine pineal gland in 1958 [1]. In recent years, melatonin has aroused a great deal of interest in plant biology due its broad distribution in the plant kingdom.

Melatonin is known to act as a biological modulator of mood, sleep, retina physiology, sexual behaviour, seasonal reproductive physiology, circadian rhythms and immunological enhancement. In mammals, melatonin is secreted from the pineal gland at night, from where it diffuses into the cerebro-spinal fluid and the blood stream, although levels quickly drop during the day. Thus, a role in photoperiodic regulation has been demonstrated because of the duration and timing of the melatonin signal [2-10].

Another recent and important action attributed to melatonin is its scavenging capacity against biological free radicals (anti-oxidant activity), such as reactive oxygen and nitrogen species (ROS/RNS), including the hydroxyl radical, singlet oxygen, peroxyl radical, hydrogen peroxide, peroxynitrite anion and nitric oxide, and protects cells, tissues from oxidative stress [11-19]. In this respect, it is interesting that melatonin shows certain peculiarities as an anti-oxidant. The melatonin molecule presents no pro-oxidative effects, while melatonin-intermediate products show antioxidant properties. There is an important synergistic action with other anti-oxidants, such as ascorbic acid, glutathione, etc. [20-22].

4.1. Melatonin Biosynthesis Pathways

The biosynthetic pathway of melatonin has been clearly determined in mammals, including humans, and other groups (amphibian, reptilian, avian) (*Figure 1*). Melatonin is synthetized from serotonin (5-hydroxytryptamine) via an acetylation reaction catalyzed by arylalkylamine N-acetyltransferase, also called serotonin N-acetyltransferase (SNAT) (EC 2.3.1.87), a regulated enzyme that controls melatonin biosynthesis. The last step is catalyzed by hydroxyindole O-methyltransferase (HIOMT) (EC 2.1.1.4). Serotonin is generated from 5-hydroxytryptophan in a reaction catalyzed by the aromatic L-aminoacid decarboxylase (EC

4.1.1.28). Tryptophan is hydroxylated by the action of the tryptophan 5-hydroxylase (EC 1.14.16.4) to form 5-hydroxytryptophan. All the enzymes mentioned have been isolated and characterized in animal tissues, but not all of them in plant material [4,11,23-31].

In plants, the melatonin biosynthesis pathway is not clearly understood, although it is known to start with tryptophan, with serotonin as an intermediate and N-acetylserotonin as direct precursor of melatonin (see *Figure 1*). This is similar to the way that it is produced in animals.

An additional pathway appears in plants, where the formation of tryptamine from tryptophan by the action of L-tryptophan decarboxylase (EC 4.1.1.28) has been described (*Figure 1*). Tryptamine is then transformed into serotonin by the action of tryptophan 5-hydroxylase (EC 1.14.16.4). Another possibility not described is that tryptamine also acts as substrate of SNAT, forming N-acetyltryptamine, which can be hydroxylated to form N-acetylserotonin. Thus, more alternatives in the biosynthesis of melatonin in plants would be possible than in mammals [32-38].

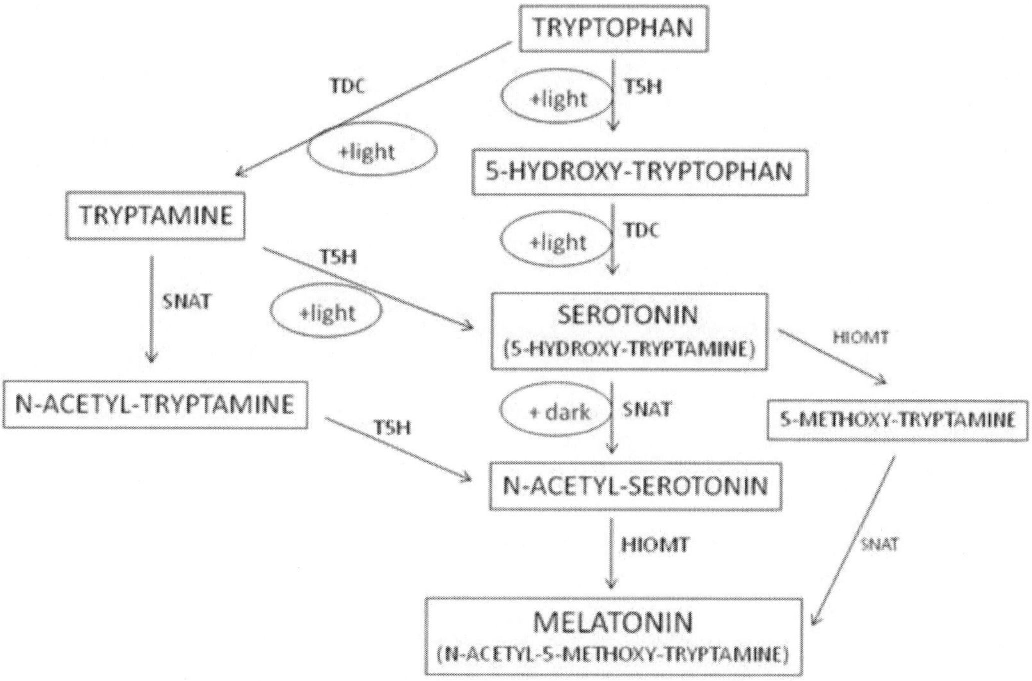

Figure 1. Biosynthetic pathway of melatonin in mammals and plants. The enzymes for the respective steps are: T5H=Tryptophan 5-hydroxylase (EC 1.14.16.4); TDC=Tryptophan decarboxylase (EC 4.1.1.28); SNAT=Serotonin N-acetyltransferase (EC 2.3.1.87); HIOMT=Hydroxyindole O-methyltransferase (EC 2.1.1.4).

4.2. Melatonin Catabolism: Enzymatic and Non-Enzymatic Pathways

The amphiphilic nature of melatonin and its anti-oxidative properties establish complex catabolic pathways (see Figure 2). In mammals, melatonin is converted into aqueous-soluble metabolites for excretion via the kidneys. The main pathway leading to melatonin excretion is its transformation into 6-hydroxymelatonin by cytochrome P450 in the liver (and other organs), where it is also conjugated with sulfate, forming 6-hydroxymelatonin sulfate, which is excreted in the urine [39,40]. Melatonin can also be degraded through the kynuric pathway, in which melatonin is transformed into several hydroxylated compounds such as 2-hydroxymelatonin, 4-hydroxymelatonin, and cyclic-3-hydroxymelatonin. All these compounds can be formed by different radical oxygen-species (ROS) and also by UV-B [41-44]. UV-B radiation provokes the accumulation of N^1-acetyl-N^2-formyl-5-methoxy-kynuramine (AFMK), a melatonin catabolic product that evolves to N^1-acetyl-5-methoxy-kynuramine (AMK) through the action of a variety of enzymes (catalase, arylamine-formamidase, peroxidase) and in the presence of UV and ROS. AFMK can also be formed directly from melatonin by the action of multiple enzymes such as arylacylamidase, myeloperoxidase, horseradisperoxidase, hemoperoxidase, indolamine dioxygenase, cytochrome c, haemoglobin or cytochrome P450. Lastly, melatonin can be enzymatically degraded in the so-called indolic pathway (see Figure 2), in which 5-methoxytryptamine is formed by a melatonin deacetylase, and 5-methoxytryptophol and 5-methoxyindolil-acetic acid appear from 5-methoxyindolil-acetaldehyde [11,45-50].

As can be seen in Figure 2, UV-B radiation (alone or together with ROS) provokes the generation of multiple melatonin metabolites, generally due to the stabilization of melatonin radical species. Surprisingly, many of the intermediate compounds have anti-oxidant activity similar to or even higher than that of melatonin itself. This effect means that the whole "melatonin metabolite pool" works cooperatively as an excellent endogenous synergistic anti-oxidant complex [12,51-55].

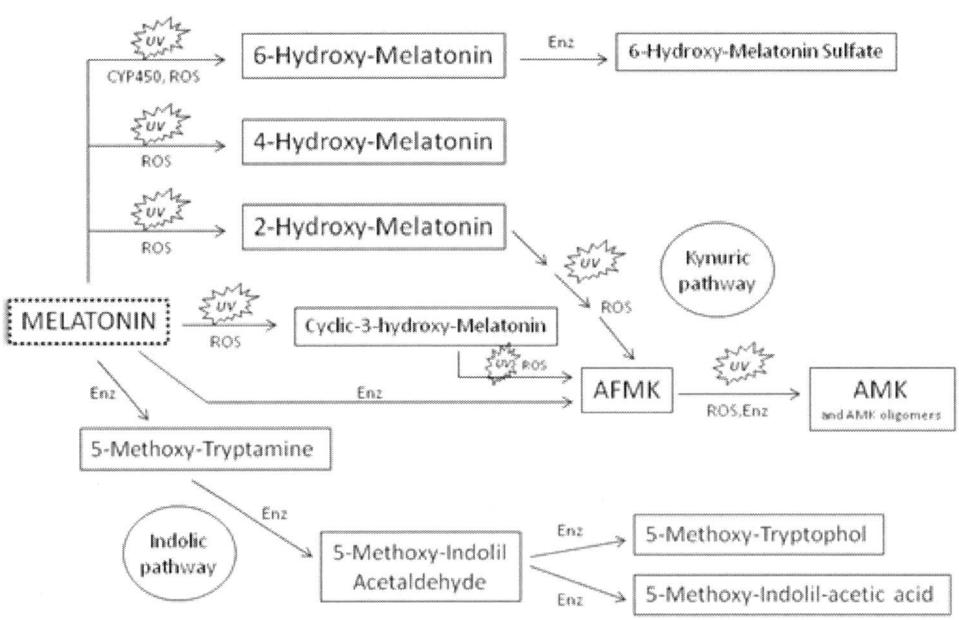

Figure 2. Catabolic pathway of melatonin with enzymatic (Enz), photocatalytic (UV) and free-radical steps (ROS). Kynuric and indolic pathway details are shown (see text). AFMK=N1-acetyl-N2-formyl-5-methoxykynuramine; AMK=N1-acetyl-5-methoxykynuramine.

4.3. MELATONIN IN PLANTS

In vascular plants, melatonin was first detected in 1995, when several groups found it, mainly in mono- and dicotyledon edible plant families [56-58]. Since then, melatonin has been detected and quantified in roots, shoots, leaves, fruits and seeds of a considerable variety of plant species. The levels of melatonin in plant organs vary considerably, from picograms to micrograms per gram of plant material. Generally, seeds and leaves present the highest level of melatonin and fruits the lowest. Variables such as the variety, the cultivation conditions and the extraction protocols used are important aspects for an accurate quantification of melatonin levels in tissues and plant organs. Until now, studies of melatonin in plants have concentrated on measuring its levels and on ascertaining its possible implications in human food, since melatonin from plant foods is absorbed from the gastrointestinal tract and incorporated in the blood stream; it also crosses the blood-brain barrier and the placenta, while at subcellular level, it is mainly incorporated in the nucleus and mitochondria. The possibility of modulating blood melatonin levels in mammals through the ingestion of plant-derived foods has led to numerous studies in which melatonin is considered as a health-related phytochemical. Also, the cultivation of plants with high endogenous melatonin levels to a best environmental adaptation is an interesting focus [59-74].

4.4. MELATONIN FUNCTIONS IN PLANTS

Melatonin has been shown to be involved in several physiological aspects in plants, acting as circadian regulator, cytoprotector, growth promotor and in rhizogenesis, cellular expansion and stress-protection. In this respect, several reviews with summarized data can be consulted [33,34,75-80].

From a functional point of view, the roles of melatonin in plants can be classified into two modes of action: i) by direct action, working through melatonin receptors, and ii) by indirect action, independent of receptors, working as an anti-oxidative molecule.

Considerable data points to the possible role of melatonin as a plant growth regulator. Melatonin stimulates the growth of lupin hypocotyls and the coleoptiles of some monocots such as wheat, barley and oat [81,82]. In a similar way to the widely distributed auxin (indolil-3-acetic acid, IAA), melatonin is able to stimulate and inhibit the expansion of roots, depending on its concentration. In relevant experiments, Arnao et al. [83] showed that melatonin provokes the generation of new roots in aerial organs (adventitious roots) or in established roots (lateral roots), confirming the great capability of melatonin as a stimulating agent of rhizogenesis. This organogenic capacity of melatonin was also showed on commercial sweet cherry rootstocks [84]. Also, melatonin was able to promote the cellular expansion of cotyledons in *Lupinus albus* [85].

Many authors have described the possible role of melatonin as a regulator of light-dark cycles in plants. This function has been clearly established in mammals, where the role in photoperiodic regulation has been demonstrated based on the duration and timing of the melatonin signal [2-4]. In plants, some oscillations in endogenous melatonin levels have been described in *Chenopodium rubrum* [86-88], hyacinth [49] and, recently, in grapes [89], pointing to its possible role as a light-dark regulator. All these data and more, indicate that a direct interaction of melatonin-receptor(s) would be necessary for melatonin to work as a plant regulator. In animals, various subtypes of melatonin membrane-bound receptors (MT1 and MT2) have been characterized, and their genes have been sequenced. These melatonin receptors, which act through coupled G proteins, are widely distributed throughout the body and their expression varies between different target organs [90-92]. So, melatonin activates several nuclear receptors (RZR/ROR) [93-95]. However, to date, no melatonin receptor has been described in plants. Another aspect that has not been studied is the possible interaction of melatonin with the intracellular auxin receptor TIR-1 [96].

Protection against free radical damage is another important role of melatonin, an indirect action-mode that has been taken on more importance in recent years. Its anti-oxidative potential has been demonstrated in numerous studies, mainly in animals but, to a lesser extent, in plants, yeasts and bacteria [7,78]. In plants it has been [97] demonstrated that melatonin treatment of *Daucus carota* L. cultured root cells attenuated cold-induced cell apoptosis. Studying the possible protective role of melatonin during flower development of *Hypericum perforatum* L., it was shown that [98] serotonin and melatonin are present at higher concentrations at various stages; the highest melatonin level coinciding with the maximum regeneration potential of isolated anthers, suggesting that melatonin acts as a stress-protecting agent, providing an adaptative mechanism to ensure reproduction. The role of melatonin as a stress-protector has been analyzed. Arnao et al. [99] demonstrated that barley plants subjected to different treatments with chemical agents, such as hydrogen peroxide, sodium chloride and zinc sulphate, at different concentrations, stimulates the endogenous level of melatonin in roots, in a time- and concentration- dependent manner. This increase in melatonin levels probably plays an important role in the anti-oxidative defense against chemical-induced stress. It has been [100] demonstrated that melatonin added to the soil enhanced the tolerance and survival of pea plants against copper contamination, and these authors proposed that plants with high levels of melatonin can be used for phytoremediation purposes. In addition, the incubation of *Brassica oleracea* L. seeds with exogenous melatonin solutions (hydropriming) had a protective effect against copper, a highly toxic metal [101]; this was also observed in cucumber seeds under chilling stress [102]. All these data point to a role for melatonin as anti-oxidative protector against harmful toxic metals, UV radiation, and/or ROS. These findings suggesting that melatonin helps ensure a proper redox status in plant cells. Also, melatonin plays a significant role in the natural senescence process, and it has been reported that barley and apple leaves treated with melatonin delay the dark-induced senescence in a concentration dependent manner, while slowing down chlorophyll loss in detached leaves [103-104].

4.5. Light As Modulator of Melatonin Levels

Light plays a relevant role in many aspects: i) as modulator of some enzymes of the biosynthetic pathway of melatonin (see *Figure 1*); ii) as activator of many steps in the melatonin catabolic pathways (see *Figure 2*); and iii) as a decisive factor in many physiological responses mediated by melatonin, mainly in light-stressing situations.

In regards the first point, unlike in animals, the enzymes tryptophan 5-hydroxylase (T5H) and aromatic L-amino acid decarboxylase (L-tryptophan decarboxylase, TDC) are regulated by light (see *Figure 1*). In experiments with detached rice leaves, it was [105] demonstrated that during the senescence process, rice leaves maintained under constant light increase their melatonin level 150-fold with respect to leaves under constant darkness. The levels of tryptamine, serotonin and N-acetylserotonin also strongly increase. The transcripts and proteins of T5H and TDC (see *Figure 1*) were high in constant light, but low in darkness, indicating that the expression of these enzymes were regulated by light. However, HIOMT transcript of rice was not shown to be induced under light. In animals, serotonin N-acetyltransferase (SNAT) is a regulated enzyme that is inhibited by light (see *Figure 1*). In the dark, the enzyme actively provides N-acetylserotonin to HIOMT, which increases the level of melatonin in blood at the onset of the night period [28-31]. Therefore, these appear to be significant differences in the regulation of the melatonin biosynthetic pathway by light between animals and plants. Nevertheless, until all the enzymes involved in the melatonin biosynthesis pathway in plants can be isolated and characterized, the regulation of the pathway may not be known exactly.

In regards to the catabolic pathway, many steps are influenced by UV radiation (see *Figure 2*). In the kynuric pathway (enzymatic, pseudoenzymatic, photocatalytic, and free-radical reactions), for example, many melatonin intermediates are generated by the joint action of ROS and UV radiation [11-17,45-48]. The secondary, tertiary and quaternary metabolites derived from melatonin are highly effective since, together, they behave as a 'free radical scavenging cascade', protecting tissues from oxidative stress [52-54].

Thirdly, a possible role for melatonin as UV light protector has been suggested by some authors. Researchers have [106] described the high melatonin content of Alpine and Mediterranean plants exposed to high UV radiation, compared with the same plants grown in low UV radiation conditions. Also, it has been reported that there is [49] a possible relation between UV exposure and melatonin levels in the water hyacinth *Eichhornia crassipes*, finding that plants grown in sunlight showed a higher melatonin content than plants grown in a controlled chamber. Similar data were presented by others [107] in *Lycopersicon esculentum*. In this case, tomato plants cultivated indoors under moderate conditions had lower melatonin content than those cultivated outdoors in more variable conditions. The leaves of plants cultivated in outdoor conditions had roughly ten times more melatonin than leaves of plants cultivated *in vitro* and seven times more than the leaves of chamber-grown plants. Recently, a 10-fold difference in melatonin levels was measured in grape berries exposed to sunlight compared with berries grown in the shade [89]. In *Glycyrrhyza uralensis* roots, the melatonin content doubled as a result of UV-B radiation treatment, reaching a surprising 80 μg/g FW [108]. Lastly, transgenic *Nicotiana sylvestris* plants expressing SNAT and HIOMT genes exposed to UV-B radiation, showed lower DNA damage due to the relevant increase in their melatonin levels [109]. In all these cases, the authors suggested that melatonin can act as an anti-oxidant compound that prevents plants from the UV injuries and the oxidative stress caused by excess radiation.

To conclude, a whole body of data suggests that stress agents or stress situations (UV radiation, solar light, chemical agents, low temperatures, water deficit, and biotic stressors) can induce the biosynthesis of melatonin in plants, possibly through the upregulation of enzyme(s) of the melatonin biosynthesis pathway. This increase in melatonin levels probably plays an important role in the anti-oxidative defense against such stressor agents, detoxifying ROS. This behavior might form a part of a central strategy whereby the plant correctly regulates the redox status of cells.

References

[1] Lerner AB, Case JD, Takahashi Y, Lee TH, Mori W, (1958), Isolation of melatonin, a pineal factor that lightens melanocytes, *J. Am. Soc.*, 80:2587.

[2] Bartness TJ, Goldman BD, (1989), Mammalian pineal melatonin: a clock for all seasons, *Experientia*, 45:939-945.

[3] Reiter RJ, (1993), The melatonin rhythm: both a clock and a calendar, *Experientia*, 49:654-664.

[4] Yu HS, Reiter RJ, (1993), Melatonin: Biosynthesis, Physiological Effects, and Clinical Applications, CRC Press, Boca Raton, FL.

[5] Reiter RJ, (1986), Normal patterns of melatonin levels in the pineal gland and body fluids of humans and experimental animals, *J. Neural Trans*, S21:35–54.

[6] Reiter RJ, (1991), Pineal melatonin: cell biology of its synthesis and of its physiologic interactions, *Endocrine Rev.*, 12:151–180.

[7] Tan DX, Hardeland R, Manchester LC, Paredes SD, Korkmaz A, Sainz RS, Mayo JC, Fuentes-Broto L, Reiter RJ, (2010), The changing biological roles of melatonin during evolution: from an antioxidant to signals of darkness, sexual selection and fitness, *Biol. Rev.*, 85:607-623.

[8] Carrillo-Vico A, Guerrero J, Lardone P, Reiter RJ, (2012), A review of the multiple actions of melatonin on the immune system, *Endocrine*, 27:189-200.

[9] Shiu SY, (2007), Towards rational and evidence-based use of melatonin in prostate cancer prevention and treatment, *J. Pineal. Res.,* 43:1-9.

[10] Humbert W, Pevet P, (1994), The decrease of pineal melatonin production with age, *Ann. NY Acad. Sci.*, 719:43-61.

[11] Tan D, Manchester L, Terron M, Flores L, Reiter RJ, (2007), One molecule, many derivates: A never-ending interaction of melatonin with reactive oxygen and nitrogen species? *J. Pineal. Res.,* 42:28-42.

[12] Galano A, Tan DX, Reiter RJ, (2011), Melatonin as a natural ally against oxidative stress: a physic-chemical examination, *J. Pineal. Res.*, 51:1-16.

[13] Tan DX, Chen LD, Poeggeler B, Manchester LC, Reiter RJ, (1993), Melatonin: a potent, endogenous hydroxyl radical scavenger, *Endocrine* J, 1:57-60.

[14] Beyer CE, Steketee JD, Saphier D, (1998), Antioxidant properties of melatonin: an emerging mystery, *Biochem. Pharmacol.,* 56:1265-1272.

[15] Reiter RJ, Tan DX, Acuña-Castroviejo D, Burkhardt S, Karbownik M, (2000), Melatonin: mechanisms and actions as an antioxidant, Curr Top Biophys, 24:171-183.

[16] Reiter RJ, Tan DX, Manchester LC, Qi W, (2001), Biochemical reactivity of melatonin with reactive oxygen and nitrogen species, *Cell Biochem. Biophys.,* 34:237-256.

[17] Tan DX, Manchester LC, Reiter RJ, Plummer BF, Limson J, Weintraub ST, Qi W, (2000), Melatonin directly scavenges hydrogen peroxide: a potentially new metabolic pathway of melatonin biotransformation, *Free Rad. Biol. Med.,* 29:1177-1185.

[18] Reiter RJ, Burkhardt S, Cabrera J, Garcia JJ, (2002), Beneficial neurobiological effects of melatonin under conditions of increased oxidative stress, *Curr. Med. Chem.,* 2:45-58.

[19] Rodriguez C, Mayo JC, Sainz RM, Antolín I, Herrera F, Martín V, Reiter RJ, (2004), Regulation of antioxidant enzymes: a significant role for melatonin, *J. Pineal. Res.*, 36:1-9.

[20] Poeggeler B, Reiter RJ, Hardeland R, Sewerynek E, Melchiorri D, Barlow-Walden LR, (1995), Melatonin, a mediator of electron transfer and repair reactions, acts synergistically with the chain-breaking antioxidants ascorbate, trolox and glutathione, *Neuroendocrinol. Lett*, 2:87-92.

[21] Cano A, Alcaraz O, Arnao MB, (2003), Free radical-scavenging activity of indolic compounds in aqueous and ethanolic media, *Anal. Bioanal. Chem.*, 376:33-37.

[22] Tan DX, Hardeland R, Manchester LC, Poeggeler B, Lopez-Burillo S, Mayo JC, Sainz RM, Reiter RJ, (2003), Mechanistic and comparative studies of melatonin and classic antioxidants in terms of their interactions with the ABTS cation radical, *J. Pineal. Res.*, 34:249-259.

[23] Abe M, Itoh M, Miyata M, Ishikawa S, Sumi Y, (1999), Detection of melatonin, its precursor and related enzyme activities in rabbit lens, *Exp. Eye Res.*, 68:255-262.

[24] Tan DX, Manchester LC, Reiter RJ, Qi WB, Zhang M, Weintraub ST, Cabrera J, Sainz RM, Mayo JC, (1999), Identification of highly elevated levels of melatonin in bone marrow: its origin and significance, *Biochim. Biophys. Acta*, 1472:206-214.

[25] Conti A, Conconi S, Hertens E, Skwarlo-Sonta K, Markowska M, Maestroni GJM, (2000), Evidence for melatonin synthesis in mouse and human bone marrow cells, *J. Pineal. Res.,* 28:193-202.

[26] Hardeland R, Reiter RJ, Poeggeler B, Tan DX, (1993), The significance of the metabolism of the neurohormone melatonin: antioxidative protection and formation of bioactive substances, *Neurosci. Biobehav. Rev.*, 17:347-357.

[27] Sugden D, (1989), Melatonin biosynthesis in the mammalian pineal gland, *Experientia*, 45:922-932.

[28] Vollrath L, Huesgen A, (1988), Response of pineal serotonin N-acetyltransferase activity in male guinea pigs exposed to light pulses at night, *J. Neural. Transm.*, 72:55-66.

[29] Foulkes NS, Borjigin J, Snyder SH, Sassone-Corsi P, (1997), Rhythmic transcription: the molecular basis of circadian melatonin synthesis, *Trends Neurosci.*, 20:487-492.

[30] Karolczak M, Kore HW, Stehle JH, (2005), The rhythm and blues of gene expression in the rodent pineal gland, *Endocrine*, 27:89-100.

[31] Liu T, Borjigin J, (2005), N-acetyltransferase is not the rate-limiting enzyme of melatonin synthesis at night, *J. Pineal. Res.*, 39:91-96.

[32] Murch S, KrishnaRaj S, Saxena PK, (2000), Tryptophan is a precursor for melatonin and serotonin biosynthesis in in vitro regenerated St. John's wort (*Hypericum perforatum L. cv. Anthos*) plants, *Plant Cell Rep.*, 19:698-704.

[33] Arnao MB, Hernandez-Ruiz, J, (2006), The physiological function of melatonin in plants, *Plant Signal Behav.,* 1:89-95.

[34] Posmyk M, Janas KM, (2009), Melatonin in plants, *Acta Physiol. Plant*, 31:1-11.

[35] Facchini PJ, Huber-Allanach KL, Tari LW, (2000), Plant aromatic L-amino acid decarboxylases: evolution, biochemistry, regulation, and metabolic engineering applications, *Phytochemistry*, 54:121–138.

[36] Fujiwara T, Maisonneuve S, Isshiki M, Mizutani M, Chen L, Wong HL, Kawasaki T, Shimamoto K, (2010), Sekiguchi lesion gene encodes a cytochrome P450 monooxygenase that catalyzes conversion of tryptamine to serotonin in rice, *J. Biol. Chem.*, 285:11308–11313.

[37] Kang K, Kong K, Park S, Natsagdorj U, Kim YS, Back K, (2011), Molecular cloning of a plant N-acetylserotonin methyltransferase and its expression characteristics in rice, *J. Pineal. Res.,* 50:304–309.

[38] Schröder P, Abele C, Gohr P, Stuhlfauth-Roisch U, Grosse W, (1999), Latest on enzymology of serotonin biosynthesis in walnut seeds, *Adv. Exp. Med. Biol.*, 467:637-644.

[39] Ma X, Idle JR, Krausz KW, Gonzalez FJ, (2005), Metabolism of melatonin by human cytochromes P450, *Drug Metab. Disp.*, 33:489-494.

[40] Lahiri DK, Ge YW, Sharman EH, Bondy SC, (2004), Age-related changes in serum melatonin in mice: higher levels of combined melatonin and 6-hydroxymelatonin sulfate in the cerebral cortex than serum, hearth, liver and kidney tissues, *J. Pineal. Res.*, 36:217-223.

[41] Tan DX, Manchester LC, Reiter RJ, Plummer BF, Limson J, Weintraub ST, Qi WB, (2000), Melatonin directly scavenges hydrogen peroxide: A potentially new metabolic pathway of melatonin biotransformation, *Free Rad. Biol. Med.*, 29:1177-1185.

[42] Horstman JA, Wrona WZ, Dryhurst G, (2002), Further insights into the reaction of melatonin with hydroxyl radical, *Bioorg. Chem.,* 30:371-382.

[43] Zhang H, Squadrito GL, Uppu R, Pryor WA, (1999), Reaction of peroxynitrite with melatonin: a mechanistic study, *Chem. Res. Toxicol.,* 12:526-534.

[44] Dellegar SM, Murphy SA, Bourne AE, DiCesare JC, Purser GH, (1999), Identification of the factors affecting the rate of deactivation of hypochlorous acid by melatonin, *Biochem. Biophys. Res. Commun.*, 257:431-439.

[45] Fischer TW, Sweatman TW, Semak I, Sayre RM, Worstman J, Slominski A, (2006), Constitutive and UV-induced metabolism of melatonin in keratinocytes and cell-free systems, *FASEB J.*, 20:1564-1566.

[46] Slominski A, Tobin DJ, Zmijewski MA, Worstman J, Paus R, (2007), Melatonin in the skin: synthesis, metabolism and functions, *Trends Endocr. Metab.*, 19:17-24.

[47] De Almeida EA, Martinez GR, Klitzke CF, De Medeiros MH, Di Mascio P, (2003), Oxidation of melatonin by singlet molecular oxygen produces N1-acetyl-N2-formyl-5-methoxykynuramine, *J. Pineal. Res.*, 35:131-137.

[48] Tan DX, Manchester LC, Burkhardt S, Sainz RM, Mayo JC, Kohen R, Shohami E, Huo YS, Hardeland R, Reiter RJ, (2001), N1-acetyl-N2-formyl-5-methoxy-kynuramine, a biogenic amine and melatonin metabolite, functions as a potent antioxidant, *FASEB J.,* 15:2294-2296.

[49] Tan DX, Manchester LC, Di Mascio P, Martinez GR, Prado FM, Reiter RJ, (2007), Novel rhythms of N1-acetyl-N2-formyl-5-methoxykynuramine and its precursor melatonin in water hyacinth: importance for phytoremediation, *FASEB J.*, 21:1724-1729.

[50] Mayo JC, Sainz RM, Tan DX, Hardeland R, Leon J, Rodriguez C, Reiter RJ, (2005), Anti-inflammatory actions of melatonin and its metabolites, N1-acetyl-N2-formyl-5-methoxykynuramine (AFMK) and N1-acetyl-5-methoxykynuramine (AMK), in macrophages, *J. Neuroimmunol.*, 165:139-149.

[51] Schaefer M, Hardeland R, (2008), The melatonin metabolite N1-acetyl-5-methoxykynuramine is a potent singlet oxygen scavenger, J Pineal Res, 46:49-52.

[52] Tan DX, Manchester LC, Reiter RJ, Qi WB, Karbownik M, Calvo JR, (2000), Significance of melatonin in antioxidative defence system: reactions and products, *Biol. Signal Recep.*, 9:137–159.

[53] Tan DX, Reiter RJ, Manchester LC, Yan MT, El-Sawi M, Sainz RM, Mayo JC, Kohen R, Allegra M, Hardeland R, (2002), Chemical and physical properties and potential mechanisms: melatonin as a broad spectrum antioxidant and free radical scavenger, *Curr. Top. Med. Chem.*, 2:181–197.

[54] Hardeland R, (2005), Antioxidative protection by melatonin: multiplicity of mechanisms from radical detoxification to radical avoidance, *Endocrinology*, 27:119-130.

[55] Hardeland R, Pandi-Perumal SR, (2005), Melatonin, a potent agent in antioxidative defense: actions as a natural food constituent, gastrointestinal factor, drug and prodrug, *Nutr. Metab.*, 2:1-15.

[56] Dubbels R, Reiter RJ, Klenke E, Goebel A, Schnakenberg E, Ehlers C, Schiwara HW, Schloot W, (1995), Melatonin in edible plants identified by radioimmunoassay and high performance liquid chromatography-mass spectrometry, *J. Pineal. Res.*, 18:28-31.

[57] Kolar J, Machackova I, Illnerova H, Prinsen E, van Dongen W, van Onckelen HA, (1995), Melatonin in higher plant determined by radioimmunoassay and high performance liquid chromatography-mass spectrometry, *Biol. Rhythm. Res.*, 26:406-409.

[58] Hattori A, Migitaka H, Iigo M, Itoh M, Yamamoto K, Ohtani-Kaneko R, Hara M, Suzuki T, Reiter RJ, (1995), Identification of melatonin in plants and its effects on plasma melatonin levels and binding to melatonin receptors in vertebrates, *Biochem. Mol. Biol. Int.*, 35:627-634.

[59] Reiter RJ, Tan DX, Burkhardt S, Manchester LC, (2001), Melatonin in plants, *Nutr. Rev.*, 59:286-290.

[60] Tettamanti C, Cerabolini B, Gerola P, Conti A, (2000), Melatonin identification in medicinal plants, *Acta Phytother.*, 3:137–144.

[61] Benot S, Goberna R, Reiter RJ, Garcia-Mauriño S, Osuna C, Guerrero JM, (1999), Physiological levels of melatonin contribute to the antioxidant capacity of human serum, *J. Pineal. Res.*, 27:59-64.

[62] Reiter RJ, Manchester LC, Tan DX, (2005), Melatonin in walnuts: influence on levels of melatonin and total antioxidant capacity of blood, *Nutrition*, 21:920-924.

[63] Chen G, Huo Y, Tan DX, Liang Z, Zhang W, Zhang Y, (2003), Melatonin in Chinese medicinal herbs, *Life Sci.*, 73:19–26.

[64] Burkhardt S, Tan DX, Manchester LC, Hardeland R, Reiter RJ, (2001), Detection and quantification of the antioxidant melatonin in Montmorency and Balaton tart cherries (*Prunus cerasus*), *J. Agr. Food Chem.*, 49:4898–4902.

[65] Arnao MB, Hernández-Ruiz J, (2009), Assessment of different sample processing procedures applied to the determination of melatonin in plants, *Phytochem. Anal*, 20:14–18.

[66] Hernández-Ruiz J, Arnao MB, (2008), Distribution of melatonin in different zones of lupin and barley plants at different ages in the presence and absence of light, *J. Agr. Food Chem.*, 56:10567–10573.

[67] Iriti M, Rossoni M, Faoro F, (2006), Melatonin content in grape: myth or panacea? *J. Sci. Food Agr.*, 86:1432-1438.

[68] Lamont KT, Somers S, Lacerda L, Opie LH, Lecour S, (2011), Is red wine a SAFE sip away from cardioprotection? Mechanisms involved in resveratrol- and melatonin-induced cardioprotection, *J. Pineal. Res.*, 50:374–380.

[69] Maldonado MD, Moreno H, Calvo JR, (2009), Melatonin present in beer contributes to increase the levels of melatonin and antioxidant capacity of the human serum, *Clinical Nutr.*, 28:188–191.

[70] Vitalini S, Gardana C, Zanzotto A, Fico G, Faoro F, Simonetti P, Iriti M, (2011), From vineyard to glass: agrochemicals enhance the melatonin content, total polyphenols and antiradical activity of red wines, *J. Pineal. Res.*, 51:278-285.

[71] Vitalini S, Gardana C, Zanzotto A, Simonetti P, Faoro F, Fico G, Iriti M, (2011), The presence of melatonin in grapevine (*Vitis vinifera* L.) berry tissues, *J. Pineal. Res.*, 51:331-337.

[72] Murch SJ, Alan AR, Cao J, Saxena PK, (2009), Melatonin and serotonin in flowers and fruits of *Datura metel* L., *J. Pineal. Res.*, 47:277–283.

[73] Okazaki M, Ezura H, (2009), Profiling of melatonin in the model tomato (*Solanum lycopersicum* L.) cultivar Micro-Tom, *J. Pineal. Res.*, 46:338–343.

[74] Reiter RJ, Tan DX, Manchester LC, Simopoulos AP, Maldonado MD, Flores LJ, Terron MP, (2007), Melatonin in edible plants (phytomelatonin): identification, concentrations, bioavailability and proposed functions, *World Rev. Nutr. Diet.*, 97:211–230.

[75] Paredes SD, Korkmaz A, Manchester LC, Tan DX, Reiter RJ, (2009), Phytomelatonin: a review, *J. Exp. Bot.*, 60:57–69.

[76] Kolar J, Machackova I, (2005), Melatonin in higher plants: occurrence and possible functions, *J. Pineal. Res.*, 39:333-341.

[77] Hardeland R, Pandi-Perumal SR, Poeggeler B, (2007), Melatonin in plants. Focus on a vertebrate night hormone with cytoprotective properties, *Funct. Plant Sci. Biotech.*, 1:32–45.

[78] Hardeland R, Poeggeler B, (2003), Non-vertebrate melatonin, *J. Pineal. Res.*, 34:233–241.

[79] Balzer I, Hardeland R, (1996), Melatonin in algae and higher plants. Possible new roles as a phytohormone and antioxidant, *Bot. Acta*, 109:180–183.

[80] Tan DX, Hardeland R, Manchester LC, Korkmaz A, Ma S, Rosales-Corral S, Reiter RJ, (2012), Functional roles of melatonin in plants, and perspectives in nutritional and agricultural science, *J. Exp. Bot.*, 63:577–597.

[81] Hernández-Ruiz J, Cano A, Arnao MB, (2004), Melatonin: a growth-stimulating compound present in lupin tissues, *Planta*, 220:140-144.

[82] Hernández-Ruiz J, Cano A, Arnao MB, (2005), Melatonin acts as a growth-stimulating compound in some monocot species, *J. Pineal. Res.*, 39:137-142.

[83] Arnao MB, Hernández-Ruiz J, (2007), Melatonin promotes adventitious- and lateral root regeneration in etiolated hypocotyls of *Lupinus albus* L., *J. Pineal. Res.*, 42:147–152.

[84] Sarropoulou VN, Therios IN, Dimassi KN, (2011), Melatonin promotes adventitious root regeneration in *in vitro* shoot tip explants of the commercial sweet cherry rootstocks CAB-6P (*Prunus cerasus* L.), Gisela 6 (*P. cerasus* x *P. canescens*) and MxM 60 (*P. avium x P. mahaleb*), *J. Pineal. Res.*, 52:38-46.

[85] Hernández-Ruiz J, Arnao MB, (2008), Melatonin stimulates the expansion of etiolated lupin cotyledons, *Plant Growth Regul.*, 55:29-34.

[86] Kolar J, Johnson CH, Machackova I, (1999), Presence and possible role of melatonin in a short-day flowering plant *Chenopodium rubrum*, *Adv. Exp. Med. Biol.*, 460:391–393.

[87] Kolar J, Johnson CH, Machackova I, (2003), Exogenously applied melatonin affects flowering of the short-day plant *Chenopodium rubrum*, *Physiol. Plant.*, 118:605-612.

[88] Kolar J, Machackova I, Eder J, Prinsen E, Van Dongen W, Van Onckelen H, Illnerova H, (1997), Melatonin occurrence and daily rhythm in *Chenopodium rubrum*, *Phytochemistry*, 44:1047–1413.

[89] Boccalandro HE, Gonzalez CV, Wunderlin DA, Fernanda M, (2011), Melatonin levels determined by LC–ESI–MS/MS, deeply fluctuate during the day in *Vitis vinifera cv Malbec*: Evidence of its antioxidant role in fruits, *J. Pineal. Res.*, 51:226–232.

[90] Dubocovich ML, Markowska M, (2005), Functional MT1 and MT2 melatonin receptors in mammals, *Endocrine*, 27:101–110.

[91] Reppert SM, Weaver DR, Godson C, (1996), Melatonin receptors step into the light: cloning and classification of subtypes, *Trends Pharm. Sci.*, 17:100-102.

[92] Von Gall Ch, Stehle JH, Weaver DR, (2002), Mammalian melatonin receptors: molecular biology and signal transduction, *Cell Tiss. Res.*, 309:151-162.

[93] Carlberg C, (2000), Gene regulation by melatonin, *Ann. New York Acad. Sci.*, 917:387–396.

[94] Carlberg C, Wiesenberg I, (2005), The orphan receptor family RZR/ROR, melatonin and 5-lipoxygenase: an unexpected relationship, *J. Pineal. Res.*, 18:171–178.

[95] Wiesenberg I, Missbach M, Carlberg C, (1998), The potential role of the transcription factor RZR/ROR as a mediator of nuclear melatonin signaling, *Rest. Neurol. Neurosci.*, 12:143–150.

[96] Dharmasiri N, Dharmasiri S, Estelle M, (2005), The F-box protein TIR1 is an auxin receptor, *Nature*, 435:441-445.

[97] Lei XY, Zhu RY, Zhang GY, Dai YR, (2004), Attenuation of cold-induced apoptosis by exogenous melatonin in carrot suspension cells: the possible involvement of polyamines, *J. Pineal. Res.*, 36:126–131.

[98] Murch S, Saxena PK, (2002), Mammalian neurohormones: potential significance in reproductive physiology of St. John`s wort (*Hypericum perforatum* L.)? *Naturwissenschaften*, 89:555-560.

[99] Arnao MB, Hernández-Ruiz J, (2009), Chemical stress by different agents affects the melatonin content of barley roots, *J. Pineal. Res.*, 46:295–299.

[100] Tan DX, Manchester LC, Helton P, Reiter RJ, (2007), Phytoremediative capacity of plants enriched with melatonin, *Plant. Signal Behav.*, 2:514–516.

[101] Posmyk MM, Kuran H, Marciniak K, Janas KM, (2008), Presowing seed treatment with melatonin protects red cabbage seedlings against toxic copper ion concentrations, *J. Pineal. Res.*, 45:24–31.

[102] Posmyk MM, Balabusta M, Wieczorek M, Sliwinska E, Janas KM, (2009), Melatonin applied to cucumber (*Cucumis sativus* L.) seeds improves germination during chilling stress, *J. Pineal. Res.*, 46:214–223.

[103] Arnao MB, Hernández-Ruiz J, (2009), Protective effect of melatonin against chlorophyll degradation during the senescence of barley leaves, *J. Pineal. Res.*, 46:58–63.

[104] Wang P, Yin L, Liang D, Li C, Ma F, Yue Z, (2012), Delayed senescence of apple leaves by exogenous melatonin: towards regulating the ascorbate–glutathione cycle, *J. Pineal. Res.*, 53:11-20.

[105] Byeon Y, Park S, Kim YS, Park DH, Lee S, Back K, (2012), Light-regulated melatonin biosynthesis in rice during the senescence process in detached leaves, *J. Pineal. Res.*, 53:107-111.

[106] Conti A, Tettamanti C, Singaravel M, Haldar C, Pandi-Perumal SR, Maestroni GJM, (2002), Melatonin: an ubiquitous and evolutionary hormone, In: Haldar C, Singaravel M, Maitra SK, Eds, Treatise on pineal gland and melatonin, Science Publishers, Enfield, NH, 105–143.

[107] Arnao MB, Hernández-Ruiz J, (2013), Growth conditions influence the melatonin content of tomato plants, *Food Chem.*, 138:1212-1214.

[108] Afreen F, Zobayed SM, Kozai T, (2006), Melatonin in *Glycyrrhiza uralensis*: response of plant roots to spectral quality of light and UV-B radiation, *J. Pineal. Res.*, 41:108–115.

[109] Zhang LG, Jia JF, Xu Y, Wang Y, Hao JG, Li TK, (2012), Production of transgenic *Nicotiana sylvestris* plants expressing melatonin synthetase genes and their effect on UV-B induced DNA damage, *In vitro Cell Dev. Biol.* (Plant), 48:275-282.

In: UV Radiation: Properties, Effects, and Applications
Editor: James A. Radosevich
ISBN: 978-1-63321-090-5

Chapter 5

ROLE OF HYPOXIA AND OXIDATIVE STRESS IN MUTAGENESIS AND METABOLIC REPROGRAMMING: DRIVING FORCES WHICH INDUCE MALIGNANCY WITH UV-RADIATION RESISTANCE

Humberto De Vitto, Ph.D.*
University of Milano-Bicocca, Biotechnology
and Biosciences Department, Milan, Italy

ABSTRACT

"The Hallmarks of Cancer" were first documented as six traits that cells acquire during their progression to malignancy: a) limitless proliferation potential, b) sustained angiogenesis, c) apoptosis resistance, d) self-sufficiency in growth signals, e) insensitivity to anti-growth signals and f) ability to tissue invasion and metastasize to distinct organs. Indeed, radiation resistance of tumor cells might be explained in part by a variety of DNA damage in oncogenes and/or onco-suppressors and metabolic reprogramming that are involved in signaling pathways related to cell survival. Herein, we focus our analysis on environmental conditions that play key role in enhanced cell proliferation, evasion from apoptosis and acquiring radiation resistance such as hypoxia and oxidative stress. The text will be divided into three main parts. The first will be a comprehensive introduction of role of hypoxia and oxidative stress in tumor cell radio-sensitization. The second part will address about how UV-radiation induces mutagenesis with emphasis on dysfunctions of oncogenes and onco-suppressors. The third part will cover the reprogramming of energy metabolism that accompanies cell transformation focusing on those aspects that entail UV-induced tumor microenvironment (TME). This preamble was included in order to facilitate the understanding of the ongoing debate concerning role of glycolytic pathway and mitochondria in the process of malignant cell transformation. Finally, we will explore how we might combine treatments to kill UV-resistant cancer cells.

* Correspondence: Piazza della Scienza 2, 20126 Milan, Italy. Email: vitto@bioqmed.ufrj.br.

Keywords: UV-radiation, hypoxia, oxidative stress, DNA damage, metabolism, radio therapy resistance

ABBREVIATIONS

2-DG	2-deoxyglucose
3-BrPA	3- bromopyruvate
A	Adenine
ACC	Acetyl CoA Carboxylase
ACLY	ATP Citrate Lyase
AkT1/2/3	Serine/threonine-specific protein kinase B 1/2/3
ANT	Adenine Nucleotide Translocator
ATM	Ataxia Telangiectasia Mutated
ATP	Adenosine Tri-Phosphate
BCL-2	B-cell lymphoma 2
BIRC1/2/5	baculoviral inhibitor of apoptosis repeat-containing 1/2/5, called Survivin
C	Cytosine
CO_2	Carbon Dioxide
ChIP	Chromatin immunoprecipitation
CPD	Cyclobutane Pyrimidine Dimers
DCA	Dichloroacetate
DNA	Deoxynucleic acid
eNOS	endothelial Nitric Oxide Synthase
ERK	Extracellular-signal-Regulated Kinases
ER	Endoplasmatic Reticulum
ETC	Electron Transport Chain
FAS	Fatty Acid Synthase
FUS-1	Fused in Sarcoma 1
G	Guanine
HBP	Hexosamine Biosynthesis Pathway
HK	Hexokinase
HIF-1α	Hypoxia Inducible Factor-1α
H2O2	*Hydrogen Peroxide*
HNO	*High Nitric Oxide*
HRE	*Hypoxia Response Element*
IR	Ionizing Radiation
MCF-7	Michigan Cancer Foundation-7 (a breast cancer cell line)
MEK/ERK	*Mitogen-activated protein kinase/Extracellular signal-regulated kinases*
mtDNA	mitochondrial Deoxynucleic acid
NADH	Nicotinamide Adenine Dinucleotide
NADPH	Nicotinamide Adenine Dinucleotide Phosphate Hydrogen
NER	Nucleotide Excision Repair

NIH3T3	Mouse Embryo Fibroblast Cell Line
NO	Nitric Oxide
NOS	Nitric Oxide Synthase
$^{\cdot}OH$	hydroxyl radical
$O2^{\cdot}$	Hydrogen superoxide radical
O2	Oxygen
OXPHOS	Oxidative Phosphorylation
PDH	Pyruvate Dehydrogenase
PFK-1	Phosphofuctokinase-1
PHD2	Prolyl Hydroxylase Domain Protein 2
PK	Pyruvate kinase
PKA	Pyruvate kinase A
PI3	Phosphatidylinositide 3 Kinase
PRXII	Peroxiredoxin II
RAS	Rat Sarcoma Virus
RNA	Ribonucleic acid
RNS	Reactive Nitrogen Species
ROS	Reactive Oxygen Species
SOD-2	Superoxide dismutase 2
T	Thymadine
TCA	*Tricarboxylic Acid* Cycle
TNF-α	*Tumor Necrosis Factor-alpha*
TME	Tumor Microenvironment
TP53	Tumor Protein p53
UDP	Uridine Di-Phosphate
UPR	Unfolded Protein Response
UTP	Uridine Tri-Phosphate
UVR	Ultra Violet Radiation
VDAC	Voltage-Dependent Anionic Channel
VEGF	Vascular Endothelial Growth Factor
VHL	Von Hippel-Lindau protein

1. INTRODUCTION

Ionizing radiation (IR) therapy is the main current cancer patient treatment; approximately 50% of patients are submitted for this type of therapy [1]. Resistance of many human solid tumors to the most currently available cancer therapies is an important clinical issue [2]. Multiples subtypes of solid tumors exhibit chemo and radiation resistant, due in part by genetic mutations in oncogenes and/or onco-suppressors, and also due a metabolic shifts induced by environmental conditions in the tumor microenvironment (TME) such as hypoxia or chronic oxidative stress including: Reactive Oxygen Species (ROS) and Reactive Nitrogen Species (RNS) [3, 4]. In the harsh TME, tumors are more than insular masses of proliferating cancer cells. Instead, they are complex tissues composed of multiple distinct cell types that participate in heterotypic interactions with one another [5]. Consequently, the primary limitation of radiation therapy is the tolerance of normal tissues and other microenvironment cell types. Nevertheless, the real challenge of cancer therapy has been to developed specific strategies to increase radiation sensitivity of tumor cells and/or increase radiation resistance of normal tissues. These strategies might be favorable as a combined treatment response with antineoplastic agents/metabolic and IR after a clear and comprehensive understanding of the genetic background and metabolic alterations between tumor and normal cells, and others TME cell types.

1.2. Tumor Cells Radiosensitization: Key Role of Environmental Conditions

The main mechanism response of cancer and normal cells to IR is strongly dependent upon oxygen (O2), as suggested in 1958 by P. Howard-Flanders ‘‘oxygen fixation hypothesis” [6]. In aerobic conditions, O2 is central to biology. This molecule serves as the final electron acceptor in the mitochondrial oxidative phosphorylation process (OXPHOS) [7]. However, at low O2 concentration and/or a partial reduction of O2 as electron acceptor might increase the risk of generating ROS, linking hypoxia and chronic oxidative stress with tumor cells radio-sensitization. As consequence, ROS and RNS react with cellular macromolecules and alter their biochemical or physical properties, resulting in many cellular dysfunctions such as DNA damage, and metabolic shifts, which alter between survival and death molecular signals [8]. According to Shinohara and co-workers, patients treated that have severely hypoxic cells require 2-3 times more radiation dose as do well-oxygenated cells to achieve similar cell killing. Apart from the tumor cells, neighboring cells such as endothelial cells and others TME cell types may influence radio sensitivity [9].

Therefore, IR can lead to the expression of molecules that may increase chemotherapy and radiation resistance such as hypoxia inducible factor-1α (HIF-1α); a master transcription factor that regulates hundreds of genes expressed in the hypoxia condition, or a vascular endothelial growth factor (VEGF), a secreted protein that regulates angiogenesis. Recently, many studies are investigating potential targeting of HIFs for therapeutic purposes, and current therapy is using VEGF receptor tyrosine kinase inhibitors and humanized VEGF antibody to treat solid tumors [10, 8]. Since hypoxia governs tumor cells to acquire chemotherapy and radiation resistant in the TME, it will be interesting to exploit the main mechanism of HIF-1α. In 2008, Kaelin and Ratcliffe suggested that, under well-oxygenated conditions, HIF-1 α is bound by the von Hippel-Lindau (VHL) protein. VHL protein recruits an ubiquitin ligase that targets HIF-1α for proteasomal degradation [11]. HIF-1α can be hydroxylated by prolyl hydroxylase domain protein 2 (PHD2) in two proline residues (Pro^{402} and Pro^{564}; a required step for the binding of the VHL), which leads to HIF-1α ubiquitination and proteasomal degradation. These hydroxylation reactions use O2 and α-ketoglutarate as substrates and enzyme activity is inhibited under hypoxic conditions, leading to increased HIF-α stability and transcriptional activity [12]. (Figure 1) HIF-α may directly coordinate greater than 800 genes (i.e., at least 1 out of every 30 human genes) [13, 14]. As consequence, HIF1α and HIF-1β play key roles in many critical hallmarks of cancer biology including: metabolic reprogramming [15, 16], angiogenesis [17-19], invasion [20, 21], metastasis [22-24], resistance to radiation therapy [25], and chemotherapy [26].

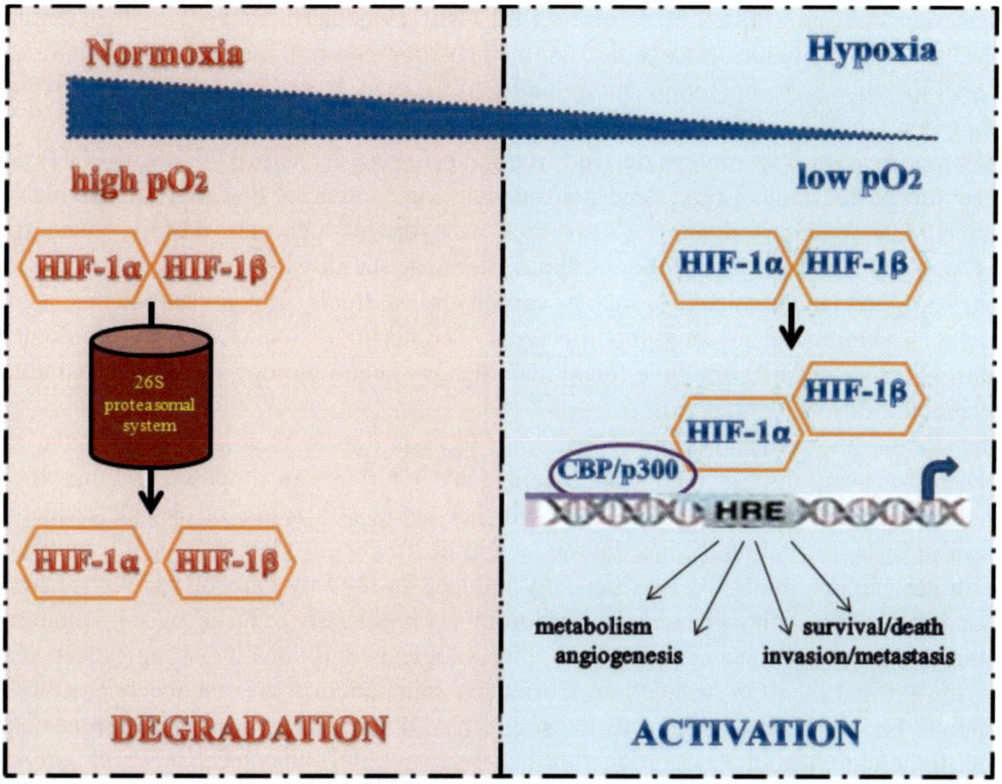

Figure 1. Molecular mechanism of degradation and activation of the HIF-1α and HIF-1β in normoxia and hypoxia conditions. Under well-oxygenated environment (normoxia) the HIF-1α and HIF-1β subunits are constitutively expressed in many tissues. HIF-1α has a short life-half and is degraded by the 26S proteasomal system while HIF-1b remains constant. On the other hand, under low concentration of O2 (hypoxia) HIF-1α associates with HIF-1 β and together both bind to hypoxia-response elements (HRE) in DNA, resulting in transcriptional activation of hundreds of target genes involved in cancer biology including: metabolism, angiogenesis, invasion/metastasis, and cell survival/death. CBP/p300 is a co-activator.

The molecular mechanism underlying cancer cell radiation resistance remains to be completely elucidated, but it has been well documented that IR produces free radicals in biological tissues including: ROS and RNS. Interestingly, many studies have been suggested that in the harsh TME, ROS and RNS also play a direct role in oncogenic transformation and/or onco-suppressors alterations and metabolism shifts, linking chronic oxidative stress with chemotherapy and radiation resistance in cancer biology [27, 28]. As an example, over-expression of RAS, a signal transduction protein, confers radiation resistance in many solid tumors [29]; while ATM, a serine/threonine protein kinase that is recruited and activated by DNA double-strand breaks, increases sensitivity to IR, leading to cell cycle arrest, DNA repair, and apoptosis [30]. In addition, oncogene inhibition can radiosensitize cells, it is therefore possible that the selective radiosensitizing effect of darinaparsin is due to inhibition of dysregulated oncogenic pathways in the tumor, but not normal cells [31]. Taken together, these data suggest that oncogenes are important for cancer cell survival after IR. Radiation resistance has been shown to be related to the expression of several other genes such as TP53 [32], BCL2 [33], and BIRC5 [34]. Interestingly, p53, an onco-suppressors protein that is mutated or silenced in many solid tumors [35-37], has been reported to protect intestinal cells from radiation damage via cell-cycle arrest [38]. This evidence demonstrates that chronic oxidative stress might induce oncogenic and/or onco-suppressors transformation, supporting tumor cells to acquire chemotherapy and radiation resistance. It will be interesting to exploit the comprehensive role of ROS and RNS in the TME.

ROS represent a set of oxygen-derived free radicals such as hydroxyl radical ($^{\cdot}OH$) and hydrogen superoxide radical ($O_2^{\cdot}$); and non-radical oxygen species that may be chemically transformed into reactive oxygen radicals such as hydrogen peroxide (H_2O_2). Since free radicals have unpaired electrons, they achieve chemical stability by donating or removing electrons from biological molecules, such as carbohydrates, lipids, proteins and nucleic acids, resulting in structural and physical modifications. The accumulation of damaged biological molecules is associated with multiple functional changes in cell biology including: mutations and metabolism shifts [39, 40].

Much of the biomedical interest concerning ROS is due to their potential role in the pathogenesis of many diseases and also in aging. In 1956, Harwan proposed the first "Free radical theory of aging." The author suggested that aging is a result of chronic oxidative modification of biological molecules, culminates in death [41]. Today, it is well established that in an aerobic condition, O_2 is completely reduced to H_2O by cytochrome *c* oxidase in mitochondria [7, 40]. Although, a small portion of O_2 is partially reduced by mitochondria, generating ROS, as consequence, many pathophysiological conditions including cancer, may increase mitochondrial ROS production. Curiously, mitochondria are not the only cellular ROS source. Endoplasmic reticulum (E.R.) stress, NADPH oxidases, and peroxissomes also represent important sites of ROS production however, the molecular mechanism underlying ROS production remains to be fully elucidated. Indeed, in the TME the imbalance between ROS generation and removal mechanisms might lead to chronic oxidative stress. It is well documented that normal, cancer cells, and others microenvironment cell types have been shown to have different molecular expression profiles of anti-oxidant defenses, to avoid oxidative stress. These include: 1) preventive (avoid ROS generation, such as UCPs and hexokinase), 2) scavengers (suppress unpaired electrons; enzymes like superoxide dismutase, catalase and glutathione peroxidase and non-enzymatic like α-tocopherol, or thioredoxin) and, 3) repair {repair molecules damaged by ROS, such as poly (ADP ribose) polymerase, PARP and aldehyde dehydrogenase} [42].

Interesting studies suggest that individuals with chronic oxidative stress may have low Fus1 levels and hence become more sensitive to the side effects of radiation therapy. This study has been shown that differential expression of FUS-1 (a mitochondrial protein) was associated with a deficiency of mitochondrial onco-suppressors Fus1 that results in dysregulation of mitochondrial homeostasis including the generation of high levels of ROS [43, 44]. Fus1 expression is epigenetically suppressed by asbestos exposure [45, 46] and tobacco smoking, [47, 48] demonstrating that inhibition of Fus1 expression can occur in normal tissues under chronic oxidative stress.

Another interesting study has been shown that an adapted model system exposed to chronic IR became resistant to radiation treatment, generating many radiation-resistant clones. The authors carried out a comparative proteomic analysis to search for expression alterations between both clones. The study suggested that protein expression level of peroxiredoxin II (PrxII) is significantly up-regulated in radio-resistant MCF-7 breast cells when compared with that in radiosensitive MCF-7 breast cells. Also, Western blot analysis of radio-resistant MCF-7 cells and radiosensitive MCF-7 cells confirmed the increased expression of PrxII protein in MCF-7 radio-resistant breast cancer cells. In addition, over-expression of PrxII in radiosensitive MCF-7 cells resulted in increased radiation resistance in breast cancer cells. As conclusion, the authors suggested that ROS contributed to the toxicity of IR, and the intracellular scavenging of these species is an important modulator of IR–induced damage [49]. The up-regulation of PrxII in radio-resistant breast cancer cells suggested that human PrxII plays a key role in eliminating the generation of ROS [50]. In this scenario, these antioxidant mechanisms, as cited above, induced by oxidative stress play a central protective role against ROS and RNS production, leading to therapy resistance [51, 52]. These findings support investigating potential targeting of antioxidant mechanisms for new therapeutic purposes in cancer. (Figure 2)

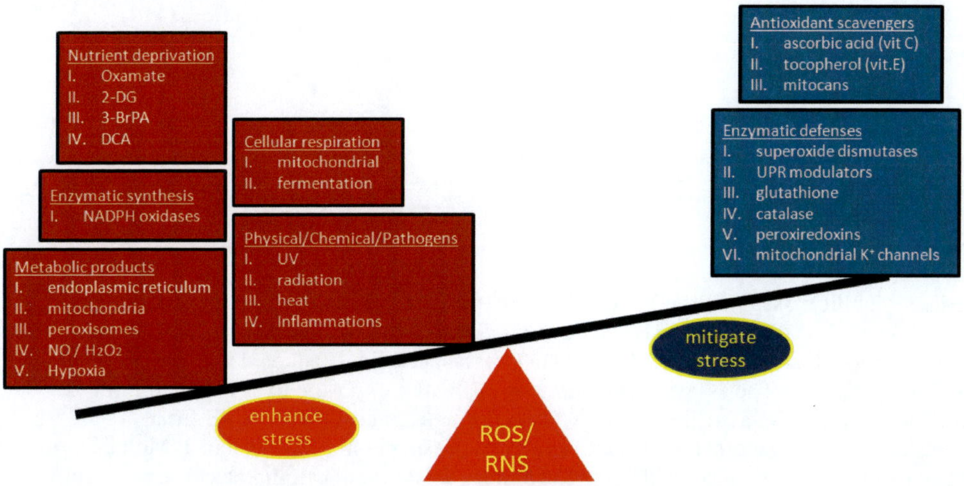

Figure 2. The main cellular mechanisms between imbalance of production and removal of ROS/RNS. This figure shows the key targets concerning ROS/RNS. Some of these mechanisms can be used for new therapeutic purposes in the pathogenesis of many diseases, including cancer and aging. In order: Nitric oxide (NO), Hydrogen peroxide (H2O2), ultraviolet light (UV), vitamin C (vit. C), vitamin E, Unfolded Protein Response (UPR), 2-deoxyglucose (2-DG), bromopyruvate (3-BrPA3), and dichloroacctatc (DCA).

Nitric oxide (NO) also plays a key role in the cancer cell radio-sensitization. NO has been mainly discussed as a potentially oncogenic molecule over the past decades [53]. Although, NO has been observed in cancer biology associated with apoptosis, proliferation, metastasis, angiogenesis, and a modulator for chemo/radio/immune therapy [10]. NO is well documented to be a robust radio-sensitizer of mammalian cells under hypoxia conditions [53]. NO donors have been used in a variety of studies and demonstrated a synergistic enhancement for radio-sensitivity, in both *in vivo* and *in vitro* settings [54-57]. It has also been shown that the cytotoxic impact of radiation is significantly increased through up-regulated nitric oxide synthases (NOS) in tumor cells [58]. Despite this, a dichotomy exists between enhanced radio-sensitization, conditions which elevate NO levels and up-regulated NOS are contributors to tumor malignancy. The relationship between NOS expression and tumor growth is evident in the literature from studies with a variety of cancer types including: human oral squamous cell carcinoma [59, 60], gastric carcinoma [61], lymph node metastases, and breast cancer, as a universal property for human solid tumors [62-68].

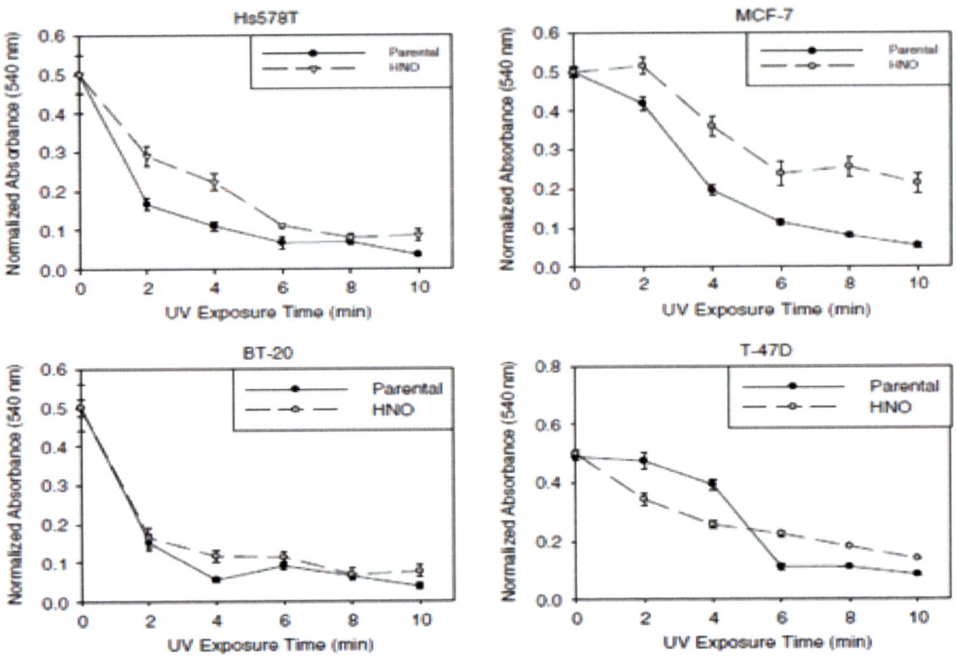

Figure 3. Cellular viability assays after UV-radiation exposure. MTT assay results for four human breast cancer cell lines (Hs578T, MCF-7, BT-20 and T-47D). Assays were performed with breast parent cell lines (without treatments with NO) and HNO breast cell lines (adapted to NO donor exposure, DETA-NONOate). The assays were measured 72 h after exposure to UV radiation for varying periods (0–10 min). Data are presented as mean normalized absorbance ± SE for a representative trial. In order: nanometers (nm). Courtesy by Vesper et al., (2013)"Part I. Molecular and cellular characterization of high nitric oxide-adapted human breast adenocarcinoma cell lines." Tumour Biol. 34 (2013): 203-214.

Studies suggest that long-term exposure to high concentration of NO applied to four human breast cancer cell line (BT-20, Hs578T, MCF-7, TD-47) that they are prone to acquire aggressiveness phenotypes, and are called HNO (High Nitric Oxide) adapted cells. When these four breast cell lines were challenged with UV-radiation for varying periods of time (0–10 min), the adapted HNO cell lines were more resistant to UV irradiation than its corresponding parent cell lines at longer exposure times [69]. (Figure 3) Curiously, the RNS adaptation mechanism for BT-20 breast cell line induced mutagenesis in p53 onco-suppressor gene and mitochondrial DNA (mtDNA), leading to mitochondrial dysfunction, and metabolic shifts (up-regulation of glycolysis genes and down-regulation of Tricarboxylic Acid cycle - TCA genes). After the NO adaptation process, it was observed that there was an induction of synergic NOS2 and NOS3 over-expression, which sustained a pro-inflammatory cellular signaling pattern (up-regulation of c-Myc, IL-8, S100A8 and Elf-2 genes). Taken together, these results suggest that chronic oxidative stress facilitated progression to malignancy and radiation resistance [70, 71]. Hence, response to cytotoxic therapy may be improved by modulating the TME [72, 9, 73]. This may opened up new avenues for interrupting hypoxia and the negative effects of chronic oxidative stress negative effects on tumor radio-sensitivity [25].

2. Potential Mutagenic DNA Damage

2.1. Mechanisms of UVR-Induced Mutations

X-ray and UV-radiation are energy sources known to induce DNA damage [69]. It is well established that X-ray radiation chemically reacts with water in the body to yield free radicals that may in part cause DNA damage (see introduction section), while UV-radiation can react directly with DNA, generating mutagenic processes and carcinogenic compounds of radiation, thus giving rise to dimeric products between adjacent pyrimidine bases. The two main types of chemical and structural DNA modifications are cyclobutane pyrimidine dimers (CPDs) and pyrimidine [6-4] pyrimidone products [74]. CPD contains a four member ring arising from the coupling of the C=C double bonds of pyrimidines [75]. The formation of dimers inhibits the enzymes DNA polymerases and interferes with base pairing during DNA replication, leading to mutagenesis. On the other hand, 6,4 pyrimidine-pyrimidone products contain a cross bond between the DNA sugar carbons 6 and 4 that occur in a low frequency (1/3) when compared with CPDs. Since 6,4 pyrimidine-pyrimidone products are more potentially mutagenic; the entire mutagenic mechanism is not completely understood [76]. (Figure 4)

UV-radiation causes mutagenesis and DNA photoproduct formation in skin cells, playing a key role in tumor malignancy. In particular, there is clear evidence for the involvement of these lesions in photo carcinogenesis or tumor radiation resistance as supported by the high rate of p53 mutations, including TC to TT or CC to TT transitions, which are detected at bi-pyrimidine sites in skin cancer [77-79]. In this regard, these photoproducts could be removed by nucleotide excision repair (NER). In the NER, repair proteins are recruited to the DNA damage site; however, cells deficient in these proteins, such as when p53 is mutated, are unable to repair the photoproducts. In eukaryotes, this complex process relies on the products of approximately 30 genes. Defects in some of these genes have been documented to cause human disease, such as skin cancer [80, 81]. Also, UV-radiation induces oxidative stress by formation of 8-oxo-7, 8-dihydro-2-deoxyguanosine, which is a classically known pathway that results in DNA strand breaks, and oxidized pyrimidine bases [82, 83].

Interestingly, a variety of studies has been shown that UVA-radiation could activate cyclobutane pyrimidine dimers in DNA in many models system such as bacteria [84], mammalian cells [85, 86], and skin [87, 88]. As a notable trend, recent studies showed that exposed to UVA can generate more CPD formation than 8-oxodGuo in rodent cell lines [89, 90] and in human skin cells [91]. UVA- radiation preferentially induces the production of CPDs at TT sites without any detectable formation of pyrimidine (6-4) pyrimidone photoproducts [90-92]. These observations clearly indicate that UVA-radiation does not induce CPDs via a direct excitation pathway, but more likely through a triplet energy transfer photosensitization mechanism. Also, exposure to UVB-radiation leads to the production of DNA damage, mainly cyclobutane pyrimidine dimers, and it was suggested that the thymidine dinucleotide (pTpT), mimicking small DNA fragments released in the course of excision repair mechanisms, could induce S-phase cell cycle arrest in addition to increased melanogenesis in human melanocytes [93]. Taken together, both mutagenesis mechanisms as described above might be associated with an increased radiation resistance in cancer biology, linking mutagenesis in oncogene and/or onco-suppressor gene with UV-radiation resistance.

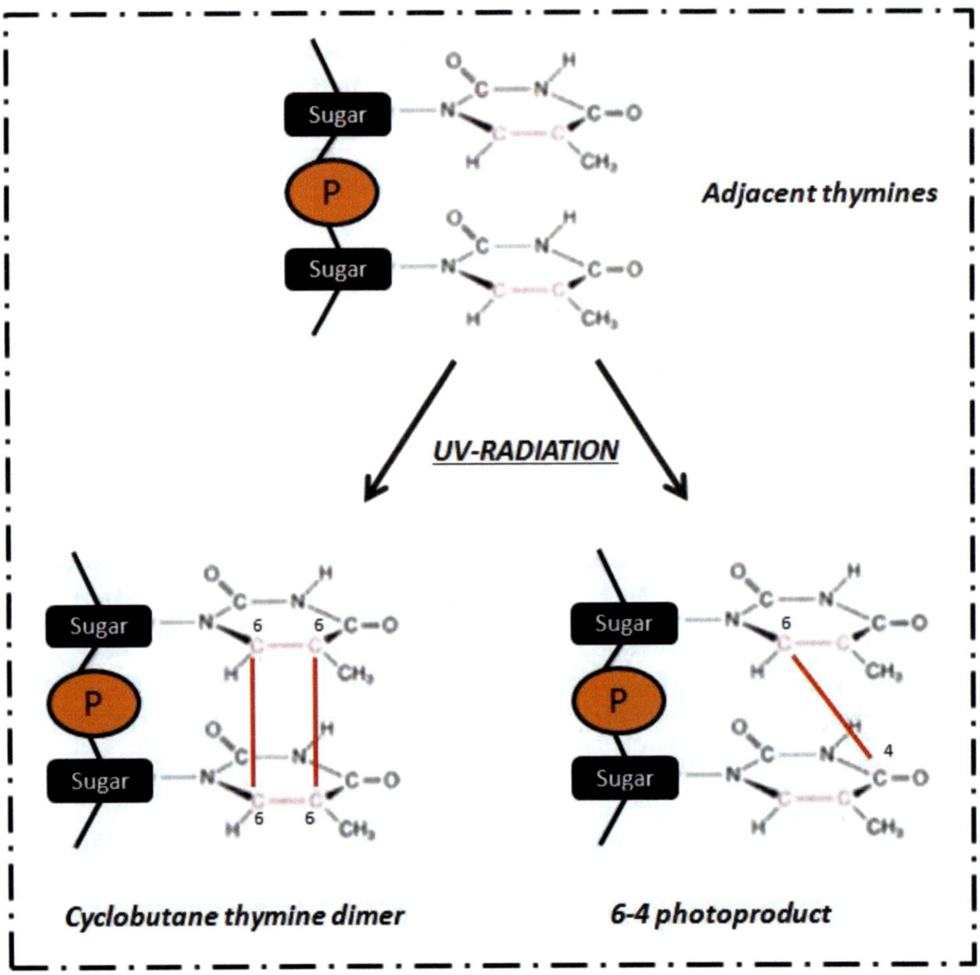

Figure 4. UV-radiation induced pyrimidine dimer pathway alterations. The two main types of structural DNA modifications: cyclobutane pyrimidine dimers (CPD) and pyrimidine (6-4) pyrimidone photoproduct. In order: carbon element (C), oxygen element (O), nitrogen element (N), hydrogen element (H), Phosphate (P), 6C (carbon number 6) and 4C (carbon number 4).

2.2. Role of Oncogenes and/or Onco-Suppressors in Radiation Resistance and in Sensitizing Tumor Cells to Radiation

Cigarette smoke, xenobiotics, diet, chronic ultraviolet radiation exposure, and cellular and tissue injuries caused by pathogens are environmental conditions that may produce ROS and RNS, which support carcinogenesis [94-96]. In the harsh TME, the low O2 concentration and the imbalance of ROS and RNS play key role in oncogenic transformation and/or onco-suppressors alterations, linking hypoxia and oxidative stress with chemotherapy and radiation resistance in the cancer biology. DNA damage of oncogenes and onco-suppressors may explain in part the mechanisms by which hypoxia and ROS/RNS may promote tumorigenesis.

Mutations in the oncogene *RAS,* and oncosuppressor *P53,* have been shown in multiple types of human cancers [97, 98]. Transition-type mutations at dipyrimidine sites and G:C to T:A transversions, in addition to being induced by the presence of 8-oxoguanine during DNA replication [95], are observed in *RAS* and *P53* genes in human solid tumors [99]. Moreover, it is well established that hypoxia and oxidative stress govern important dysfunctional elements in the glycolysis pathway, TCA cycle and OXPHOS. In addition, glutamine addiction and fatty acid metabolism (see energy metabolic reprogramming section) are governed by these stressors. They result in mutations in the RAS oncogene, and in the onco-suppressor P53, which can result in an advantage for tumor survival, enhanced proliferation, and evasion from apoptosis [100]. In this regard, the role of intracellular redox state in regulating growth, cell signaling, and/or gene expression is becoming recognized as an important issue in the radiotherapy.

Notably, it's estimated that approximately 20% of all human tumors have functional mutations in one of three RAS genes. The HRAS gene is mutated in approximately 4% of samples analyzed with the highest frequency in salivary gland (20%), urinary tract (12%) and cervix (9%). The KRAS gene is mutated in approximately 21.7% of samples analyzed with the highest frequency in pancreas (59%), biliary tract (32%), large intestine (30%), small intestine (27%) and lung (19%). Finally, the N-RAS gene is mutated in approximately 8% of samples analyzed with the highest frequency in skin (17%), nervous system (16%) and hematopoietic and lymphoid tissues (11%). Therefore, multiple approaches have been undertaken to develop tumor therapies efficiently targeting RAS and RAS downstream pathway gene products [97]. In a study which targeted mutated H-RAS proteins in NIH 3T3 cell line has been shown that 213 protein spots were changed in H-RAS transformed cell lines. Upon further characterization, finally 64 polypeptides were identified to be consistently deregulated by mutated H-RAS (36 up-/28 down-regulated). Many H-RAS targets identified were associated with the cytoskeleton or molecular chaperones [101, 102]. Also, large-scale genomic expression profiling approaches for the analysis of RAS function has been performed. In one study, G12V KRAS transformed ovarian epithelial cells showed that more than 200 genes were identified with altered expression; 79 targets being sensitive to inhibition of MEK/ERK or PI3K [103]. Apart from the cascades identified, the RAF/MEK/ERK pathway as well as the PI3K pathway still happened to be the most valuable targets for cancer therapy. Finally, the use of large-scale RNA interference libraries showed considerable promise in quickly uncovering the crucial components of RAS induced signaling cascades. The recent discovery that RAS oncogene expression is modulated by microRNAs has added a new facet to RAS research. In the days to come, microRNAs regulating other downstream targets of the RAS signaling cascades will be deciphered as well. There is no doubt that these studies will keep up the momentum and excitement of this ever growing field of RAS biology, thereby retaining RAS at the heart of cancer research for decades [97].

Out of these signatures for various tumor types it will be important to identify those targets that are relevant for oncogene addiction, a phenomenon describing the acquired dependence of tumor cells on an activated oncogene for their survival and/or proliferation [104-105]. Analyses of this type may provide molecular targets for future therapeutic approaches by targeting mutation-specific events in radio-resistant tumors.

Inactivation of onco-suppressor genes is one of the key hallmarks of tumor progression. Unlike other onco-suppressor genes, p53 is inactivated by mis-sense mutations in approximately 50% of all human cancers. It has become clear that the resulting mutant p53 proteins do not represent only the mere loss of wild-type p53 onco-suppressor activity, but gain new oncogenic properties favoring the insurgence, maintenance, increased metastatic potential, along with chemo-resistance, and radio-resistance properties of the tumors. Accordingly, the real challenge is the fine deciphering of the molecular mechanisms underlying the gain of function of mutant p53 proteins [98].

The p53 structure comprises an N-terminal transcriptional activation domain, a central DNA-binding domain and a C-terminal site with oligomeric and regulatory activities [106]. The transcriptional activity of p53 is mainly regulated at the post-translational level. Indeed, phosphorylation, acetylation, prolyl isomerization and sumoylation are the most characterized post-translation modifications on p53 that impact on their transcriptional activity [107, 108]. Mutant p53 proteins are characterized by an enhanced half-life compared with the wild type-p53 protein and by the inability to recognize wild type-p53 DNA-binding sites [109]. The molecular basis of the enhanced half-life of mutant p53 proteins still needs to be understood. The mechanism(s) may depend on the inefficient degradation exerted by the E3 ubiquitin ligase Mdm2; a gene that is a direct transcriptional target of wild type-p53. As a consequence, the activation is impaired in cells carrying mutant p53 proteins [110]. This suggests the hypothesis that the tumor cell context contains additional physical or chemical factors contributing to the increased half-life of mutant p53 proteins. (Figure 5)

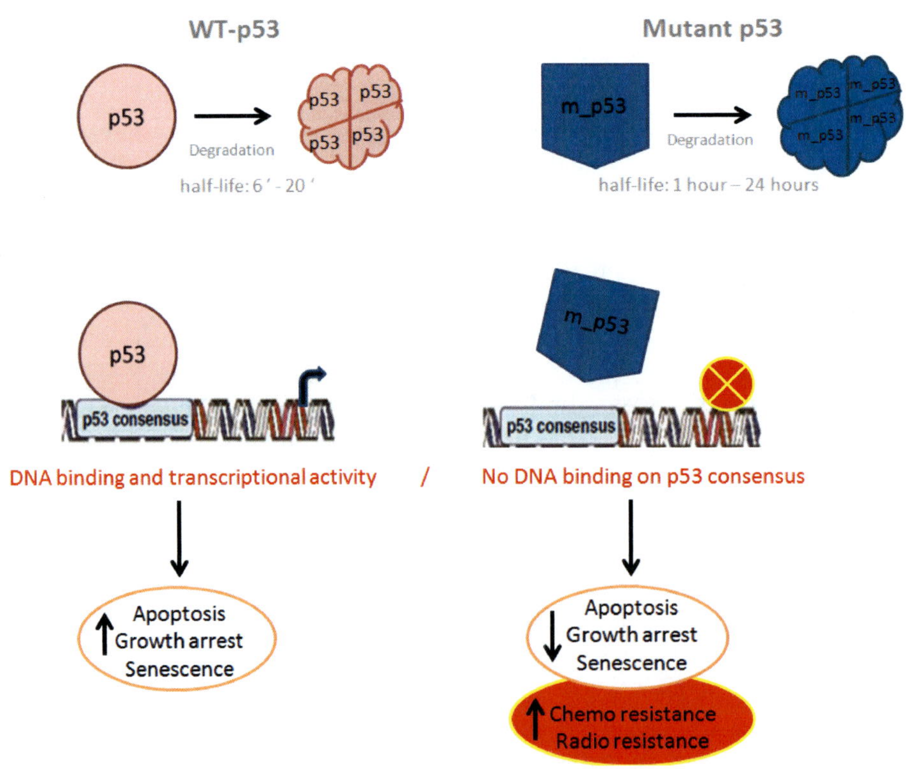

Figure 5. The half-life and the ability to bind specifically to DNA are hallmarks which distinguish wild-type from mutant p53 proteins. The main biological p53 functions in cancer biology are: apoptosis signals, cell growth, senescence and chemo/radio resistance. Figure modified from Strano et al., "Mutant p53: an oncogenic transcription factor." Oncogene. 26 (2007): 2212-2219.

Recently, genome-wide approaches have been showed that mutant p53 is capable of modulate the expression of large patterns of transcripts. These findings support the idea that mutant p53 might function as a transcription factor capable to turn on and/or off specific clutters of genes in response to different stimuli [111, 112]. As consequence, mutant p53 might trigger different pathways that represent the molecular basis of a large spectrum of gain-of-function activities, which range from the resistance to anti-cancer treatments to genomic instability and increased metastasis. Also, Chromatin immunoprecipitation (ChIP) experiments have shown that mutant p53 can be recruited *in vivo* onto regulatory regions of diverse genes. Further, recent evidence indicates that mutant p53 is present on promoter regions together with DNA-binding proteins, acetylases and/or deacetylases [113].

3. Energy Metabolic Reprogramming

The hypothesis, called "The Hallmarks of Cancer" was proposed by Douglas Hanahan and Robert A. Weinberg, in 2000 [5]. This hypothesis counted six traits that cells acquire during their progression to malignancy: a) limitless proliferation potential, b) sustained angiogenesis, c) resistance to apoptosis, d) self-sufficiency in growth signals, e) insensitivity to anti-growth signals and, f) the ability to invade tissue and metastasize to distant organs. Ten years later, other parameters incremented the new generation of hallmarks of cancer, such as: g) reprogramming of cellular energy metabolism, h) evasion of cancer cells from attack and elimination by the immune system, i) genomic instability, j) tumor-promoting inflammation, k) presence of a stress phenotype [114]. In particular, cancer cells display high proliferation rates and evasion from apoptosis supported by metabolic features that appear to be activated with malignancy. In this metabolic section, we will address the central role of enhanced glycolysis and mitochondrial dysfunctions as an essential stimulus for survival and proliferation of cancer cells in the TME. We will also explore insights about how nutrient deprivation pathways and mitochondria provide excellent targets to challenge UV-resistance cancer cells. Since metabolism is a prerequisite for cellular growth, one may wonder whether a new strategy for developing novel, specific, and more effective antineoplastic agents may be derived from a comprehensive understanding of cancer cell metabolism. Chemotherapeutics, and several biological antineoplastic agents have been designed to act on cell proliferation (by targeting signaling or cell cycle events); a property that is not peculiar to cancer but also to fast growing normal cells [115]. If a difference in metabolism could be found that is strictly specific for cancer cells, the enzyme(s) that catalyze this step should represent convenient targets and/or biomarkers to be used in neoplastic diseases.

3.1. Cancer Metabolic Alterations: From Warburg Effect to Reverse Warburg Effect

In the TME with a restricted supply of O_2 (hypoxia condition) cancer cells may acquire a universal property which drives them toward malignancy; an increase in glucose uptake and consumption, and consequent reduction of converting pyruvate to lactate, in order to recycle NADH back to NAD+. This sustains the metabolic flux via glycolysis (Pasteur Effect). As a consequence, lactate, can feed as building blocks to support catabolism in the TME, or it needs to be expelled from the tumor to the extracellular microenvironment in order for the tumor to survive by continuing its aberrant glycolytic metabolism (glucose to lactate) [116]. In a normoxic environment, cells oxidize pyruvate to CO2, in the mitochondria. This is first done by the activity of pyruvate dehydrogenase (PDH), following by the activity of the TCA cycle enzymes, and finally, by the mitochondrial activity which drives the electrons flux into the enzymatic mitochondrial complexes (OXPHOS). Thus, there is an electrochemical gradient formed that is able to generate energy, as ATP production.

In an overview glucose metabolism, glucose generates enough energy for macromolecules biosynthesis required for cell proliferation. Also, glucose participates in the pentose phosphate pathway which generates ribose 5-Phosphate for nucleic acid synthesis and NADPH. NADPH is essential in the process of scavenging ROS, anabolic metabolism, fatty acid metabolism, and amino acid metabolism [117]. The TCA cycle plays a central role for cellular metabolic regulation, generally in pathways for the catabolism and anabolism. TCA cycle intermediates such as oxalacetate and α-ketoglutarate, which produce aspartate and glutamate respectively, are important molecules in the production of purines and pyrimidines. Others studies have shown that citrate plays a role in the fatty and steroid production. Glutamine addiction plays a crucial role to generate oxalacetate which participates in the carbon skeleton maintenance, and provides enhanced cell proliferation properties [118]. (Figure 6)

As a universal property of cancer, cancer cells are prone to use glucose as the main substrate to drive ATP production. Even in normoxia, there is a shift of the metabolism from oxidative to fermentative ATP production known as the "*Warburg Effect*." Otto Warburg described how cancer cells displayed a decreased respiration and increased lactate production, suggesting that cancer cells were dependent of fermentative metabolism for ATP production [119].

One plausible explanation for the Warburg effect in cancer cell is a result of impaired mitochondrial OXPHOS machinery. Over the past 70 years after Warburg discovery, the cancer cell metabolism remains a puzzle. Other concepts about glucose metabolism in cancer cells have been proposed; for example, the reversible repressive effect of glucose over respiration, which is known as the "*Crabtree effect*" [120]. The *Crabtree effect* has the same metabolic signature as observed in the Warburg effect, although the shift from oxidative to fermentative metabolism is observed in a reversible fashion, even in normoxia, with functional oxidative phosphorylation machinery. In this scenario, the OXPHOS is rapidly and reversibly down-regulated by the repressive effect of glucose, whereas in the Warburg effect, there is a long irreversible effect favoring fermentation due to the increased expression of proteins involved in glucose transport and metabolism.

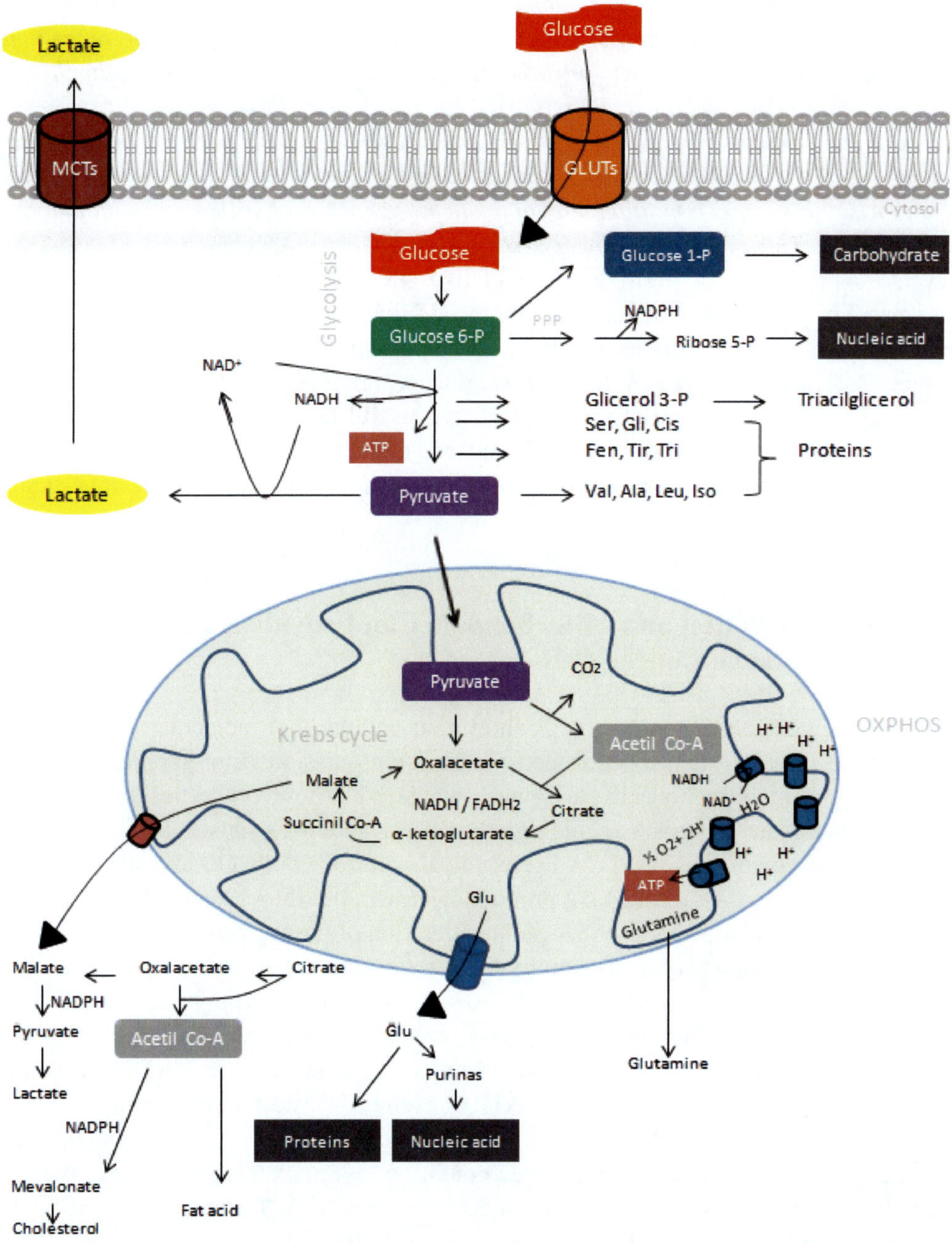

Figure 6 Biologic pathways of cell energy metabolism. Glycolysis, TCA cycle, OXPHOS, glutamine and fatty acid metabolism, sustained catabolism reactions for cellular energy production. In order: (GLUTs), (MCTs), (PPP), (NADPH), (ATP), (FADH2), (Ser), (Gly), (Cys), (Phe), (Tyr), (Val), (Ala), (Iso), (Leu) and (Glu).

An interestingly point that should be addressed will be the seemingly paradoxical role of mitochondria in cancer cells. The Warburg hypothesis says that all cancer cells derive most of their energy from aerobic glycolysis [121]. However, recent studies have shown the additional dependence on oncogene-stimulated mitochondrial biogenesis for energy generation [122, 42*]. According to this hypothesis, lactate could be an energetic substrate for tumor cells. The lactate shuttle between tumor cell and stromal cells, feeds the other tumor cells via lactate metabolism, which only the tumor cells utilize to generate mitochondria energy. As a consequence, tumor cells might signal the tumor clonal expansion process by induced oxidative stress in adjacent tumor cells. It has been demonstrated that an increase in nitric oxide in the stromal fibroblasts can generate mitochondrial dysfunction by ROS production. Also, this stress condition may activate autophagy and mitophagy with increased aerobic glycolysis. This process has been described by* Martinez-Outschoorn and co-workers as a reverse Warburg effect [123] when oxidative stress in cancer associated fibroblasts functions as a "metabolic" and "mutagenic" motor, driving epithelial cancer cell evolution. In this line of thought, mutagenic and metabolic reprogramming has been proposed to support carcinogenesis. Thus, any strategy to deprive glucose, increase TCA cycle activity, with impairment of mitochondria activity, may provide a useful target to kill cancer cells while preserving normal tissues [124, 125]. Curiously, nutrient deprivation of glucose and mitochondria could provide excellent targets to treat solid tumor cells resistant to radiotherapy.

3.2. Nutrient Deprivation and Mitochondria Can Provide Excellent Targets to Treat Uv-Resistance Cancer Cells

It is well documented that cancer cells show four cellular and molecular properties that can be measured *in vitro*, such as enhanced proliferation, evasion from apoptosis, genomic instability, and the inability to activate senescence [125]. These properties act as driving forces toward malignant processes including: chemo and radio-resistance. In fact, enhanced glycolysis, down-regulation of the TCA cycle, impairment of OXPHOS and the decoupling of glucose and glutamine addiction (as a product of growth building blocks and glutathione), have been proposed as hallmarks of enhanced proliferation of human solid tumor.

In generally, mitochondrial dysfunctions have been reported to occur in several cancer cells [126, 127]. In K-*ras* transformed cells a substantial down-regulation of OXPHOS, accompanied by a strong reduction of Complex I activity, has been reported together with an increase of mitochondrial membrane potential and a reduction of the NADH ubiquinone reductase activity [128]. The decrease of ATP forming ability of cancer mitochondria is compensated by the ATP produced by glycolysis. Calculations predict that the energy flux in the time required by a cancer cell in normoxic condition to completely oxidize one molecule of glucose through the TCA cycle (to produce 36 molecules of ATP), ten more molecules of glucose are converted to lactic acid to make an additional 20 molecules of ATP [129]. Thus in the same time period in which a normal cell produces 36 ATP from a molecule of glucose, a cancer cell appears able to produce 56 of them. These calculations suggest a very large and fast glycolytic flux, as compared to the slow rate of the TCA cycle, and may mediate a large pyruvate utilization that sustains the energy requirement for cancer cell growth. At this stage, it is possible to conclude that the enhanced growth of cancer cells is accompanied by a profound rewiring of glycolytic and mitochondrial metabolism. Recently, studies have investigated the potential targeting of these main metabolic shifts for therapeutic purposes and current therapy [130]. (See the next section)

As cited above, cancer cells are prone to use glucose as the main substrate to drive ATP production (Warburg Effect). Interestingly, glucose limitation can induce cell cycle arrest in normal and in transformed cells however, it specifically enhances cell death in transformed cells [131]. Many biological processes have been associated with the death of transformed cells during glucose restriction, such as ROS/RNS accumulation, ATP level decrease, and mitochondrial dysfunction [128, 131, 132]. For example, one of these processes might govern cell death signals by activating a mitochondrial apoptotic course, as regulation of BH3-only proteins, or by activating a caspase-8-dependent apoptotic process, or by inducing necrosis [133-135]. A notable observation by Palorini and co-workers was that the glucose-dependent mechanism of cancer cell death may provide a suitable target through which apoptosis may be induced in tumors while shielding healthy normal tissues. They studied four different cell lines including: 1) NIH-3T3 fibroblasts harboring an oncogenic *K-RAS* gene (G12V), 2) parental immortalized NIH-3T3 cells, 3) a human breast cancer cell line, MDA-MB-231, carrying an oncogenic *K-RAS* gene (G13D) and, 4) a parental MDA-MB-231 cell line. In this study, they proposed that glucose restriction supports cell death in K-ras transformed cells by interposing the Hexosamine Biosynthesis Pathway (HBP) and fomenting the Unfolded Protein Response (UPR) [132]. UPR is a cellular response to protein folding alteration induced by a variety of effectors [136] that may lead either to cell survival or to cell death depending of the power and time of stimulus duration [137]. It is well documented that under physiological conditions, 1–3% of intracellular glucose is derived from glycolysis due to the HBP [138] and therefore glucose availability drives the flux through the HBP. The main product of HBP is UDP-N-acetylglucosamine, which is a key donor molecule for post-translational modifications, such as N- and O-glycosylation [139].

The potential role of the HBP in glucose shortage-induced ER stress in cancer cells has not been fully elucidated. Current ongoing studies are directed at studying the UPR mechanism in p53 mutated cancer cell lines in a microenvironment of glucose restriction. These studies will be directed at understanding the cell death mechanism activated by glucose deprivation, and how these transformed cancer cells try to survive this type of stress. These studies will also include challenging these cell lines with combined treatments targeting tumor metabolic features associated with survival and cell death. These studies could open up new avenues to investigate metabolic shifts and chemo and radio resistance.

4. Implications for Cancer Therapy and Why Combined Treatments Might Kill UV-Resistance Cancer Cells

Human solid tumors often display chemo and radiation resistant due in part by genetic mutations in oncogenes and/or onco-suppressors and also environmental conditions such as hypoxia or chronic oxidative stress, which cooperate to acquiring the malignant transformation. It is well established that metabolism is a prerequisite for cellular growth, suggesting that a clear and comprehensive understanding of cancer cell metabolism could be one strategy for developing novel, specific and more effective antineoplastic agents [140]. Multiples antineoplastic agents have been designed and combined to inhibit cell proliferation, specially by acting on targeted signaling or cell cycle events; a property that is not specific to cancer but, also to fast growing normal cells including: atrocities and testicular cells [115].

Therefore, if a difference in metabolism could be found between normal and cancer cells, such as these enzyme(s) that catalyze these steps; this should represent suitable targets and/or biomarkers to be used in cancer treatment.

The peculiar metabolic pathways modified in cancer cells are tightly correlated to altered glycolysis with changes in regulatory steps and alterations in transcriptional activity, rerouting of glutamine addiction, fatty acid metabolism, and mitochondria, which play a key role in the regulation of cellular metabolism and apoptosis [140]. Three specific, irreversible enzyme steps of glycolysis have been considered as possible therapeutic targets for cancer disease, including: 1) hexokinase (HK), 2) Phophofructokinase-1 (PFK-1) and, 3) Pyruvate kinase (PK). A variety of compounds target these glycolysis steps, such as: 1) 2-deoxyglucose (2-DG), 2) 3- bromopyruvate (3-BrPA) and, 3) dichloroacetate (DCA). Furthermore, there is a set of compounds targeting mitochondria, named mitocans including: 1) Bcl-2 homology-3 (BH3) mimetics, 2) thiol redox inhibitors, 3) drugs targeting the voltage-dependent anionic channel (VDAC) or the adenine nucleotide translocator (ANT), 4) agents as metformin and phenformin that interfering with the electron transport chain (ETC), 5) lipophilic cations targeting the inner membrane, 6) agents interfering with the mitochondrial DNA and, 7) drugs acting on not well-defined sites. They often lead to specific cancer cell death through mitochondria destabilization, by inducing ROS and a reduction of total intracellular ATP, leading to activation of apoptotic signals [72, 141].

Table 1: Examples of promising metabolic targets for cancer therapy

Targets	Pathways	Agents or approaches (company)
Bioenergetic metabolism		
CPT1	β-oxidation	Etomoxir, Oxfenicine, Perhexiline and RNAi
Complex I	Mitochondrial respiration	Metformin and Phenformin
GLUT1	Glycolysis	WZB117 and RNAi
GLS1	Glutamine metabolism	968, BPTES and RNAi
Hexokinases	Glycolysis	2-DG, 3-BP, Oxamate, Lonidamine, Methyl jasmonate and RNAi
MCT1	Krebs cycle	AR-C155858, AR-C117977, AZD3965 (AstraZeneca), CHC and RNAi
PDK1	Krebs cycle	DCA
PKM2	Glycolysis	TLN-232 (Thallion) and RNAi
Anabolic metabolism		
Choline kinase	Lipid biosynthesis	CK37, TCD-717 (TCD Pharma) and RNAi
HMGCR	Mevalonate pathway	Statins
IDHs	Lipid biosynthesis	AGI-5198 (Xcessbio), AGI-6780 (Xcessbio) and RNAi
MGLL	Lipid biosynthesis	JZL184 and RNAi
PGAM1	Pentose phosphate pathway	PGMI-004A and RNAi
PHGDH	Anaplerosis	RNAi
PKM2	Pentose phosphate pathway	TEPP-46, SAICAR and Serine
Other metabolic circuitries		
HIF1	Hypoxic responses	Acriflavine and PX-478
IDO	Tryptophan metabolism	RNAi
mTOR	Cell growth, autophagy	Rapalogues and Torins
PTGS2, AMPK?	Cell growth, autophagy	Aspirin

Recent studies suggest that combined treatments targeting both glycolysis and mitochondria, or other specific tumor features, may be lethal for cancer cells. As an example, it has been shown that treatment of cancer cells with DCA, a TCA cycle inducer, is able to redirect their metabolism from glycolysis to OXPHOS and hence to lead them towards apoptosis [142]. DCA is an inhibitor of pyruvate dehydrogenase kinase-1 (PDK-1), a kinase that regulates the flux of pyruvate into mitochondria by inhibiting Pyruvate dehydrogenase (PDH). By inhibiting PDK-1, DCA redirects pyruvate away from the production of lactate, inducing pyruvate entry into the mitochondrial TCA cycle. Therefore, it has been proposed that induction of a reversion of the Warburg effect coupled to a treatment able to interfere in the mitochondrial activity (rotenone) and/or in the glycolysis flux (oxamate) could specifically kill cancer cells.

Increased *de novo* FA synthesis has been recognized as a hallmark of cancer and recent advances in this field have further confirmed the importance of the lipogenic enzymes in cancer cell survival (FAS, ACC and ACLY) [143]. These findings suggested *de novo* FA synthesis as a rational therapeutic target. Great interest has been given to FAS proteins as targets in several oncogenic pathways [144]. The metabolic rewiring of cancer cells has been viewed as a promising source of novel drug targets for cancer therapy [145] (Table 1).

CONCLUSION AND OUTLOOK

This chapter sought to explain a variety of cancer hallmarks, substantiate the complexity of the biology of cancer, and present a comprehensive discussion concerning the role of hypoxia and chronic oxidative stress as driving forces which lead malignant cells to become UV-radiation resistant. Among the ten acquired capabilities (the new generation of hallmarks of cancer) [114] presented herein, are cellular and molecular properties that can be measured *in vitro*, such as oncogene and/or onco-suppressors transformations and deregulating cellular energetic. In particular, the refinement of these organizing principles will surely come in the future, and will represent remarkable conceptual progress in this regard.

Notably, there are significant advances in our understanding of the role of genetic transformation by oncogenic RAS and onco-suppressors like p53. More work needs to be done on how these genes act on aerobic glycolysis and mitochondrial cellular metabolism, which are the hallmarks of chronically sustained proliferation and evasion from apoptosis. Similarly, recent advances in the identification of metabolic targeting compounds are able to combine cytotoxic and cytostatic effects to kill cancer cells.

It remains perplexing and hard to understand that tumors are complex tissues composed of multiple distinct cell types and how they participate in heterotypic interactions with one another. These cell types show multiples genetic background and different metabolic features, which facilitate chemo and radiation resistance. As suggested by Douglas Hanahan and Robert A. Weinberg: "We continue to foresee cancer research as an increasingly logical science, in which myriad phenotypic complexities are manifestations of a small set of underlying organizing principles".

REFERENCES

[1] Delaney, G., Jacob, S., Featherstone, C., Barton, M. (2005) "The role of radiotherapy in cancer treatment: estimating optimal utilization from a review of evidence-based clinical guidelines." *Cancer* 104: 1129–1137.

[2] Tian, J., Zhao, H., Nolley, R., Reese, SW., Young, SR., Li, X., Peehl, DM., Knox, SJ. (2012) "Darinaparsin: Solid Tumor Hypoxic Cytotoxin and Radiosensitizer." *Clin. Cancer Res.* 18: 3366-3376.

[3] Ruan, K., Song,G., Ouyang, G. (2009) "Role of hypoxia in the hallmarks of human cancer." *J. Cell Biochem.* 107: 1053-1062.

[4] De Ridder, M., Verellenm, D., Verovski, V., Storme, G. (2008) "Hypoxic tumor cell radiosensitization through nitric oxide." Nitric Oxide 19: 164-169.

[5] Hanahan, D., Weinberg, RA. (2000) "The hallmarks of cancer." *Cell.* 100: 57-70.

[6] Howard-Flanders, P., Moore. D. (1958) "The time interval after pulsed irradiation within which injury to bacteria can be modified by dissolved oxygen. I. A search for an effect of oxygen 0.02 second after pulsed irradiation." *Radiat. Res.* 9: 422-437.

[7] Mitchell, P. (1961) "Coupling of phosphorylation to electron and hydrogen transfer by a chemi-osmotic type of mechanism." *Nature* 191: 144-148.

[8] Semenza, GL. (2012) "Hypoxia-Inducible Factors in Physiology and Medicine." *Cell* 148: 399-408.

[9] Shinohara, ET., Maity, A. (2009) "Increasing sensitivity to radiotherapy and chemotherapy by using novel biological agents that alter the tumor microenvironment." *Curr. Mol. Med.* 9: 1034-1045.

[10] Yasuda, H. (2008) "Solid tumor physiology and hypoxia-induced chemo/radio-resistance: novel strategy for cancer therapy: nitric oxide donor as a therapeutic enhancer." *Nitric Oxide* 19: 205-216.

[11] Kaelin, WG. Jr., Ratcliffe, PJ. (2008) "Oxygen sensing by metazoans: the central role of the HIF hydroxylase pathway." *Mol. Cell.* 30: 393-402.

[12] Semenza, GL. (2010) "Oxygen homeostasis." *Wiley Interdiscip. Rev. Syst. Biol. Med.* 2: 336-361.

[13] Xia, X., Lemieux, ME., Li, W., Carroll, JS., Brown, M., Liu, XS., Kung, AL. (2009) "Integrative analysis of HIF binding and transactivation reveals its role in maintaining histone methylation homeostasis." *Proc. Natl. Acad. Sci. U S A.*, no. 106: 4260-4265.

[14] Schödel, J., Oikonomopoulos, S., Ragoussis, J., Pugh, CW., Ratcliffe, PJ., Mole, DR. (2011) "High-resolution genome-wide mapping of HIF-binding sites by ChIP-seq." Blood. 117: 207-217.

[15] Zhang, H., Gao, P., Fukuda, R., Kumar, G., Krishnamachary, B., Zeller, KI., Dang, CV., Semenza, GL. (2007) "HIF-1 inhibits mitochondrial biogenesis and cellular respiration in VHL-deficient renal cell carcinoma by repression of C-MYC activity." *Cancer Cell.* 11: 407-420.

[16] Luo, W., Hu, H., Chang, R., Zhong, J., Knabe,l M., O'Meally, R., Cole, RN., Pandey, A., Semenza, GL. (2011) "Pyruvate kinase M2 is a PHD3-stimulated coactivator for hypoxia-inducible factor 1." *Cell* 145: 732-744.

[17] Liao, D., Johnson, RS. (2007) "Hypoxia: a key regulator of angiogenesis in cancer." *Cancer Metastasis Rev.* 26: 281-290.

[18] Lee, K., Qian, DZ., Rey, S., Wei, H., Liu, JO., Semenza, GL. (2009a) "Anthracycline chemotherapy inhibits HIF-1 transcriptional activity and tumor-induced mobilization of circulating angiogenic cells." *Proc. Natl. Acad. Sci. U S A.* 17: 2353-2358.

[19] Lee, K., Zhang, H., Qian, DZ., Rey, S., Liu, JO., Semenza, GL. (2009b) "Acriflavine inhibits HIF-1 dimerization, tumor growth, and vascularization." *Proc. Natl. Acad. Sci. U S A.* 106: 17910-17915.

[20] Sullivan, R., Graham, CH. (2007) "Hypoxia-driven selection of the metastatic phenotype." *Cancer Metastasis Rev.* 26: 319-331.

[21] Krishnamachary, B., Semenza, GL. (2007) "Analysis of hypoxia-inducible factor 1α expression and its effects on invasion and metastasis." *Methods Enzymol.* 435: 347-354.

[22] Erler, JT., Bennewith, KL., Nicolau, M., Dornhöfer, N., Kong, C., Le, QT., Chi, JT., Jeffrey, SS., Giaccia, AJ. (2006) "Lysyl oxidase is essential for hypoxia-induced metastasis." *Nature* 440: 1222-1226.

[23] Wong, CC., Gilkes, DM., Zhang, H., Chen, J., Wei, H., Chaturvedi, P., Fraley, SI., Wong, CM., Khoo, US., Ng, IO., Wirtz, D., Semenza, GL. (2011) "Hypoxia-inducible factor 1 is a master regulator of breast cancer metastatic niche formation." *Proc. Natl. Acad. Sci. U S A.* 108: 16369-16374.

[24] Zhang, H., Wong, CC., Wei, H., Gilkes, DM., Korangath, P., Chaturvedi, P., Schito, L., Chen, J., Krishnamachary, B., Winnard, PT Jr., Raman, V., Zhen, L., Mitzner, WA., Sukumar, S., Semenza, GL. (2012) "HIF-1-dependent expression of angiopoietin-like 4 and L1CAM mediates vascular metastasis of hypoxic breast cancer cells to the lungs." *Oncogene.* 31: 1757-1770.

[25] Moeller, BJ., Richardson, RA., Dewhirst, MW. (2007) "Hypoxia and radiotherapy: opportunities for improved outcomes in cancer treatment." *Cancer Metastasis Rev.* 26: 241-248.

[26] Rohwer, N., Cramer, T. (2011) "Hypoxia-mediated drug resistance: novel insights on the functional interaction of HIFs and cell death pathways." *Drug Resist. Updat.* 14: 191-201.

[27] Griffin, RJ., Makepeace, CM., Hur, WJ., Song, CW. (1996) "Radiosensitization of hypoxic tumor cells in vitro by nitric oxide." *Int. J. Radiat. Oncol. Biol. Phys.* 36: 377-383.

[28] Yazlovitskaya, EM., Uzhachenko, R., Voziyan, PA., Yarbrough, WG., Ivanova, AV. (2013) "A novel radioprotective function for the mitochondrial tumor suppressor protein Fus1EM." *Cell Death Dis.* 4: e687.

[29] Sklar, MD. (1988) "The ras oncogenes increase the intrinsic resistance of NIH 3T3 cells to ionizing radiation." *Science.* 239: 645-647.

[30] Westphal, CH., Rowan, S., Schmaltz, C., Elson, A., Fisher, DE., Leder, P. (1997) "ATM and p53 cooperate in apoptosis and suppression of tumorigenesis, but not in resistance to acute radiation toxicity." *Nat. Genet.* 16: 397-401.

[31] McKenna, WG., Muschel, RJ., Gupta, AK., Hahn, SM., Bernhard, EJ. (2003) "The RAS signal transduction pathway and its role in radiation sensitivity." *Oncogene* 22: 5866-5875.

[32] Biard, DS., Martin, M., Rhun, YL., Duthu, A., Lefaix, JL., May, E., May, P. (1994) "Concomitant p53 gene mutation and increased radiosensitivity in rat lung embryo epithelial cells during neoplastic." *Cancer Res.* 54: 3361-3364.

[33] Lee, JU., Hosotani, R., Wada, M., Doi, R., Kosiba, T., Fujimoto, K., Miyamoto, Y., Tsuji, S., Nakajima, S., Nishimura, Y., Imamura, M. (1999) "Role of Bcl-2 family proteins (Bax, Bcl-2 and Bcl-X) on cellular susceptibility to radiation in pancreatic cancer cells." *Eur. J. Cancer.* 35: 1374-1380.

[34] Asanuma, K., Moriai, R., Yajima, T., Yagihashi, A., Yamada, M., Kobayashi, D., Watanabe, N. (2000) "Survivin as a radioresistance factor in pancreatic cancer." *Jpn. J. Cancer Res.* 91: 1204-1209.

[35] Olivier, M., Hainaut, P. (2001) "TP53 mutation patterns in breast cancers: searching for clues of environmental carcinogenesis." *Semin. Cancer Biol.* 11: 353-360.

[36] van Bokhoven, A., Varella-Garcia ,M., Korch, C., Johannes, WU., Smith, EE., Miller, HL., Nordeen, SK., Miller, GJ., Lucia, MS. (2003) "Molecular characterization of human prostate carcinoma cell lines." *Prostate.* 57: 205-225.

[37] Deer, EL., González-Hernández, J., Coursen, JD., Shea, JE., Ngatia, J., Scaife, CL., Firpo, MA., Mulvihill, SJ. (2010) "Phenotype and genotype of pancreatic cancer cell lines." *Pancreas.* 39: 425-435.

[38] Kirsch, DG., Santiago, PM., di Tomaso, E., Sullivan, JM., Hou, WS., Dayton, T., Jeffords, LB., Sodha, P., Mercer, KL., Cohen, R., Takeuchi, O., Korsmeyer, SJ., Bronson, RT., Kim, CF., Haigis, KM., Jain, RK., Jacks, T. (2010) "p53 controls radiation-induced gastrointestinal syndromein mice independent of apoptosis." *Science* 327: 593-596.

[39] Halliwell, B., Gutteridge, JMC. (2007) Free Radicals in Biology and Medicine. Oxford: Oxford University Press.

[40] Ralph, SJ., Rodríguez-Enríquez, S., Neuzil, J., Saavedra, E., Moreno-Sánchez, R. (2010) "The causes of cancer revisited: ''Mitochondrial malignancy" and ROS-induced oncogenic transformation – Why mitochondria are targets for cancer therapy." *Mol. Aspects Med.* 31: 145-170.

[41] Harman, D. (1956) "Aging: a theory based on free radical and radiation chemistry." *J. Gerontol.* 11 (1956): 298-300.

[42] de Oliveira, MF., Amoêdo, ND., Rumjanek, FD. (2012) "Energy and redox homeostasis in tumor cells." *Int. J. Cell Biol. doi*: 10.1155/2012/593838.

[43] Ivanova, AV., Ivanov, SV., Pascal, V., Lumsden, JM., Ward, JM., Morris, N., Tessarolo, L., Anderson, SK., Lerman, MI. (2007) "Autoimmunity, spontaneous tumourigenesis, and IL-15 insufficiency in mice with a targeted disruption of the tumour suppressor gene Fus1." *J.Pathol.* 211: 591-601.

[44] Uzhachenko, R., Issaeva, N., Boyd, K., Ivanov, SV., Carbone, DP., Ivanova, AV. (2012) "Tumour suppressor Fus1 provides a molecular link between inflammatory response and mitochondrial homeostasis." *J. Pathol.* 227: 456-469.

[45] Ivanova, AV., Ivanov, SV., Prudkin, L., Nonaka, D., Liu, Z., Tsao, A., Wistuba, I., Roth, J., Pass, HI. (2009) "Mechanisms of FUS1/ TUSC2 deficiency in mesothelioma and its tumorigenic transcriptional effects." *Mol. Cancer.* 8: doi: 10.1186/1476-4598-8-91.

[46] Lerman MI, Minna JD. (2000) "The 630-kb lung cancer homozygous deletion region on human chromosome 3p21.3: identification and evaluation of the resident candidate tumor suppressor genes. The International Lung Cancer Chromosome 3p21.3 Tumor Suppressor Gene Consortium." *Cancer Res.* 60: 6116-6133.

[47] Prudkin, L., Behrens, C., Liu, DD., Zhou, X., Ozburn, NC., Bekele, BN., Minna, JD., Moran, C., Roth, JA., Ji, L., Wistuba, II. (2008) "Loss and reduction of FUS1 protein expression is a frequent phenomenon in the pathogenesis of lung cancer." Clin Cancer Res. 14: 41-47.

[48] Nagasawa, H., Keng, P., Harley, R., Dahlberg, W., Little, JB. (1994) "Relationship between gamma-ray-induced G2/M delay and cellular radiosensitivity." *Int. J. Radiat. Biol.* 66: 373-379.

[49] Li, Z., Xia, L., Lee, LM., Khaletskiy, A., Wang, J., Wong, JY., Li, JJ. (2001) "Effector genes altered in MCF-7 human breast cancer cells after exposure to ractionated ionizing radiation." *Radiat. Res.* 155: 543-553.

[50] Rhee, SG., Chang, TS., Bae, YS., Lee, SR., Kang, SW. (2003) "Cellular regulation by hydrogen peroxide." *J. Am. Soc. Nephrol.* 8: 211-215.

[51] Kowaltowski, AJ., Netto, LE., Vercesi, AE. (1998) "The thiolspecific antioxidant enzyme prevents mitochondrial permeability transition. Evidence for the participation of reactive oxygen species in this mechanism." *J. Biol. Chem.* 273: 12766-12769.

[52] Netto, LE., Kowaltowski, AJ., Castilho, RF., Vercesi, AE. (2002) "Thiol enzymes protecting mitochondria against oxidative damage." *Methods Enzymol.* 348: 260-270.

[53] Paradise, WA., Vesper, BJ., Goel, A., Waltonen, JD., Altman, KW., Haines, GK., Radosevich, JA. (2010) "Nitric oxide: perspectives and emerging studies of a well known cytotoxin." *Int. J. Mol. Sci.* 11: 2715-2745.

[54] Blackett, NM., Wooliscroft, WE., Fielden, EM., Lillicrap, SC. (1974) "Radiation modifying effect of the free radical norpseudopelletierene-N-oxyl on normal bone marrow stem cells in vitro and in vivo." *Radiat. Res.* 58: 361-372.

[55] Cooke, BC., Fielden, EM., Johnson, M. (1976) "Polyfunctional radiosensitizers: I. Effects of a nitroxyl biradical on the survival of mammalian cells in vitro." *Radiat. Res.* 65: 152-162.

[56] George, R., Chung, TD., Vedam, SS., Ramakrishnan, V., Mohan, R., Weiss, E., Keall, PJ. (2006) "Audio-visual biofeedback for respiratory-gated radiotherapy: impact of audio instruction and audio-visual biofeedback on respiratory-gated radiotherapy." *Int. J. Radiat. Oncol. Biol. Phys.* 65: 924-933.

[57] Hogle, WP. (2007) "Radiation therapy in the treatment of breast cancer." Semin Oncol Nurs. 23: 20-28.

[58] Chen, HH., Su, WC., Chou, CY., Guo, HR., Ho, SY., Que, J., Lee, WY. (2005) "Increased expression of nitric oxide synthase and cyclooxygenase-2 is associated with poor survival in cervical cancer treated with radiotherapy." *Int. J. Radiat. Oncol. Biol. Phys.* 63: 1093-1100.

[59] Bentz, BG., Haines, GK 3rd., Hanson, DG., Radosevich, JA. (1998) "Endothelial constitutive nitric oxide synthase (ecNOS) localization in normal and neoplastic salivary tissue." *Head Neck.* 20: 304-309.

[60] American Cancer Society. (2008) http://www.cancer.org/ downloads/PRO/Oral Cancer.pdf. (Accessed July 04, 2013).

[61] Wang, YZ., Cao, YQ., Wu, JN., Chen, M., Cha, XY. (2005) "Expression of nitric oxide synthase in human gastric carcinoma and its relation to p53, PCNA." *World J. Gastroenterol.* 11: 46-50.

[62] Vesper, BJ., Elseth, KM., Tarjan, G., Haines, GK 3rd., Radosevich, JA. (2010) "Long-term adaptation of breast tumor cell lines to high concentrations of nitric oxide." *Tumor Biol.* 31: 267-275.

[63] Loibl, S., von Minckwitz, G., Weber, S., Sinn, HP., Schini-Kerth, VB., Lobysheva, I., Nepveu, F., Wolf, G., Strebhardt, K., Kaufmann, M. (2002) "Expression of endothelial and inducible nitric oxide synthase in benign and malignant lesions of the breast and measurement of nitric oxide using electron paramagnetic resonance spectroscopy." *Cancer.* 95: 1191-1198.

[64] Dueñas-Gonzalez, A., Isales, CM., del Mar Abad-Hernandez, M., Gonzalez-Sarmiento, R., Sangueza, O., Rodriguez-Commes, J. (1997) "Expression of inducible nitric oxide synthase in breast cancer correlates with metastatic disease." *Mod. Pathol.* 10: 645-649.

[65] Tschugguel, W., Schneeberger, C., Unfried, G., Czerwenka, K., Weninger, W., Mildner, M., Gruber, DM., Sator, MO., Waldhör, T., Huber, JC. (1999) "Expression of inducible nitric oxide synthase in human breast cancer depends on tumor grade." *Breast Cancer Res. Treat. 56*: 145-151.

[66] Glynn, CU., Boersma, B., Martin, D., Howe, T., Ridnour, L., Wink, D., Yi, M., Stephens, R., Yfantis, H., Ambs, S. 2007 "iNOS expression is associated with basal-like breast cancer phenotype and predicts poor survival in ERneg breast cancer." *Nitric Oxide*, 17(Supp. 1): S17.

[67] Nakamura, Y., Yasuoka, H., Tsujimoto, M., Yoshidome, K., Nakahara, M., Nakao, K., Nakamura, M., Kakudo, K. (2006) "Nitric oxide in breast cancer: induction of vascular endothelial growth factor-C and correlation with metastasis and poor prognosis." *Clin. Cancer Res.* 12: 1201-1207.

[68] Bulut, AS., Erden, E., Sak, SD., Doruk, H., Kursun, N., Dincol, D. (2005) "Significance of inducible nitric oxide synthase expression in benign and malignant breast epithelium: an immunohistochemical study of 151 cases." *Virchows Arch.* 447: 24-30.

[69] Vesper, BJ., Onul, A., Haines, GK 3rd., Tarjan, G., Xue, J., Elseth, KM., Aydogan, B., Altman, MB., Roeske, JC., Paradise, WA., De Vitto, H., Radosevich, JA. (2013) "Part I. Molecular and cellular characterization of high nitric oxide-adapted human breast adenocarcinoma cell lines." *Tumor Biol.* 34: 203-214.

[70] De Vitto, H., Mendonça, BS., Elseth, KM., Vesper, BJ., Portari, EA., Gallo, CV., Paradise, WA., Rumjanek, FD., Radosevich, JA. (2013a) "Part II. Mitochondrial mutational status of high nitric oxide adapted cell line BT-20 (BT-20- HNO) as it relates to human primary breast tumors." *Tumor Biol.* 34: 337-347.

[71] De Vitto, H., Mendonça, BS., Elseth, KM., Onul, A., Xue, J., Vesper, BJ., Gallo, CV., Rumjanek, FD., Paradise, WA., Radosevich, JA. (2013b) "Part III. Molecular changes induced by high nitric oxide adaptation in human breast cancer cell line BT-20 (BT-20-HNO): a switch from aerobic to anaerobic metabolism." *Tumor Biol.* 34: 403-413.

[72] Ralph, SJ., Low, P., Dong, L., Lawen, A., Neuzil J. (2006) "Mitocans: mitochondrial targeted anti-cancer drugs as improved therapies and related patent documents." *Recent Pat. Anticancer Drug Discov.* 1: 327-346.

[73] Palorini, R., De Rasmo, D., Gaviraghi, M., Sala Danna, L., Signorile, A., Cirulli, C., Chiaradonna, F., Alberghina, L., Papa, S. (2013a) Oncogenic K-ras expression is associated with derangement of the cAMP/PKA pathway and forskolin-reversible alterations of mitochondrial dynamics and respiration. Translated by 32: 352-362., *Oncogene.*

[74] Cadet, J., Sage, E., Douki, T. (2005) "Ultraviolet radiation-mediated damage to cellular DNA." *Mutat. Res.* 571: 3-17.

[75] Setlow, RB. (1966) "Cyclobutane-type pyrimidine dimers in polynucleotides." *Science.* 153: 379-386.

[76] Van Holde, K. E., Mathews Christopher, K. (1990) Biochemistry. Menlo Park, Calif: Benjamin / Cummings Pub. Co.

[77] Brash DE, Rudolph JA, Simon JA, Lin A, McKenna GJ, Baden HP, Halperin AJ, Pontén J. (1991) "A role for sunlight in skin cancer: UV-induced p53 mutations in squamous cell carcinoma." *Proc. Natl. Acad. Sci. U S A.* 88: 10124-10128.

[78] Ziegler, A., Leffell, DJ., Kunala, S., Sharma, HW., Gailani, M., Simon, JA., Halperin, AJ., Baden, HP., Shapiro, PE., Bale, AE. (1993) "Mutation hotspots due to sunlight in the p53 gene of nonmelanoma skin cancers." *Proc. Natl. Acad. Sci. U S A.* 90: 4216-4220.

[79] Dumaz, N., Drougard, C., Sarasin, A., Daya-Grosjean, L. (1993) "Specific UV-induced mutation spectrum in the p53 gene of skin tumors from DNA-repair-deficient xeroderma pigmentosum patients." *Proc. Natl. Acad. Sci. U S A.* 90: 10529-10533.

[80] Clancy, S. (2008) DNA Damage & Repair: Mechanisms for Maintaining DNA Integrity. *Nature.*

[81] Cortat, B., Garcia, CC., Quinet, A., Schuch, AP., de Lima-Bessa, KM., Menck, CF. (2013) "The relative roles of DNA damage induced by UVA irradiation in human cells." *Photochem. Photobiol. Sci.* [Epub ahead of print].

[82] Kielbassa, C., Roza, L., Epe, B. (1997) "Wavelength dependence of oxidative DNA damage induced by UV and visible light." *Carcinogenesis.* 18: 811-816.

[83] Pouget, JP., Douki, T., Richard, MJ., Cadet, J. (2000) "DNA damage induced in cells by gamma and UVA radiation as measured by HPLC/GC-MS and HPLC-EC and Comet assay." *Chem. Res. Toxicol.* 13: 541-549.

[84] Tyrrell, RM. (1973) "Induction of pyrimidine dimers in bacterial DNA by 365 nm radiation." *Photochem. Photobiol.* 17: 69-73.

[85] Freeman, SE., Ryan, SL. (1990) "Wavelength dependence for UV-induced pyrimidine dimer formation in DNA of human peripheral blood lymphocytes." *Mutat. Res.* 235(3):181-6.

[86] Perdiz, D., Grof, P., Mezzina, M., Nikaido, O., Moustacchi, E., Sage, E. (2000) "Distribution and repair of bipyrimidine photoproducts in solar UV-irradiated mammalian cells. Possible role of Dewar photoproducts in solar mutagenesis." *J. Biol. Chem.* 275: 26732-26742.

[87] Young, AR., Chadwick, CA., Harrison, GI., Nikaido, O., Ramsden, J., Potten, CS. (1998) "The similarity of action spectra for thymine dimers in human epidermis and erythema suggests that DNA is the chromophore for erythema." *J. Invest. Dermatol.* 1998 Dec;111(6):982-8.

[88] Freeman, SE., Hacham, H., Gange, RW., Maytum, DJ., Sutherland, JC., Sutherland, BM. (1989) "Wavelength dependence of pyrimidine dimer formation in DNA of human skin irradiated in situ with ultraviolet light." *Proc. Natl. Acad. Sci. U S A* 86: 5605-5609.

[89] Douki, T., Perdiz, D., Gróf, P., Kuluncsics, Z., Moustacchi, E., Cadet, J., Sage, E. (1999) "Oxidation of guanine in cellular DNA by solar UV radiation: biological role." *Photochem Photobiol.* 70: 184-190.

[90] Douki, T., Reynaud-Angelin, A., Cadet, J., Sage, E. (2003) "Bipyrimidine photoproducts rather than oxidative lesions are the main type of DNA damage involved in the genotoxic effect of solar UVA radiation." *Biochemistry.* 42(30):9221-6.

[91] Courdavault, S., Baudouin, C., Charveron, M., Favier, A., Cadet, J., Douki, T. (2004) "Larger yield of cyclobutane dimers than 8-oxo-7,8-dihydroguanine in the DNA of UVA-irradiated human skin cells." *Mutat. Res.* 556: 135-142.

[92] Rochette, PJ., Therrien, JP., Drouin, R., Perdiz, D., Bastien, N., Drobetsky, EA., Sage, E. (2003) "UVA-induced cyclobutane pyrimidine dimers form predominantly at thymine-thymine dipyrimidines and correlate with the mutation spectrum in rodent cells." *Nucleic. Acids Res.* 31: 2786-2794.

[93] Pedeux, R., Al-Irani, N., Marteau, C., Pellicier, F., Branche, R., Ozturk, M., Franchi, J., Doré, JF. (1998) "Thymidine dinucleotides induce S phase cell cycle arrest in addition to increased melanogenesis in human melanocytes." *J. Invest. Dermatol.* 111: 472-477.

[94] Goetz, ME., Luch, A. (2008) "Reactive species: a cell damaging rout assisting to chemical carcinogens." Cancer Lett. 266: 73-83.

[95] Halliday, GM. (2005) "Inflammation, gene mutation and photoimmunosuppression in response to UVR-induced oxidative damage contributes to photocarcinogenesis." *Mutat. Res.* 571: 107-120.

[96] Liu, Y., Gao, W., Zhang, D. (2010) "Effects of cigarette smoke extract on A549 cells and human lung fibroblasts treated with transforming growth factor-beta1 in a coculture system." *Clin. Exp. Med.* 10: 159-167.

[97] Rajalingam, K., Schreck, R., Rapp, UR., Albert, S. (2007) "Ras oncogenes and their downstream targets." *Biochim. Biophys. Acta.* 1773: 1177-1195.

[98] Strano, S., Dell'Orso, S., Di Agostino, S., Fontemaggi, G., Sacchi, A., Blandino, G. (2007) "Mutant p53: an oncogenic transcription factor." *Oncogene.* 26: 2212-2219.

[99] Nishigori, C., Hattori, Y., Toyokuni, S. (2004) "Role of reactive oxygen species in skin carcinogenesis." *Antioxid. Redox. Signal.* 6: 561-70.

[100] Ortega, A., Carretero, J., Obrador, E., Estrela, JM. (2008) "Tumoricidal activity of endothelium-derived NO and the survival of metastatic cells with high GSH and Bcl-2 levels." *Nitric Oxide.* 19: 107-114.

[101] Park, JW., Kim, S., Bahk, YY. (2006a) "A proteomic approach for dissecting H-Ras signaling networks in NIH/3T3 mouse embryonic fibroblast cells." Proteomics. 6: 2433-2443.

[102] Park, JW., Kim, S., Lim, KJ., Simpson, RJ., Kim, YS., Bahk, YY. (2006b) "A proteomic approach for unraveling the oncogenic H-Ras protein networks in NIH/3T3 mouse embryonic fibroblast cells." *Proteomics.* 2006 Feb;6(4):1175-86. 6: 1175-1186.

[103] Tchernitsa, OI., Sers, C., Zuber, J., Hinzmann, B., Grips, M., Schramme, A., Lund, P., Schwendel, A., Rosenthal, A., Schäfer, R. (2004) "Transcriptional basis of KRAS oncogene-mediated cellular transformation in ovarian epithelial cells." *Oncogene.* 23: 4536-4555.

[104] Jonkers, J., Berns, A. (2004) "Oncogene addiction: sometimes a temporary slavery." *Cancer Cell.* 6: 535-538.

[105] Weinstein, IB. (2002) "Cancer. Addiction to oncogenes--the Achilles heal of cancer." *Science.* 297: 63-64.

[106] McKinney, C., Prives, C. (2005) Regulation of p53 DNA Binding. 25 Years of p53 Research. Dordrecht: *Springer.*

[107] Costanzo, A., Merlo, P., Pediconi, N., Fulco, M., Sartorelli, V., Cole, PA., Fontemaggi, G., Fanciulli, M., Schiltz, L., Blandino, G., Balsano, C., Levrero, M. (2002) "DNA damage-dependent acetylation of p73 dictates the selective activation of apoptotic target genes." *Mol. Cell.* 9: 175-186.

[108] Xu, Y. (2003) "Regulation of p53 responses by post-translational modifications." *Cell Death Differ.* 10: 400-403.

[109] Sigal, A., Rotter, V. (2000) "Oncogenic mutations of the p53 tumor suppressor: the demons of the guardian of the genome." *Cancer Res.* 60: 6788-6793.

[110] Lain, S., Lane, D. (2005) Novel p53-based Therapies: Strategies and Future Prospect, 25 Years of p53 Research. Dordrecht: *Springer*.

[111] Scian, MJ., Stagliano, KE., Ellis, MA., Hassan, S., Bowman, M., Miles, MF., Deb, SP., Deb, S. (2004) "Modulation of gene expression by tumor-derived p53 mutants." *Cancer* Res. 64: 7447-7454.

[112] Mizuarai, S., Yamanaka, K., Kotani, H. (2006) "Mutant p53 induces the GEF-H1 oncogene, a guanine nucleotide exchange factor-H1 for RhoA, resulting in accelerated cell proliferation in tumor cells." *Cancer Res.* 66: 6319-6326.

[113] Di Agostino, S., Strano, S., Emiliozzi, V., Zerbini, V., Mottolese, M., Sacchi, A., Blandino, G., Piaggio, G. (2006)"Gain of function of mutant p53: the mutant p53/NF-Y protein complex reveals an aberrant transcriptional mechanism of cell cycle regulation." *Cancer Cell.* 10: 191-202.

[114] Hanahan, D., Weinberg, RA. (2011) "Hallmarks of cancer: the next generation." *Cell* 144: 646-674.

[115] Lee, C., Safdie, FM., Raffaghello, L., Wei, M., Madia, F., Parrella, E., Hwang, D., Cohen, P., Bianchi, G., Longo, VD. (2010) "Reduced levels of IGF-Imediate differential protection of normal and cancer cells in response to fasting and improve chemotherapeutic index." *Cancer Res.* 70: 1564-1572.

[116] Mathupala, SP., Colenm CB., Parajuli, P., Sloan, AE. (2007) "Lactate and malignant tumors: a therapeutic target at the end stage of glycolysis." *J. Bioenerg. Biomembr.* 39: 73-77.

[117] Ortega, AD., Sánchez-Aragó, M., Giner-Sánchez, D., Sánchez-Cenizo, L., Willers, I., Cuezva, JM. (2009) "Glucose avidity of carcinomas." *Cancer Lett.* 276. 125-135.

[118] DeBerardinis RJ, Lum JJ, Hatzivassiliou G, Thompson CB. (2008) "The biology of cancer: metabolic reprogramming fuels cell growth and proliferation." *Cell Metab.* 7: 11-20.

[119] Warburg, O. (1930) Metabolism of tumors. London:: Arnold Constable.

[120] Smolková, K., Plecitá-Hlavatá, L., Bellance, N., Benard, G., Rossignol, R., Ježek, P. (2011) "Waves of gene regulation suppress and then restore oxidative phosphorylation in cancer cells." *Int. J. Biochem. Cell Biol.* 43: 950-968

[121] Warburg, O. (1956) "On the origin of cancer cells." Science 123: 309-314.

[122] Gordan, JD., Thompson, CB., Simon, MC. (2007) "HIF and c-Myc: sibling rivals for control of cancer cell metabolism and proliferation." *Cancer Cell.* 12: 108-13.

[123] Martinez-Outschoorn, UE., Balliet, RM., Rivadeneira, DB., Chiavarina, B., Pavlides, S., Wang, C., Whitaker-Menezes, D., Daumer, KM., Lin, Z., Witkiewicz, AK., Flomenberg, N., Howell, A., Pestell, RG., Knudsen, ES., Sotgia, F., Lisanti, MP. (2010) "Oxidative stress in cancer associated fibroblasts drives tumor-stroma co-evolution: A

new paradigm for understanding tumor metabolism, the field effect and genomic instability in cancer cells." *Cell Cycle.* 9: 3256-3276.

[124] Palorini, R., Simonetto, T., Cirulli, C., Chiaradonna, F. (2013b) "Mitochondrial complex I inhibitors and forced oxidative phosphorylation synergize in inducing cancer cell death." *Int. J. Cell Biol.*, doi: 10.1155/2013/243876.

[125] Alberghina, L., Gaglio, D., Gelfi, C., Moresco, RM., Mauri, G., Bertolazzi, P., Messa, C., Gilardi, MC., Chiaradonna, F., Vanoni, M. (2012) "Cancer cell growth and survival as a system-level property sustained by enhanced glycolysis and mitochondrial metabolic remodeling." *Front Physiol.*, doi: 10.3389/fphys.2012.00362.

[126] Modica-Napolitano, JS., Singh, KK. (2004) "Mitochondrial dysfunction in cancer." *Mitochondrion.* 4: 755-762.

[127] Wallace, DC. (2005) "A mitochondrial paradigm of metabolic and degenerative diseases, aging, and cancer: a dawn for evolutionary medicine." *Annu. Rev. Genet.* 39: 359-407.

[128] Baracca, A., Chiaradonna, F., Sgarbi, G., Solaini, G., Alberghina, L., Lenaz, G. (2010) "Mitochondrial Complex I decrease is responsible for bioenergetic dysfunction in K-ras transformed cells." *Biochim. Biophys. Acta.* 1797: 314-323.

[129] Koppenol, WH., Bounds, PL., Dang, CV. (2011) "Otto Warburg's contributions to current concepts of cancer metabolism." *Nat. Rev. Cancer.* 11: 325-337.

[130] Alberghina, L. Westerhoff, H. (2005) Systems Biology: Definitions and Perspectives. Heidelberg:: *Springer.*

[131] Chiaradonna, F., Sacco, E., Manzoni, R., Giorgio, M., Vanoni, M., Alberghina, L. (2006) "Ras-dependent carbon metabolism and transformation in mouse fibroblasts." *Oncogene.* 25: 5391-5404.

[132] Palorini, R., Cammarata, F., Balestrieri, C., Monestiroli, A., Vasso, M., Gelfi, C., Alberghina, L., Chiaradonna, F. (2013c) "Glucose starvation induces cell death in K-ras-transformed cells by interfering with the hexosamine biosynthesis pathway and activating the unfolded protein response." *Cell Death Dis.* doi: 10.1038/cddis.2013.257.

[133] Caro-Maldonado, A., Tait, SW., Ramírez-Peinado, S., Ricci, JE., Fabregat, I., Green, DR., Muñoz-Pinedo, C. (2010) "Glucose deprivation induces an atypical form of apoptosis mediated by caspase-8 in Bax-, Bak-deficient cells." *Cell Death Differ.* 17: 1335-1344.

[134] El Mjiyad, N., Caro-Maldonado, A., Ramírez-Peinado, S., Muñoz-Pinedo, C. (2011) "Sugar-free approaches to cancer cell killing." *Oncogene.* 30: 253-264.

[135] Yuneva, M., Zamboni, N., Oefner, P., Sachidanandam, R., Lazebnik, Y. (2007) "Deficiency in glutamine but not glucose induces MYC-dependent apoptosis in human cells." *J. Cell Biol.* 178: 93-105.

[136] Ron, D., Walter, P. (2007) "Signal integration in the endoplasmic reticulum unfolded protein response." *Nat. Rev. Mol. Cell Biol.* 8: 519-529.

[137] Tabas, I., Ron, D. (2011) "Integrating the mechanisms of apoptosis induced by endoplasmic reticulum stress." *Nat. Cell Biol.* 13: 184-90.

[138] McClain, DA. (2002) "Hexosamines as mediators of nutrient sensing and regulation in diabetes." *J. Diabetes Complications.* 16: 72-80.

[139] Slawson, C., Copeland, RJ., Hart, GW. (2010) "O-GlcNAc signaling: a metabolic link between diabetes and cancer?" *Trends Biochem. Sci.* 35: 547-555.

[140] Chiaradonna, F., Moresco, RM., Airoldi, C., Gaglio, D., Palorini, R., Nicotra, F., Messa, C., Alberghina, L. (2012) "From cancer metabolism to new biomarkers and drug targets." *Biotechnol. Adv.* 30: 30-51.

[141] Biasutto, L., Dong, LF., Zoratti, M., Neuzil, J. (2010) "Mitochondrially targeted anti-cancer agents." *Mitochondrion.* 10: 670-681.

[142] Bonnet, S., Archer, SL., Allalunis-Turner, J., Haromy, A., Beaulieu, C., Thompson, R., Lee, CT., Lopaschuk, GD., Puttagunta, L., Bonnet, S., Harry, G., Hashimoto, K., Porter, CJ., Andrade, MA., Thebaud, B., Michelakis, ED. (2007) "A mitochondria-K+ channel axis is suppressed in cancer and its normalization promotes apoptosis and inhibits cancer growth." *Cancer Cell.* 11: 37-51.

[143] Menendez, JA., Lupu, R. (2007) "Fatty acid synthase and the lipogenic phenotype in cancer pathogenesis." *Nat. Rev. Cancer.* 7: 763-777.

[144] Mashima, T., Seimiya, H., Tsuruo, T. (2009) "De novo fatty-acid synthesis and related pathways as molecular targets for cancer therapy." *Br. J. Cancer.* 100: 1369-1372.

[145] Galluzzi L., Kepp O., Vander Heiden M.G., Kroemer G. (2014) "Metabolic targets for cancer therapy" *Nature Reviews* 12: 829-846.

In: UV Radiation: Properties, Effects, and Applications
Editor: James A. Radosevich
ISBN: 978-1-63321-090-5

Chapter 6

CO-CARCINOGEN AND BIOENERGETIC MODULATION OF UV-INDUCED DNA DAMAGE

Benjamin C. Thompson, Diona L. Damian and Gary M. Halliday*

Discipline of Dermatology, Royal Prince Alfred Hospital, Bosch Institute, University of Sydney, Sydney, Australia

ABSTRACT

Both arsenic and bioenergetics modulate cellular responses to UV radiation. Exposure of our skin to UV radiation leads to DNA damage. Cellular responses to DNA damage are crucial for the stability of the genome and prevention of skin cancer. Non-melanoma skin cancer is the most common cancer in humans. The main risk factor is exposure to UV radiation, although on a global level arsenic exposure, usually through contamination of drinking water, is also a significant cause. UV and arsenic act as co-carcinogens in the skin. Arsenic has a range of cellular effects that are not fully understood. It increases oxidative stress and also modulates the cell's response to DNA damage. The cell's response to UV-induced DNA damage is also highly dependent on bioenergetics. Cellular ATP and NAD+ are depleted following exposure to UV radiation, subsequently compromising repair pathways and leading to inefficient DNA repair. Nicotinamide, the amide form of vitamin B3, replenishes ATP and NAD+. It ensures that the cell has adequate energy following UV radiation, maximising the capacity for DNA repair. In summary, arsenic and bioenergetics are two clinically relevant factors that alter the cells capabilities to respond appropriately to DNA damage and can ultimately affect genomic stability and skin carcinogenesis.

Keywords: UV Radiation, Arsenic, DNA damage, Nicotinamide

* Address for Correspondence: Prof. Gary Halliday, Dermatology, Blackburn Building D06, University of Sydney, Sydney, NSW, Australia 2006. Email: gary.halliday@sydney.edu.au.

ABBREVIATIONS

8oxoG	8-oxo-2'-deoxyguanosine
AS-TM	arsenite transformed
ATP	adenosine-5'-triphosphate
BCC	Basal cell carcinoma
BER	base excision repair
CPD	cyclobutane pyrimidine dimers
DMS	dimethyl sulphate
ENU	ethylnitrosourea
ETO	etoposide
GGR	global genome repair
MNNG	N-methyl-N'-nitro-N-nitrosoguanidine
NAD+	nicotinamide adenine dinucleotide
NER	nucleotide excision repair
NMSC	non-melanoma skin cancer
PARP	poly (ADP-ribose) polymerase
SCC	squamous cell carcinoma
TCR	transcription-coupled repair
UV	Ultraviolet
WHO	World Health Organisation

6.0. INTRODUCTION

DNA damage and repair are key mechanisms that occur in response to ultraviolet (UV) radiation. DNA damage that fails to be repaired or to lead to apoptosis can lead to skin carcinogenesis. Efficient repair of DNA lesions correlates with a reduced rate of genomic mutations and is a key protective mechanism. Inefficient repair of DNA lesions increases the rate of cancer development [1]. The genome is constantly threatened by exogenous and endogenous agents that damage DNA. The major environmental carcinogen and cause of DNA damage in skin is UV radiation from sunlight. This is divided into two wavebands, UVA (320-400nm) and UVB (290-320nm), with lower wavelengths (UVC) being filtered by the atmosphere and ozone layer so that they do not contribute to sunlight-induced DNA damage in humans. Ninety five percent of the UV radiation that reaches the surface of the Earth consists of UVA, with the remaining 5% being UVB [2]. There is significant evidence that both wavebands contribute to sunlight-induced DNA damage and carcinogenesis [3]. Different types of DNA damage are formed depending on the inciting agent, and include DNA cross-links, strand breaks and adducts [1].

Fortunately, eukaryotic cells are equipped with the ability to identify DNA damage and activate the appropriate response; either repair of the DNA damage or activation of apoptotic pathways if the damage is too great to accurately repair. These pathways can be further modified by various environmental and endogenous factors that alter the cell's ability to effectively deal with the DNA damage. These include co-carcinogens and modifiers of bioenergetics, the energy status of the cell. UV itself also influences these pathways.

UVA predominantly induces oxidative DNA damage, often quantified through the by-product 8-oxo-2'-deoxyguanosine (8oxoG) as this is the most frequent oxidative photolesion in DNA (1). UVB predominantly induces direct DNA damage, leading to the formation of the UV signature photoproducts cyclobutane pyrimidine dimers (CPD)s and 6,4-pyrimidine-pyrimidone dimmers [4]. Repair of DNA depends on the type of damage suffered; the major pathways for UV-induced DNA damage are through nucleotide excision repair (NER) and base excision repair (BER) [5].

DNA strand breaks are detected by the enzyme poly (ADP-ribose) polymerase (PARP), which is also involved in activation of DNA repair and apoptotic pathways [6]. Arsenic, a contaminator of water sources in certain populations worldwide, has been shown to be a co-carcinogen in mouse models [7]. In addition, it synergistically increases UV-induced DNA damage but does not substantially damage DNA itself at clinically relevant concentrations. It is thought to be an inhibitor of PARP and consequently to inhibit DNA repair and promote the survival of DNA-damaged cells [8, 9].

Nicotinamide (vitamin B3) can inhibit PARP at high concentrations, but is also a key substrate of PARP [10]. Nicotinamide is also the sole substrate for nicotinamide adenine dinucleotide (NAD+), an essential co-enzyme for adenosine-5'-triphosphate (ATP) production via the Krebs cycle. Nicotinamide has been show to enhance DNA repair in lymphocytes [11]. Nicotinamide replenishes cellular ATP levels after UV exposure [12], reduces the immune suppressive effects of both UVA and UVB in human skin [13, 14] and, is a promising chemo-preventive agent. Therefore the bioenergetic state of the cell is critical for enabling optimal repair of UV damaged DNA and this can be modulated by components of oxidative phosphorylation such as nicotinamide.

In summary, UV induced DNA damage and repair is a central component of carcinogenesis pathways. Modulation of repair pathways, for better or worse, with co-carcinogens such as arsenic, or by influencing energy production in cells, can significantly affect DNA damage and repair.

6.1. Skin Cancer

Skin cancer is the most common cancer in Caucasians. In Australia, by age 70, more than 50% of people will develop at least one non-melanoma skin cancer (NMSC). The incidence of skin cancer in Australia continues to rise, and is five times that of all other cancers combined [15]. The total cost of NMSC in Australia was $511 million in 2010 [16] - the most expensive of all cancer types [17].

Skin cancer is divided into melanoma and NMSC due to the different cells of origin, melanocytes or keratinocytes respectively. Basal cell carcinoma (BCC) and squamous cell carcinoma (SCC) are the most common types of NMSC, accounting for 95% of all skin cancers. Although there are other types of skin cancer, the term NMSC is generally used to describe only these two. BCC are four times more common than SCC [18].

6.1.1. Skin Cancer Carcinogens

Human skin is exposed to UV on a daily basis. UV exposure is by far the major skin carcinogen. Skin cancer risk is influenced by skin type, immunosuppression, genetic DNA repair deficiency syndromes such as xeroderma pigmentosum and environmental exposures including ionizing radiation. Metals such as chromium and arsenic act as skin cancer co-carcinogens [19]. Other chemical agents associated with skin cancer include mineral oil, coal tar, soot, polychlorinated biphenyl, 4,4'-Bipyridyl, psoralens (combined with UVA for psoriasis treatment) and, nitrogen mustard [20].

6.1.2. Electromagnetic Radiation

The sun emits a broad spectrum of electromagnetic radiation. The light that reaches the earth's surface comprises visible light (400-700nm), UV radiation and, infrared radiation [21]. At sea level, the total radiation dose is made up of 52% visible light, 42% infrared and 6% UV radiation [22]. Ultraviolet radiation is subdivided into UVA (320-400nm), UVB (290-320nm) and UVC (200-290nm). The stratosphere absorbs UVC and some (mainly shortwave) UVB [22]. The UV radiation that reaches the earth's surface consists of 95% UVA and 5% UVB [2]. UV radiation has the potential to cause immunosuppression, DNA damage, cell death, inflammation, premature aging and skin cancer [23, 24]. UV suppression of the immune system inhibits immune-mediated rejection of skin cancer and infectious agents, increasing these diseases. It may also reduce the incidence or severity of autoimmunity [25]. UV-induced DNA damage is one of the mechanisms that lead to immunosuppression [26, 27]. UV-induced inflammation results in the production of a large number of cytokines and growth factors that disrupt skin homeostasis. Chronic inflammation and accumulation of DNA damage lead to structural and functional changes. These effects contribute to skin aging and photocarcinogenesis [28].

6.1.3. Arsenic-Induced Skin Cancer

Inorganic arsenic, through contaminated drinking water, is an important contributor to skin cancer incidence globally. Arsenic is present throughout the environment, in water, soil and food. Deep-water tube wells in districts of Bangladesh and West Bengal contain elevated levels of arsenic. The World Health Organisation (WHO) pre-1993 recommended a maximum arsenic concentration for drinking water of 50μg/L. Thirty seven percent of Bangladeshi and 25% of West Bengal wells contain concentrations above these levels. The USA Environmental Protection Agency and the WHO currently recommend a maximum concentration of 10μg/L [29, 30]. More than 150 million people live in these districts and are therefore at risk. To indicate the extent of this problem, a 2001 study found that 20.6% of people from arsenic affected districts of Bangladesh had dermatological signs of chronic arsenic toxicity [31]. Arsenic is present in drinking water in other countries including Taiwan, Argentina and Chile [32, 33]. Skin lesions are the most common health outcome in chronic arsenic exposure. Arsenic accumulates in skin, hair and nails. Other organs commonly affected include lung, liver and, bladder. The most common skin cancers induced are

Bowen's disease (SCC in-situ), BCC and SCC. Non-malignant changes include hyperkeratosis and hyperpigmentation [34] (Figure 1). Arsenic is also associated with internal malignancies. A dose response relationship for arsenic contaminated well water concentration and lung, bladder and, renal cancer in males and females were observed in Taiwan, in addition to prostate and liver cancer in males. Furthermore, a dose response relationship was also noted with peripheral vascular and cardiovascular disease [35].

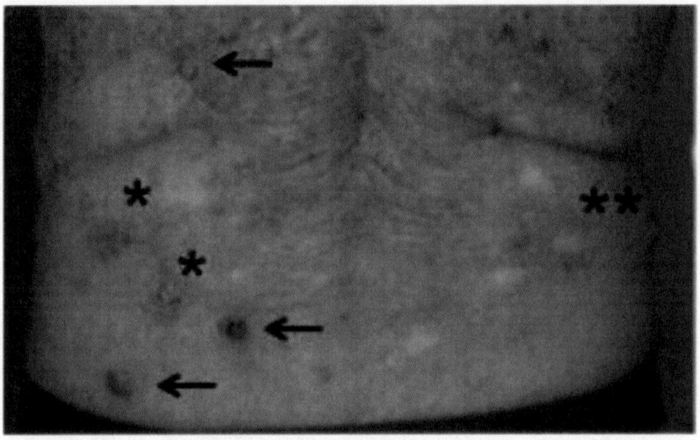

Figure 1. An elderly gentleman presenting with multiple nodular and superficial BCCs and a history of iatrogenic arsenic exposure. He has had hundreds of previous BCCs, many of which were treated with curettage leaving broad scars. Arrows represent nodular BCCs, * represents superficial BCCs, ** represents scarring.

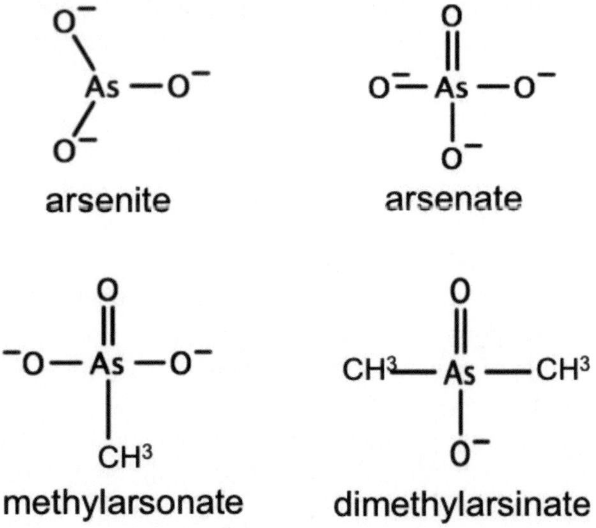

Figure 2. Arsenic compounds and their metabolites.

Arsenic is present in inorganic and organic forms. Organic arsenic compounds contain carbon and hydrogen. Inorganic arsenic is the most abundant form in the environment. It is compounded with another element, usually oxygen, chlorine or, sulfur, but not carbon. In deep water tube wells it is predominantly in the form arsenite (valency of 3). Surface water contains mostly arsenate (valency of 5), however, this is reduced to arsenite by liver metabolism [36]. Further metabolism results in the production of 2 methylated metabolites, methylarsonate and dimethylarsinate [37] (Figure 2). Arsenite is the most active form and enters cells more readily, although all forms are present in humans and so must be considered in studies of arsenic [36].

6.1.4. UV and Arsenic As Skin Cancer Co-carcinogens

Although arsenic has long been linked with cancer in humans, the mechanisms are not fully understood. There appears to be a broad range of cellular effects secondary to arsenic. Interference with the cell's response to DNA damage and subsequent genomic instability are likely to be significant co-carcinogenic mechanisms of arsenic. In mouse models, animals fed an arsenic containing diet did not develop tumours. Tumours developed only in transgenic mice containing the H-ras oncogene and painted with the tumour promoter 12-O-tetradecanoyl phorbol-13-acetate [38] or in F344/DuCrj rats pre-treated with multiple carcinogens (combined treatment with diethyl nitrosamine, N-methyl-N-nitrosourea, 1,2-dimethylhydrazine, N-butyl-N-(4-hydroxybutyl)nitrosamine and N-bis(2-hydroxylpropyl)nitrosamine) [39]. This demonstrates that arsenic acts as a cancer promoter but is not a complete carcinogen. Rossman et al. in 2001 found that mice fed sodium arsenite (10mg/L) through their drinking water for 26 weeks and exposed to UV radiation developed 2.4 times the number of skin tumours compared with UV radiation alone. The skin tumours in the arsenic-treated mice arose at an earlier time following commencement of UV irradiation. The time to 100% tumour incidence was 19 versus 25 weeks, with arsenic-induced tumours noted to be larger and more invasive. There were no tumours in the unirradiated mice. Hence UV and arsenic can act as skin cancer co-carcinogens [7].

6.2. Effect of Arsenic on DNA Repair and Apoptosis

One way that arsenic and UV may interact to augment UV-carcinogenesis is through arsenic inhibiting repair of UV-induced DNA damage. The photolesions induced by UV inhibit replication and transcription of DNA. If the lesions are not repaired efficiently, they can result in the incorporation of mutations into the cells' DNA [40]. Arsenic may exacerbate this by inhibiting the repair process [41, 8].

Different types of inciting agents induce distinct types of DNA damage that result in diverse processes being triggered. When sunlight reaches the skin surface, the majority is reflected. UVB penetrates into the outer living layer of the skin, called the epidermis, and most skin cancers in humans arise from within this layer. In the epidermis, UVB is directly absorbed by DNA, leading to the formation of dimeric photoproducts between adjacent pyrimidines on each DNA strand. The main signature photolesions formed when UVB is absorbed by DNA are CPDs and pyrimidine (6-4) pyrimidone photoproducts. This damage can result in GC-AT mutations if the cell divides before it manages to repair this damage. UVA has also been demonstrated to form CPDs, however a much greater dose is required to induced the same level of CPDs as UVB [42]. UVA penetrates deeper into the skin than UVB but has insufficient energy per photon to be absorbed by DNA. Instead, it is absorbed by a number of undefined chromophores, leading to the formation of reactive oxygen species (ROS). When ROS formation exceeds anti-oxidant defences, the cells undergo oxidative stress leading to the oxidation of nucleotide bases. Oxidation of deoxyguanine moieties to form 8oxoG is the most common marker of DNA oxidation as guanine has the lowest oxidation potential of the nucleotides that comprise DNA [4]. UVA and UVB mutations are present in similar numbers in the p53 gene sequences in benign and malignant skin tumours [1].

Fortunately, eukaryotic cells have evolved to live under conditions of continual exposure to DNA damaging agents, and have developed various mechanisms to maintain homeostasis and integrity of the genome. More than 130 genes are involved in DNA repair. Cells either tolerate DNA damage, or activate pathways leading to apoptosis and therefore removal of the damaged cell from the organism, or pathways leading to repair of the DNA. These pathways sense damage to DNA leading to cell cycle arrest. Molecules are recruited to the area of damaged DNA and form a complex that repairs the DNA. DNA is normally packaged into nucleosomes, the subunits of chromatin. The unravelling of chromosomes is required to allow for access of the enzymes to the lesion [43]. Depending on the lesion, nucleotide excision repair (NER) or base excision repair (BER) enzymes begin to repair the damage [44]. These processes are collectively termed the DNA damage response. PARP is central in this response [6]. The DNA damage response is highly dependent on supplies of ATP and therefore substances that influence the bioenergetic state of the cell have a large effect on this response.

In humans, there is evidence that arsenic-induced Bowen's disease and arsenic keratoses (premalignant skin lesions confined to the epidermis) contain significantly higher levels of 8oxoG than non-arsenic-induced Bowen's disease [45]. This suggests that arsenic increases oxidative stress, disrupts the anti-oxidant defence, and/or reduces the DNA repair capacity of cells. Consistent with this observation, HaCaT cells (human keratinocyte cell line) incubated with 10-30 μM sodium arsenite have been shown to have significantly increased levels of 8oxoG [46]. This concentration range is also known to affect cell viability [47]. At clinically relevant, non-cytotoxic levels (<5μM), incubation with sodium arsenite for 24 hours alone did not increase 8oxoG levels, suggesting that at clinically relevant concentrations arsenite does not cause sufficient oxidative stress to overcome cellular defence mechanisms. However it is likely that these clinically relevant concentrations of arsenite would compromise oxidative defence systems making the cell more vulnerable to another oxidative event such as UV radiation. Exposure of HaCaT keratinocytes to 8 J/cm^2 UV and 2 μM arsenite synergistically increased 8oxoG damage to significantly greater levels than UV irradiation alone [9]. This is consistent with the suggestion that clinically relevant doses of arsenite have a deleterious effect on anti-oxidant defence and/or the DNA damage response, but not sufficient to have a detectable affect on DNA damage until a co-carcinogen is introduced into the system. Another study demonstrated that levels of 8oxoG are raised in cells that are able to methylate arsenic compared to those unable to do so (in a comparison of epithelial cell lines) [48]. Arsenic's methylated products increase ROS and generate DNA strand breaks at low concentrations [49]. Arsenic appears to alter BER through the binding of zinc finger proteins, namely PARP-1, at low concentrations, as discussed below, but also via binding of formamidopyrimidine DNA glycosylase, a glycosylase enzyme that initiates BER, at high concentrations [49]. Therefore, it appears that arsenic acts on multiple mechanisms including inhibition of DNA repair and decreasing ROS defence mechanisms.

BER is the cell's repair pathway that removes small base lesions. The majority of oxidative DNA lesions, such as 8oxoG produced from UV, are repaired through BER. ROS are continually being produced by cellular respiration due to inefficiencies in ATP production by oxidative metabolism. UVA, UVB, and other agents increase the level of ROS and lead to the formation of DNA damage such as DNA base modifications and base loss [4]. The first step is recognition and removal of the damaged base by DNA N-glycosylases. This type of enzyme continuously monitors and scans DNA for structural abnormalities caused by mismatched or damaged bases. 8oxoG removal is undertaken by the human 8-oxoguanine DNA glyocsylase-1 enzyme. DNA glycosylases are responsible for repairing different subsets of bases, even though some have quite a broad spectrum of activity. The site is then processed by apurinic/ apyrimidinic specific endonuclease or lyase activity resulting in single strand interruptions. These interruptions are then replaced by DNA polymerase followed by ligation [5]. All of these processes requite large amounts of cellular energy.

The repair of CPDs is also affected by arsenic exposure. HaCaT cells exposed for 24 hours to sodium arsenite (0.5μM – 5μM) have defective repair of UV radiation-induced CPDs in a concentration dependent manner. This effect was reversed by specific inhibition of inducible nitric oxide synthase (iNOS) [50], suggesting that nitric oxide (NO) is involved in this inhibition of CPD repair. It is possible that arsenic-induced NO may result in formation of reactive nitrogen intermediates, which damage these repair enzymes. Reactive nitrogen intermediates are much more long-lived than ROS and therefore are considerably more damaging than ROS. Some DNA photolesions can be formed initially by the reaction of nitric oxide with a superoxide anion (associated with inflammation or other forms of oxidative stress) to form peroxynitrite. This then reacts with a guanine base to form 8-nitroguanine [51]. Sodium arsenite (1μM) also decreased NER of UV ($5J/cm^2$)-induced DNA damage as detected by the comet assay in human lymphocytes. This effect subsided 2 hours post irradiation [41]. This is consistent with 2μM sodium arsenite being shown to inhibit repair of UV-induced DNA damage in HaCaT cells [52].

CPDs are repaired via the nucleotide excision repair pathway. Nucleotide excision repair is a process that is based upon the ability to detect abnormal structures in damaged regions of DNA, to excise these, and replace them with normal DNA using the opposite strand as a template [53]. It is a flexible pathway that can repair many different structural forms of DNA damage. It may also help to remove oxidative DNA lesions such as 8oxoG. Nucleotide excision repair consists of two different pathways. These are global genome repair (GGR) and transcription-coupled repair (TCR). GGR eliminates lesions from anywhere in the genome. TCR is specific for repair of DNA on transcribed strands of active genes. They have in common the same repair mechanics, however, their ability to recognise DNA damage and the proteins involved are different [53]. The first step is recognition of the DNA repair site before construction of a multi-protein complex and unwinding or remodelling of the DNA at the repair site. The lesion is then incised bilaterally by endonuclease activity on the 5' and 3' sides. The incision is completed just wide of the structurally abnormal section. The oligonucleotide between the two incision sites is subsequently removed. A DNA polymerase replaces the cleared area and the new strand is sealed by a DNA ligase [5]. Many of the genes involved are named after the inherited condition xeroderma pigmentosum [54] and this process has a large requirement for energy.

NER appears to be compromised by arsenic exposure. DNA incision frequency is reduced in UV exposed fibroblasts that are also exposed to 2.5 μM arsenite. In addition, at higher, cytotoxic doses, ligation of repair patches is also inhibited [55]. Moreover, DNA polymerase β, DNA ligase I and DNA ligase III are actually quite insensitive to arsenite, indicating that a direct effect on these enzymes is unlikely [56]. Xeroderma pigmentosum A, another zinc finger protein, is also sensitive to arsenite and its metabolites [49].

Long-term exposure of HaCaT cells to low level inorganic arsenite (100 nM, 28 weeks), termed arsenite transformed (As-TM) cells, induces malignant transformation. HaCaT cells are pre-malignant and do not form tumours when transplanted into athymic mice, but have a large number of genetic defects including mutations in the p53 gene [57]. Hence arsenite treatment appears to complete malignant transformation of these cells. This transformation was demonstrated by the development of aggressive SCCs after inoculation into nude (athymic) mice [58]. As-TM cells (100 nM, 30 weeks) also develop a generalised resistance to apoptosis and a reduced oxidative damage response. Exposure to arsenite alone did not induce oxidative DNA damage; however, it synergistically increased acute UVA-induced (25 J/cm^2) oxidative DNA damage by greater than 50%. There was an associated increase in the Bcl2/Bax ratio and reduced caspase-3, Nrf2 and Keap1 expression, which could explain the resistance to apoptosis. There was also increased cyclin D1 and decreased p16 expression [59]. Exposure of HaCaTs for 24 hours with 2 μM arsenite also reduced UV-induced apoptosis [9]. This would enable the survival of UV-irradiated cells that have sustained high levels of UV-induced genetic damage, contributing to carcinogenesis. Nrf2, a key player in the cells anti-oxidant response, is constitutively increased in As-TM cells [60]. This is consistent with other reports that anti-oxidant response genes are upregulated in response to arsenic exposure [61]. There is growing evidence that this is significant in arsenic-induced skin cancer and involves nrf2 [60, 58, 62]. The increase in cellular anti-oxidant defence is likely to be a response to arsenic causing oxidative stress to the cells, and this, along with long-term inhibition of the apoptotic pathway could account for the ability of arsenic to push the pre-malignant HaCaT cells further along the process of malignant transformation.

Low level chronic exposure to arsenic (0.05 ppm, 18 weeks) also increases inducible nitric oxide synthase, nuclear factor-κB expression, and 8-nitroguanine [63]. This nitrosylated damage to guanine is a prevalent form of UV-induced damage which is likely to contribute to malignant transformation [64]. This lesion was also found in the brains of mice sub-chronically exposed to arsenic [65].

6.3. Effect of Arsenic on Poly(ADP ribose) Polymerase-1(PARP-1)

PARP-1 plays an important role in the cell's response to DNA strand breaks. Low levels of DNA damage lead to PARP-1-mediated DNA repair through subsequent poly(ADP-ribosyl)ation of targets. High levels of DNA damage can lead to PARP-1 over-activation, cellular energy loss and cell death while moderate DNA damage leads to activation of apoptosis pathways [66]. Arsenite at concentrations as low as 0.1 μM inhibits PARP-1 in mammalian (HeLa) cells [67]. Arsenite also inhibits PARP-1 at 0.2 nM in HaCaT cells following UV radiation leading to subsequent inefficient repair of DNA strand breaks (comet assay) and oxidative photo lesions [52, 8]. This is due to selective binding of arsenite to zinc fingers in PARP-1 [68]. Addition of zinc and subsequent competitive binding to PARP-1 reversed the arsenite-mediated inhibition of DNA repair [52, 8]. Arsenite's inhibition of UV induced apoptotic pathways also appears to involve PARP-1 inhibition [9].

PARP-1 is the most studied of the 18-member PARP family [69]. It is an abundant zinc finger nuclear enzyme that has long been linked with DNA repair and maintenance of a stable genome. It is critical in the regulation of the cell's response to DNA damage and is one of the first proteins to be activated following exposure to genetic stress. PARP-1 has three functional units; a DNA binding domain (consisting of 2 zinc fingers), a central fragment containing auto modification sites and an NAD binding domain [70]. PARP-1 binds tightly with single and double stranded breaks and catalyses the transfer of ADP-ribose from nicotinamide adenine dinucleotide (NAD+). The level of NAD+ within cells, as the main substrate of PARP-1, is critical in the ability of PARP-1 to maintain and increase activity following DNA damage. The ADP ribose is transferred to protein acceptors with nicotinamide also released. After the attachment of one ADP-ribose, further ADP-ribose moieties can be added to form branched ADP ribose polymers, depending on the enzymes involved. These are involved in chromatin assembly and DNA metabolism [71]. The main protein acceptor is PARP-1 itself [72].

Genotoxic stress-induced strand breaks are retained when PARP-1 activity is inhibited. PARP-1 inhibition also reduces DNA repair [73]. In addition, PARP-1 contributes to chromatin structure. It prevents inadvertent transcription and helps to keep the chromatin structure compact [74]. Poly (ADP-ribose) binding motifs are present in many proteins involved in the DNA damage response including p53, histones, BER protein XRCC1, and NER protein Xeroderma Pigmentosum A. The BER adaptor protein XRCC1 is dependent on poly (ADP-ribose) for its involvement in DNA strand break repair. It also activates chromatin re-modellers and subsequently helps nucleosome disassembly and reassembly before and after repair [74, 73]. PARP knockout mice are extremely sensitive to DNA damaging agents such as ionising radiation and alkylating agents. They are also very sensitive to nitrosamine-induced carcinogenesis [68].

6.4. Effect of Arsenic on p53

The p53 gene is regarded as the "guardian of the genome" and mutations in this gene are present in skin cancer [1]. It is the most commonly mutated gene in human cancers including those of the skin [75]. One of its important functions is in the regulation of cell cycle arrest following DNA damage. Cell cycle arrest allows the cells additional time to undertake DNA repair or activate programmed cell death pathways to avoid the progression of genetic damage into mutations in daughter cells arising from cell division. P53 activation increases p21 expression, an integral transcription factor required for cell cycle arrest following exposure to a genotoxic agent [76, 77]. It is an important factor in DNA repair and apoptosis [78]. Exposure of human fibroblast cells to 0.1 uM of arsenite for 14 days suppresses the ionising radiation-induced p53-dependent increase in p21, despite increased abundance of p53 protein [79]. Arsenite also alters UVB-induced p53 function in mice with reduced phosphorylation and DNA binding [80]. In HaCaT cells, exposure for 14 days to 0.1 uM arsenite decreases constitutive p21 mRNA and protein. This was associated with poly(ADP-ribosyl)ation of p53 despite a global reduction in poly(ADP-ribosyl)ation. There were also higher levels of PARP-1 [81]. It is important to note that in HaCaT cells the basal level of p53 is high with no change when exposed to stress, as opposed to normal cells where exposure to genotoxic stress leads to a change from a relatively low level of activity to high level activity [82]. This is probably because the p53 gene is mutated and the protein is non-functional in HaCaT cells, making it resistant to the normal protein degradation cellular processes and therefore more sensitive to poly(ADP-ribosyl)ation. Nevertheless these studies show that low-dose, physiological levels of arsenite, interfere with the apoptosis pathway at multiple levels. Most likely they increase overall stress on cells and by inhibiting specific cellular pathways make many regulatory proteins sensitive to oxidative damage from ROS or reactive nitrogen intermediates.

In summary, arsenic is a significant cause of skin cancer globally. UV and arsenic are co-carcinogens and the cancer-promoting activity of arsenic requires a primary carcinogen such as UV for its effect to be clinically obvious. Arsenic has a range of cellular effects that are not fully understood. These effects include enhancement of oxidative stress in addition to inhibition of repair of UV induced DNA damage and UV-induced apoptosis and most likely cell cycle control.

6.5. Energy Production and Regulation of the DNA Damage Response to UV

Cellular production of chemical energy, ATP, is critical to repair of DNA damage, apoptosis and regulation of the cell cycle. Chromatin remodelling, which gives the DNA repair process access to damaged DNA is also very energy dependent. Therefore, bioenergetics, or the energy state of the cell, is a prime regulator of how efficiently a cell responds to UV-induced DNA damage. Strategies to increase ATP levels in cells during stressful conditions such as UV irradiation and arsenic exposure are critical to how a cell responds to these challenges.

6.5.1. Nicotinamide in Health

Niacin and relative NAD+ status are important in cellular responses to DNA damage including DNA repair, immunity, and cell aging. Niacin, or vitamin B3, is a water-soluble vitamin that has 2 forms: the amide form nicotinamide and nicotinic acid (Figure 3). It consists of a pyridine ring with an amide group at position 3 [83]. Nicotinamide and nicotinic acid are dietary precursors of NAD+, its phosphorylated (NADP+) and reduced (NADH) forms. Niacin and nicotinamide can be synthesised from the *de novo* precursor tryptophan [84], and are also readily available in the diet from vegetable and animal products. Nicotinamide is safe and relatively side-effect free while nicotinic acid causes vasodilation and flushing in humans [85].

Nicotinamide is a promising skin cancer chemo-preventive agent. In UV exposed BALB/cAnNTacfBR mice, niacin supplementation decreased skin tumour incidence in a dose dependent manner; 0.1%, 0.5% and 1% supplementation decreased the incidence from 68% to: 60%, 48% or 28% repectively. This is also associated with elevated NAD+ levels in the skin [86].

Oral nicotinamide also reduces actinic keratoses (premalignant lesions) in humans. This was demonstrated in double-blinded randomised control trials comparing oral nicotinamide 500 mg with placebo. Nicotinamide once or twice daily for four months significantly reduced the number of actinic keratosis by 29% and 35% respectively in two Phase II studies [87].

Pellagra is the clinical condition that results from severe niacin deficiency in man. Pellagra does not occur if there is sufficient tryptophan, riboflavin and pyridoxine to synthesise niacin [88]. Tryptophan is obtained from the diet most commonly through the consumption of milk. Pellagra is usually associated with poverty and a diet based on maize or jowar, a type of millet commonly used in India and Egypt.

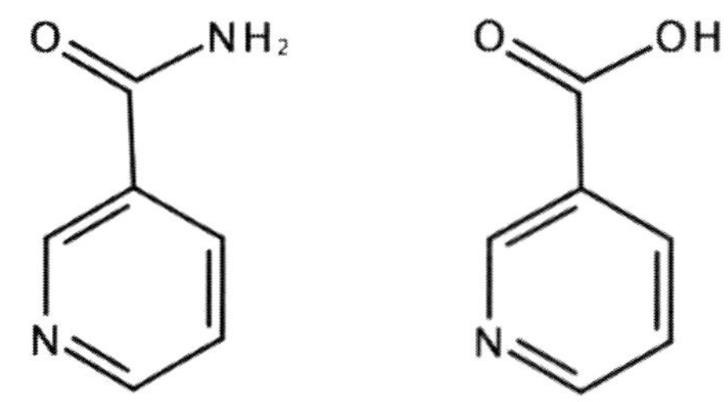

Figure 3. The chemical forms of vitamin B3, nicotinamide and nicotinic acid.

Maize contains niacin however it is in the bound form. It therefore does not contribute to availability of niacin for human nutrition. The addition of lime, a common practice historically used in Mexico, increases its bioavailability [89, 88]. There was an outbreak of pellagra in a Mozambican refugee camp in Malawi in 1990 due to unavailability of niacin containing food and a maize based diet [90]. Pellagra was first described in the eighteenth century by Gaspar Casal in Spain and referred to as mal del la rosa (sickness of the rose) and eventually termed pellagra, meaning rough skin [91]. It is characterised by dermatitis (photosensitivity), diarrhoea, dementia, and eventually death [92]. The deficiency in niacin is thought to impact cellular NAD+ levels required for cellular function, including DNA damage response pathways [93].

6.5.2. NAD+ Levels, Vitamin B3 and DNA Damage

In *ex vivo*, freshly isolated, normal human lymphocytes, nicotinamide stimulated repair of DNA damage induced by UV radiation or the genotoxic agents dimethyl sulphate (DMS) and N-methyl-N'-nitro-N-nitrosoguanidine (MNNG). It increased the level of unscheduled DNA synthesis in treated cells but not in untreated controls [11]. Unscheduled DNA synthesis describes the replication of DNA outside of the normal S-phase. This mostly occurs during NER and is therefore a measure of this process. In addition, when lymphocytes were exposed to UV radiation or MNNG, NAD+ levels decreased significantly, most likely due to consumption during energy production by the cell. This lowering of NAD+ levels was reduced by the addition of 2 mM nicotinamide [94]. By maintaining NAD+ cellular energy levels, nicotinamide appears to enhance unscheduled DNA synthesis following DNA damage. Reduction of cellular energy during the DNA damage response utilises nicotinamide and supplementation can reduce NAD+ loss. All of this data shows that the energy status of cells, and its regulation, is critical to efficient repair of damaged DNA.

In HaCaT cells, removal of nicotinamide from the cell culture media results in a greater than 90% reduction of cellular NAD+ after seven days and greater than 99% after fourteen days. Nicotinamide / NAD+ deficient cells have decreased growth rates due to an increase in the number of apoptotic cells and cell cycle arrest in the G2/M phase. ROS are also increased in these cells secondary to up-regulation of NADPH oxidase activity and probably due to less efficient production of ATP from glucose. As a result, there is an increased level of DNA damage. These effects are reversed with the addition of nicotinamide [95]. It is also known that niacin starved rats develop increased levels of oxidative DNA damage in their bone marrow in the form of 8oxoG [96]. NAD+ depletion also results in decreased p53 expression in breast, lung and skin cells [97]. Therefore NAD+ status is crucial for many cellular functions. Without it, cells cannot maintain homeostasis and efficient DNA protective pathways.

Supplementation with nicotinamide and maintenance of NAD+ levels affect how cells respond to DNA damaging agents. Nicotinamide supplementation (50-500 μM) increases the level of NAD+ following exposure to MNNG in cultured human mammary epithelial cells [97]. In addition, in murine L1210 leukaemia lymphoblast cells, both MNNG and DMS not only lower NAD+ levels but also increase PARP-1 activity which would in turn decrease ATP levels even further. Addition of nicotinamide to the media led to increased rates of strand break re-joining. Inhibition of PARP-1 by 3-aminobenzamide prevented NAD depletion and strand break re-joining [98] indicating the importance of PARP-1 over-activation in causing energy loss in cells. In summary, addition of nicotinamide and subsequent increased NAD+ levels facilitated increased DNA repair efficiency. Nicotinamide supplementation (0.1-33 μM) dose dependently increased the survival of cultured human skin fibroblasts 7 days post irradiation with UV [99]. This is most likely due, at least in part, to more efficient ATP production upon supplementation with nicotinamide.

In human *in vivo* studies, supplementation with 100 mg of nicotinic acid daily for 8 weeks increased cellular NAD+ concentrations of lymphocytes almost five-fold. ROS were increased in isolated lymphocytes from this study by exposure to hypoxanthine and xanthine oxidase. Levels of double stranded DNA expressed as a percentage of control subsequently fell over the following 24 hours as a measure of ROS-induced strand breakage. Lymphocytes from the nicotinic acid treated patients showed a smaller reduction of double stranded DNA levels thereby indicating less strand breakage in response to the ROS inducing agents. This was detected through fluorometric analysis of DNA unwinding [100].

In rats, niacin deficiency has a range of effects related to DNA repair. Kostecki et al. fed rats a niacin-deficient diet or one containing nicotinic acid (30 mg/kg representing a control diet). The rats were then exposed to ethylnitrosourea (ENU) (30 mg/kg), an alkylating agent, or etoposide (ETO), a topoisomerase II inhibitor. The effects on the bone marrow were then determined. Niacin deficiency caused a significant decrease in the level of NAD+ in the bone marrow of control and ENU treated rats. In control diet rats, NAD+ levels decreased in the first 6 hours, before increasing after 24-36 hours. Repair of DNA strand breaks, as detected through the comet assay, were delayed in the niacin-deficient rats compared to the supplemented diet. The niacin-deficient rats also had an increased frequency of micronuclei, a marker of chromosomal damage, in the bone marrow polychromatic erythrocytes [101]. Basal poly(ADP-ribose) levels were also reduced by niacin deficiency in this model. ENU increased the poly(ADP-ribose) levels five-fold in the control diet rats but the levels in the niacin-deficient rats remained 90% lower [102]. Supplementation of the rats' diet with nicotinic acid or nicotinamide at 4 g/kg, to represent pharmaceutical supplementation, increased bone marrow NAD+ levels by 24%. Basal poly(ADP-ribose) and ENU-induced poly(ADP-ribose) levels were also significantly increased. Supplementation also increased the ENU-induced mortality latency period [103]. The effect of niacin on chromosomal stability was further studied in this model. Niacin deficiency without ETO treatment caused an increase in micronuclei formation and sister chromatin exchange. ETO treatment increased the levels of micronuclei and sister chromatin exchange in all rats, indicating an increase in DNA damage. The increase in niacin-deficient rats was significantly greater than in control diet rats [104]. P53 levels were also lower in the niacin deficient rats after ETO treatment. There was also decreased apoptosis and cell cycle arrest [105]. All of these studies are consistent with an important role of vitamin B3 in health. Vitamin B3's influence over cellular NAD+ levels is crucial for many cellular functions. It allows the cell to efficiently respond to DNA damage though repair and apoptotic pathways. Deficiency of vitamin B3 removes the cells' ability to cope with genotoxic stress and subsequent carcinogenesis.

6.6. Energy Levels and Skin Immunosuppression

DNA damage activates various pathways that lead to the consumption of NAD+, and subsequent depletion of cellular ATP. PARP-1 activation leads to NAD+ depletion and glycolytic blockade [99, 106]. NAD+ is also an essential co-enzyme in ATP production via the citric acid cycle. When HaCaT cells were exposed to low levels of solar-simulated UV radiation, there was a dose-dependent depletion of cellular ATP, which was prevented by nicotinamide [12]. In addition, nicotinamide normalises the UV-induced reduction in the glycolytic rate and upregulated succinate dehydrogenase, an energy producing enzyme [12]. In a microarray study of UV-regulated genes in human skin, gene set enrichment analysis has shown that pathways involved in energy metabolism are one of the most frequently down-regulated processes following UV irradiation. This downregulation of energy metabolism could be prevented with nicotinamide [107]. This supplementation with nicotinamide and subsequent maintenances of ATP levels allows the cell to avoid the UV-induced energy crisis. Therefore, highly energy dependent pathways, including apoptosis, DNA repair, and chromatin remodelling can maintain their efficiency. In this way, vitamin B3 is photoprotective.

In addition, topical and oral nicotinamide prevent UV-induced suppression of skin immune responses. Five percent nicotinamide applied to the skin reduced solar-simulated UV-induced immunosuppression as measured by Mantoux tests (delayed type hypersensitivity responses) in humans. Interestingly, immunosuppression was 40% greater in men [107]. Oral supplementation with nicotinamide at 500 mg or 1500 mg daily for one week also reduced solar-simulated UV-induced skin immunosuppression in the same model [14]. As nicotinamide prevents the UV-induced energy crisis in cells and reverses UV down-regulation of genes involved in energy production [107], it appears likely that maintenance of energy levels is important to minimise UV immunosuppression. Photodynamic therapy, comprising irradiation of skin lesions to intense visible light following application of a photosensitiser precursor, is used as a nonsurgical treatment for some forms of NMSC. Delivery of visible light irradiation at high irradiance rates causes profound skin immunosuppression that may contribute to treatment failure. Topical 5% nicotinamide and oral nicotinamide both reduced the observed levels of immunosuppression by 59% and 66% respectively in a Mantoux model [108]. Other B group vitamins also affect skin photoimmunosuppression. Topical riboflavin (vitamin B2), which has a role in aerobic metabolism, reduced UV-induced immunosuppression as measured through contact hypersensitivity responses to nickel in nickel allergic patients [109], thus reinforcing the importance of energy levels in skin for minimising UV-induced damage (Figure 4).

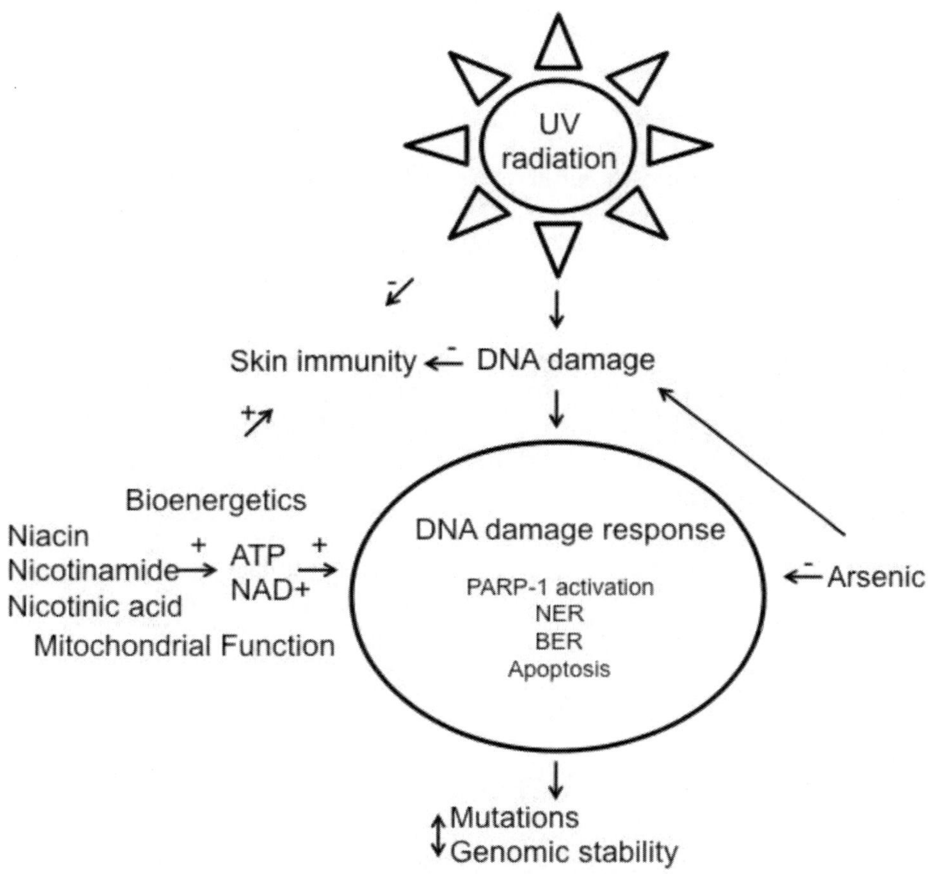

Figure 4. The effect of bioenergetics and arsenic on the DNA damage response to UV radiation. UV radiation damages our skin cells' DNA. Subsequently, the cells respond, leading to activation of PARP-1, DNA repair (by NER or BER), and / or apoptotic pathways. These energy dependent mechanisms aim to maintain a stable genome. This is therefore dependent on ATP, NAD+ and its precursors (nicotinamide, niacin, nicotinic acid). UV also causes immunosuppression directly and through DNA damage. Arsenic is a co-carcinogen in the skin, which increases oxidative stress, reducing the threshold for DNA damage. It also disrupts the DNA damage response.

CONCLUSION

UV radiation is the prime carcinogen of the skin. Its effects are wide ranging and include DNA damage, immunosuppression, and inflammation with subsequent genetic instability, which leads to cancer. Arsenic exposure, particularly through contamination of drinking water in areas of Bangladesh and West Bengal, is a significant skin cancer co-carcinogen.

The mechanisms by which arsenic contributes to carcinogenicity are not fully understood, however, there is evidence in mouse models that arsenic can act as a co-carcinogen with UV, but not independently. The majority of arsenic's effects are associated with the DNA damage response to UV radiation, a function that is energy dependent and crucial to the maintenance of a stable genome.

Energy status is an important determinant of how well a cell responds to UV-induced DNA damage. ATP and NAD+ levels can be raised or lowered by an increase or decrease in niacin (nicotinamide or nicotinic acid). Clinically, nicotinamide supplementation reduces actinic keratosis in heavily sun damaged Australians, while deficiency is characterised by dermatitis, diarrhoea, dementia, and eventually death.

Removal of nicotinamide from culture media reduces growth rates, while increasing apoptosis and cell cycle arrest. Furthermore, when lymphocytes are exposed to UV, nicotinamide supplementation increases the NAD+ and ATP depleting DNA repair pathways. Supplementation with nicotinamide leads to maintenance of ATP levels and avoids the cells depletion of energy levels, thereby allowing the cell to complete efficient DNA damage responses.

The importance of the maintenance of cellular energy is demonstrated in a wide range of cellular functions. When cells are exposed to an acute stressor such as UV, energy requirements are drastically increased, with subsequent reduction of energy reserves. Supplementation with nicotinamide allows the cells to gain an extra advantage when they are faced with such harsh treatment.

Arsenic appears to be a co-carcinogen because it interferes with DNA repair processes and a range of other DNA damage responses. This is all controlled by availability of energy to the cell, which is regulated at least partially by nicotinamide. This simple strategy may be useful not only for reducing UV-induced DNA damage but also damage where co-carcinogens such as arsenic interfere with the process (Figure 4).

REFERENCES

[1] Halliday, G. M., et al. (2005). "UV-A fingerprint mutations in human skin cancer." *Photochemistry and Photobiology* 81(1): 3-8.

[2] National Institute of Environmental Health Services (2011). "Ultraviolet radiation related exposures: solar radiation, exposure to sunlamps or sunbeds, broad-spectrum UVR, UVA, UVB, UVC." *Carcinogen Profiles* 12: 429-433.

[3] Halliday, G. M. (2005). "Inflammation, gene mutation and photoimmunosuppression in response to UVR-induced oxidative damage contributes to photocarcinogenesis." *Mutation Research* 571(1-2): 107-120.

[4] Pfeifer, G. P., et al. (2005). "Mutations induced by ultraviolet light." *Mutation Research-Fundamental and Molecular Mechanisms of Mutagenesis* 571(1-2): 19-31.

[5] Wood, R. D. (1996). "DNA repair in eukaryotes." *Annual Review of Biochemistry* 65: 135-167.

[6] Kirkland, J. B. (2010). "Poly ADP-ribose polymerase-1 and health." *Experimental Biology and Medicine* 235(5): 561-568.

[7] Rossman, T. G., et al. (2001). "Arsenite is a cocarcinogen with solar ultraviolet radiation for mouse skin: An animal model for arsenic carcinogenesis." *Toxicology and Applied Pharmacology* 176(1): 64-71.

[8] Ding, W., et al. (2009). "Inhibition of poly(ADP-ribose) polymerase-1 by arsenite interferes with repair of oxidative DNA damage." *Journal of Biological Chemistry* 284(11): 6809-6817.

[9] Qin, X. J., et al. (2012). "Poly(ADP-Ribose) Polymerase-1 Inhibition by Arsenite Promotes the Survival of Cells With Unrepaired DNA Lesions Induced by UV Exposure."*Toxicological Sciences* 127(1): 120-129.

[10] Surjana, D., et al. (2010). "Role of nicotinamide in DNA damage, mutagenesis, and DNA repair." *Journal of nucleic acids* 2010.

[11] Berger, N. A. and G. W. Sikorski (1980). "Nicotinamide stimulates repair of DNA damage in human-lymphocytes." *Biochemical and Biophysical Research Communications* 95(1): 67-72.

[12] Park, J., et al. (2010). "Nicotinamide Prevents Ultraviolet Radiation-induced Cellular Energy Loss." *Photochemistry and Photobiology* 86(4): 942-948.

[13] Sivapirabu, G., et al. (2009). "Topical nicotinamide modulates cellular energy metabolism and provides broad-spectrum protection against ultraviolet radiation-induced immunosuppression in humans." *British Journal of Dermatology* 161(6): 1357-1364.

[14] Yiasemides, E., et al. (2009). "Oral nicotinamide protects against ultraviolet radiation-induced immunosuppression in humans." *Carcinogenesis* 30(1): 101-105.

[15] Staples, M. P., et al. (2006). "Non-melanoma skin cancer in Australia: the 2002 national survey and trends since 1985." *Medical Journal of Australia* 184(1): 6-10.

[16] Fransen, M., et al. (2012). "Non-melanoma skin cancer in Australia." *Medical Journal of Australia* 197(10): 565-568.

[17] (2008). Non-melanoma skin cancer: general practice consultations, hospitalisation and mortality. C. A. AIHW. Canberra, AIHW.

[18] Staples, M., et al. (1998). "Trends in the incidence of non-melanocytic skin cancer (NMSC) treated in Australia 1985-1995: Are primary prevention programs starting to have an effect?" *International Journal of Cancer* 78(2): 144-148.

[19] Salnikow, K. and A. Zhitkovich (2008). "Genetic and epigenetic mechanisms in metal carcinogenesis and cocarcinogenesis: Nickel, arsenic, and chromium." *Chemical Research in Toxicology* 21(1): 28-44.

[20] Yuspa, S. H. (1986). "Cutaneous chemical carcinogenesis." *Journal of the American Academy of Dermatology* 15(5): 1031-1044.

[21] Diffey B, K. I. (2007). Basic principles of photobiology. Photodermatology. H. H. H Lim, J. Hawk. New York & London, Informa Healthcare: pp. 15–27.

[22] Zastrow, L., et al. (2009). "The Missing Link - Light-Induced (280-1,600 nm) Free Radical Formation in Human Skin." *Skin Pharmacology and Physiology* 22(1): 31-44.

[23] Schwarz, T. (2002). "Photoimmunosuppression." *Photodermatology Photoimmunology & Photomedicine* 18(3): 141-145.
[24] Schwarz, T. and A. Schwarz (2011). "Molecular mechanisms of ultraviolet radiation-induced immunosuppression." *European Journal of Cell Biology* 90(6-7): 560-564.
[25] Norval, M. and G. M. Halliday (2011). "The consequences of UV-induced immunosuppression for human health." *Photochem. Photobiol.* 87(5): 965-977.
[26] Kripke, M. L., et al. (1992). "Pyrimidine dimers in DNA initiate systemic immunosuppression in UV-irradiated mice." *Proceedings of the National Academy of Sciences of the United States of America* 89(16): 7516-7520.
[27] Kuchel, J. M., et al. (2005). "Cyclobutane pyrimidine dimer formation is a molecular trigger for solar-simulated ultraviolet radiation-induced suppression of memory immunity in humans." *Photochemical & Photobiological Sciences* 4(8): 577-582.
[28] Matsumura, Y. and H. N. Ananthaswamy (2004). "Toxic effects of ultraviolet radiation on the skin." *Toxicology and Applied Pharmacology* 195(3): 298-308.
[29] EPA (2002). Implementation Guidance for the Arsenic Rule. O. o. Water.
[30] WHO (2012). "Arsenic." Retrieved 31/10/12, 2012, from http://www.who.int/ipcs/assessment/public_health/arsenic/en/index.html.
[31] Rahman, M. M., et al. (2001). "Chronic arsenic toxicity in Bangladesh and West Bengal, India - A review and commentary." *Journal of Toxicology-Clinical Toxicology* 39(7): 683-700.
[32] Wu, M. M., et al. (2001). "Association of blood arsenic levels with increased reactive oxidants and decreased antioxidant capacity in a human population of northeastern Taiwan." *Environmental Health Perspectives* 109(10): 1011-1017.
[33] Bundschuh, J., et al. (2012). "One century of arsenic exposure in Latin America: A review of history and occurrence from 14 countries." *Science of the Total Environment* 429: 2-35.
[34] Centeno, J. A., et al. (2002). "Pathology related to chronic arsenic exposure." *Environmental Health Perspectives* 110 Suppl 5: 883-886.
[35] Wu, M. M., et al. (1989). "Dose-response relation between arsenic concentration in well water and mortality from cancers and vascular diseases." *American Journal of Epidemiology* 130(6): 1123-1132.
[36] Rossman, T. G. and C. B. Klein (2011). "Genetic and epigenetic effects of environmental arsenicals." *Metallomics* 3(11): 1135-1141.
[37] Thomas, D. J., et al. (2001). "The cellular metabolism and systemic toxicity of arsenic." *Toxicology and Applied Pharmacology* 176(2): 127-144.
[38] Germolec, D. R., et al. (1996). "Arsenic induces overexpression of growth factors in human keratinocytes." *Toxicology and Applied Pharmacology* 141(1): 308-318.
[39] Yamamoto, S., et al. (1995). "Cancer induction by an organic arsenic compound, dimethylarsinic acid (cacodylic acid), in F344/DURCJ rats after pretreatment with 5 carcinogens." *Cancer Research* 55(6): 1271-1276.
[40] Halliday, G. M., et al. (2011). "Ultraviolet A Radiation: Its Role in Immunosuppression and Carcinogenesis." *Seminars in Cutaneous Medicine and Surgery* 30(4): 214-221.
[41] Danaee, H., et al. (2004). "Low dose exposure to sodium arsenite synergistically interacts with UV radiation to induce mutations and alter DNA repair in human cells." *Mutagenesis* 19(2): 143-148.

[42] Javeri, A., et al. (2011). "Downregulation of Cockayne syndrome B protein reduces human 8-oxoguanine DNA glycosylase-1 expression and repair of UV radiation-induced 8-oxo-7,8-dihydro-2 '-deoxyguanine." *Cancer Science* 102(9): 1651-1658.

[43] Farrell, A. W., et al. (2011). "Chromatin Structure Following UV-Induced DNA Damage-Repair or Death?" *International Journal of Molecular Sciences* 12(11): 8063-8085.

[44] Altieri, F., et al. (2008). "DNA damage and repair: From molecular mechanisms to health implications." *Antioxidants & Redox Signaling* 10(5): 891-937.

[45] Matsui, M., et al. (1999). "The role of oxidative DNA damage in human arsenic carcinogenesis: Detection of 8-hydroxy-2 '-deoxyguanosine in arsenic-related Bowen's disease." *Journal of Investigative Dermatology* 113(1): 26-31.

[46] Ding, W., et al. (2005). "Inorganic arsenic compounds cause oxidative damage to DNA and protein by inducing ROS and RNS generation in human keratinocytes." *Molecular and Cellular Biochemistry* 279(1-2): 105-112.

[47] Klein, C. B., et al. (2007). "Further evidence against a direct genotoxic mode of action for arsenic-induced cancer." *Toxicology and Applied Pharmacology* 222(3): 289-297.

[48] Kojima, C., et al. (2009). "Requirement of Arsenic Biomethylation for Oxidative DNA Damage." *Journal of the National Cancer Institute* 101(24): 1670-1681.

[49] Schwerdtle, T., et al. (2003). "Induction of oxidative DNA damage by arsenite and its trivalent and pentavalent methylated metabolites in cultured human cells and isolated DNA." *Carcinogenesis* 24(5): 967-974.

[50] Ding, W., et al. (2008). "As(III) inhibits ultraviolet radiation-induced cyclobutane pyrimidine dimer repair via generation of nitric oxide in human keratinocytes." *Free Radical Biology and Medicine* 45(8): 1065-1072.

[51] Yermilov, V., et al. (1995). "Formation of 8-nitroguanine by the reaction of guanine with peroxynitrite in-vitro." *Carcinogenesis* 16(9): 2045-2050.

[52] Qin, X. J., et al. (2008). "Low concentration of arsenite exacerbates UVR-induced DNA strand breaks by inhibiting PARP-1 activity." *Toxicology and Applied Pharmacology* 232(1): 41-50.

[53] Ichihashi, M., et al. (2003). "UV-induced skin damage." *Toxicology* 189(1-2): 21-39.

[54] Goyal, J. L., et al. (1994). "Oculocutaneous manifestations in Xeroderma Pigmentosa." *British Journal of Ophthalmology* 78(4): 295-297.

[55] Hartwig, A., et al. (1997). "Interaction of arsenic(III) with nucleotide excision repair in UV-irradiated human fibroblasts." *Carcinogenesis* 18(2): 399-405.

[56] Hu, Y., et al. (1998). "Arsenic toxicity is enzyme specific and its affects on ligation are not caused by the direct inhibition of DNA repair enzymes." *Mutation Research-DNA Repair* 408(3): 203-218.

[57] Boukamp, P., et al. (1997). "Sustained nontumorigenic phenotype correlates with a largely stable chromosome content during long-term culture of the human keratinocyte line HaCaT." *Genes Chromosomes & Cancer* 19(4): 201-214.

[58] Pi, J., et al. (2008). "Arsenic-induced malignant transformation of human keratinocytes: Involvement of Nrf2." *Free Radical Biology and Medicine* 45(5): 651-658.

[59] Sun, Y., et al. (2011). "Arsenic transformation predisposes human skin keratinocytes to UV-induced DNA damage yet enhances their survival apparently by diminishing oxidant response." *Toxicology and Applied Pharmacology* 255(3): 242-250.

[60] Pi, J. B., et al. (2003). "Transcription factor Nrf2 activation by inorganic arsenic in cultured keratinocytes: involvement of hydrogen peroxide." *Experimental Cell Research* 290(2): 234-245.
[61] Hamadeh, H. K., et al. (2002). "Coordination of altered DNA repair and damage pathways in arsenite-exposed keratinocytes." *Toxicological Sciences* 69(2): 306-316.
[62] Zhao, R., et al. (2012). "Cross-Regulations among NRFs and KEAP1 and Effects of their Silencing on Arsenic-Induced Antioxidant Response and Cytotoxicity in Human Keratinocytes." *Environmental Health Perspectives* 120(4): 583-589.
[63] Ma, N., et al. (2012). "Arsenic-induced nitrative DNA damage in human HaCaT keratinocytes." *Nitric Oxide-Biology and Chemistry* 27: S39-S39.
[64] Song, E. J., Gordon-Thomson, C., Cole, L., Stern, H., Halliday, G.M., Damian, D.L., Reeve, V.E. and Mason, R.S. (2012). "1a,25-Dihydroxyvitamin D3 reduces several types of UV-induced DNA damage and contributes to photoprotection." *Journal of Steroid Biochemistry and Molecular Biology.*
[65] Piao, F., et al. (2011). "Abnormal Expression of 8-Nitroguanine in the Brain of Mice Exposed to Arsenic Subchronically." *Industrial Health* 49(2): 151-157.
[66] Virag, L. and C. Szabo (2002). "The therapeutic potential of poly(ADP-ribose) polymerase inhibitors." *Pharmacological Reviews* 54(3): 375-429.
[67] Hartwig, A., et al. (2003). "Very low concentrations of arsenite suppress poly(adp-ribosyl)ation in mammalian cells." *International Journal of Cancer* 104(1): 1-6.
[68] Zhou, X. X., et al. (2011). "Arsenite Interacts Selectively with Zinc Finger Proteins Containing C3H1 or C4 Motifs." *Journal of Biological Chemistry* 286(26): 22855-22863.
[69] Ame, J. C., et al. (2004). "The PARP superfamily." *Bioessays* 26(8): 882-893.
[70] Nguewa, P. A., et al. (2005). "Poly(ADP-Ribose) polymerases: Homology, structural domains and functions. Novel therapeutical applications." *Progress in Biophysics & Molecular Biology* 88(1): 143-172.
[71] Kim, M. Y., et al. (2005). "Poly(ADP-ribosyl)ation by PARP-1: 'PAR-laying' NAD(+) into a nuclear signal." *Genes & Development* 19(17): 1951-1967.
[72] de Murcia, G., et al. (1994). "Structure and function of poly(ADP-ribose) polymerase." *Molecular and Cellular Biochemistry* 138(1-2): 15-24.
[73] Heale, J. T., et al. (2006). "Condensin I interacts with the PARP-1-XRCC1 complex and functions in DNA single-strand break repair." *Molecular Cell* 21(6): 837-848.
[74] Shall, S. and G. de Murcia (2000). "Poly(ADP-ribose) polymerase-1: what have we learned from the deficient mouse model?" *Mutation Research-DNA Repair* 460(1): 1-15.
[75] Pfeifer, G. P. and A. Besaratinia (2009). "Mutational spectra of human cancer." *Human Genetics* 125(5-6): 493-506.
[76] Gartel, A. L. and S. K. Radhakrishnan (2005). "Lost in transcription: p21 repression, mechanisms, and consequences." *Cancer Research* 65(10): 3980-3985.
[77] Abbas, T. and A. Dutta (2009). "p21 in cancer: intricate networks and multiple activities." *Nature Reviews Cancer* 9(6): 400-414.
[78] Meek, D. W. (2009). "Tumour suppression by p53: a role for the DNA damage response?" *Nature Reviews Cancer* 9(10): 714-723.
[79] Vogt, B. L. and T. G. Rossman (2001). "Effects of arsenite on p53, p21 and cyclin D expression in normal human fibroblasts - a possible mechanism for arsenite's

comutagenicity." *Mutation Research-Fundamental and Molecular Mechanisms of Mutagenesis* 478(1-2): 159-168.

[80] Tang, F. Q., et al. (2006). "Arsenite inhibits p53 phosphorylation, DNA binding activity, and p53 target gene p21 expression in mouse epidermal JB6 cells." *Molecular Carcinogenesis* 45(11): 861-870.

[81] Komissarova, E. V. and T. G. Rossman (2010). "Arsenite induced poly(ADP-ribosyl)ation of tumor suppressor P53 in human skin keratinocytes as a possible mechanism for carcinogenesis associated with arsenic exposure." *Toxicology and Applied Pharmacology* 243(3): 399-404.

[82] Shaked, H., et al. (2008). "Chromatin Immunoprecipitation-on-Chip Reveals Stress-Dependent p53 Occupancy in Primary Normal Cells but Not in Established Cell Lines." *Cancer Research* 68(23): 9671-9677.

[83] Knip, M., et al. (2000). "Safety of high dose nicotinamide: a review." *Diabetologia* 43(11): 1337-1345.

[84] Belenky, P., et al. (2007). "NAD(+) metabolism in health and disease." *Trends in Biochemical Sciences* 32(1): 12-19.

[85] Gille, A., et al. (2008). Nicotinic acid: Pharmacological effects and mechanisms of action. *Annual Review of Pharmacology and Toxicology*. 48: 79-106.

[86] Gensler, H. L., et al. (1999). "Oral niacin prevents photocarcinogenesis and photoimmunosuppression in mice." *Nutrition and Cancer-an International Journal* 34(1): 36-41.

[87] Surjana, D., et al. (2012). "Oral Nicotinamide Reduces Actinic Keratoses in Phase II Double-Blinded Randomized Controlled Trials." *Journal of Investigative Dermatology* 132(5): 1497-1500.

[88] Etheridge, E. W. (1993). VII. 103: Pellagra. The Cambridge World History of Human Disease. K. F. Kiple. Cambridge Histories Online, Cambridge University Press. 1.

[89] Funk, C. (1913). "Studies on pellagra. I. The influence of the milling of maize on the chemical composition and the nutritive value of maize-meal." *Journal of Physiology-London* 47(4/5): 389-392.

[90] Malfait, P., et al. (1993). "An outbreak of pellagra related to changes in dietary niacin among Mozambican Refugees in Malawi." *International Journal of Epidemiology* 22(3): 504-511.

[91] Roberts, C. S. (1990). Goldberger and the Mal de la Rosa. Clinical Methods: The History, Physical, and Laboratory Examinations. H. K. Walker, W. D. Hall and J. W. Hurst. Boston.

[92] Williams, A. C. and D. B. Ramsden (2007). "Pellagra: A clue as to why energy failure causes diseases?" *Medical Hypotheses* 69(3): 618-628.

[93] Wan, P., et al. (2011). "Pellagra: a review with emphasis on photosensitivity." *British Journal of Dermatology* 164(6): 1188-1200.

[94] Sims, J. L., et al. (1981). "Effects of Nicotinamide on NAD and Poly (ADP-ribose) metabolism in DNA-damaged human-lymphocytes." *Journal of Supramolecular Structure and Cellular Biochemistry* 16(3): 281-288.

[95] Benavente, C. A. and E. L. Jacobson (2008). "Niacin restriction upregulates NADPH oxidase and reactive oxygen species (ROS) in human keratinocytes." *Free Radical Biology and Medicine* 44(4): 527-537.

[96] Tang, K., et al. (2008). "Niacin deficiency causes oxidative stress in rat bone marrow cells but not through decreased NADPH or glutathione status." *Journal of Nutritional Biochemistry* 19(11): 746-753.

[97] Jacobson, E. L., et al. (1999). "Mapping the role of NAD metabolism in prevention and treatment of carcinogenesis." *Molecular and Cellular Biochemistry* 193(1-2): 69-74.

[98] Durkacz, B. W., et al. (1980). "(ADP-ribose)n participates in DNA excision repair." *Nature* 283(5747): 593-596.

[99] Jacobson, E. L., et al. (2001). "Optimizing the energy status of skin cells during solar radiation." *Journal of Photochemistry and Photobiology B-Biology* 63(1-3): 141-147.

[100] Weitberg, A. B. (1989). "Effect of nicotinic-acid supplementation invivo on oxygen radical - induced genetic-damage in human-lymphocytes." *Mutation Research* 216(4): 197-201.

[101] Kostecki, L. M., et al. (2007). "Niacin deficiency delays DNA excision repair and increases spontaneous and nitrosourea-induced chromosomal instability in rat bone marrow." *Mutation Research-Fundamental and Molecular Mechanisms of Mutagenesis* 625(1-2): 50-61.

[102] Boyonoski, A. C., et al. (2002). "Niacin deficiency decreases bone marrow poly(ADP-ribose) and the latency of ethylnitrosourea-induced carcinogenesis in rats." *Journal of Nutrition* 132(1): 108-114.

[103] Boyonoski, A. C., et al. (2002). "Pharmacological intakes of niacin increase bone marrow poly(ADP-ribose) and the latency of ethylnitrosourea-induced carcinogenesis in rats." *Journal of Nutrition* 132(1): 115-120.

[104] Spronck, J. C. and J. B. Kirkland (2002). "Niacin deficiency increases spontaneous and etoposide-induced chromosomal instability in rat bone marrow cells in vivo." *Mutation Research-Fundamental and Molecular Mechanisms of Mutagenesis* 508(1-2): 83-97.

[105] Spronck, J. C., et al. (2007). "Niacin deficiency alters p53 expression and impairs etoposide-induced cell cycle arrest and apoptosis in rat bone marrow cells." *Nutrition and Cancer-an International Journal* 57(1): 88-99.

[106] Ying, W. H., et al. (2003). "NAD(+) repletion prevents PARP-1-induced glycolytic blockade and cell death in cultured mouse astrocytes." *Biochemical and Biophysical Research Communications* 308(4): 809-813.

[107] Damian, D. L., et al. (2008). "UV radiation-induced immunosuppression is greater in men and prevented by topical nicotinamide." *Journal of Investigative Dermatology* 128(2): 447-454.

[108] Thanos, S. M., et al. (2012). "Nicotinamide reduces photodynamic therapy-induced immunosuppression in humans." *British Journal of Dermatology* 167(3): 631-636.

[109] Damian, D. L., et al. (2010). "Topical riboflavin attenuates ultraviolet B- and ultraviolet A-induced immunosuppression in humans." *Photodermatology Photoimmunology & Photomedicine* 26(2): 66-69.

In: UV Radiation: Properties, Effects, and Applications
Editor: James A. Radosevich
ISBN: 978-1-63321-090-5

Chapter 7

ULTRAVIOLET RADIATION EFFECTS ON SKIN CELLS AND IMMUNE SYSTEM

Mariela L. Paz, Eliana M. Cela, Adrián D. Friedrich, Federico S. Weill, Juliana Leoni and Daniel H. González Maglio
Immunology Department, Pharmacy and Biochemistry School,
University of Buenos Aires, Buenos Aires, Argentina

ABSTRACT

Human skin is constantly challenged by environmental stimuli, like microorganisms, chemicals and sunlight-contained ultraviolet radiation (UVR). Skin is composed of a stratified epithelium -epidermis- and the underlying connective and adipose tissue - dermis and hypodermis respectively. There are several structures contained in the dermis like: 1) blood vessels, 2) lymphatic vessels, 3) sebaceous glands, 4) sweat glands, 5) hair follicles and, 6) sensitive nerve terminals. Among the different cells belonging to this important tissue, keratinocytes are the main epidermal cell type (almost 95% of all epidermal cells). As a consequence, these cells are more frequently exposed to the harmful effects of UVR (especially those induced by UVB), since this radiation has a low penetration capacity.

UVR affects keratinocytes by producing direct DNA damage (leading to cell cycle arrest and apoptosis) and also alterations in the cellular metabolism, including mitochondrial dysfunction. However, direct cellular alterations are not the unique harmful effect produced on keratinocytes. These cells are capable of producing large quantities of different cytokines, growth factors, and lipid-derived inflammatory mediators in response to UVR exposure, leading to both local inflammation and systemic immunosuppression. Taken together, all of these effects can induce abnormal transformation of epidermal cells. Malignant transformed keratinocytes lead to the development of the most common kind of human tumor: non-melanoma skin cancer (being basal cell carcinoma -BCC- and squamous cell carcinoma -SCC- the more frequent types).

Cutaneous photosensitivity (CP) is a condition where there is an abnormal skin reaction to UVR. The mechanisms involved in CP development are mostly related to autoimmunity, and there has been great controversy about whether anti-nuclear antibodies produce photosensitive skin lesions, specially, anti-Ro/SS-A auto-antibodies.

In our laboratory, during the last decade, we have been studying UVR effects on the skin and immune system using two different models: a human keratinocyte cell line (HaCaT cells) and a immunocompetitive hairless mice strain (SKH:1). Using these models we have investigated cellular alterations, mitochondrial dysfunction and, acute vs chronic inflammation; as well as SCC development and treatment. Treatment strategies included a non-steroidal anti-inflammatory drug, naproxen, and a probiotic-isolated molecule, lipoteichoic acid from *Lactobacillus rhamnosus*. We have also studied anti-Ro/SS-A auto-antibodies in both models, and their relationship with CP.

Keywords: Keratinocytes, Inflammation, Oxidative Stress, Squamous Cell Carcinoma, Probiotics, Naproxen, Cutaneous Photosensitivity, Autoimmunity

Abbreviations

autoAbs:	Auto-antibodies
COX:	Cyclooxygenase
CP:	Cutaneous Photosensitivity
DCs:	Dendritic Cells
EGFR:	Epidermal Growth Factor Receptor
H&E:	Hematoxylin and Eosin
IFN:	Interferon
IL:	Interleukin
LCs:	Langerhans Cells
LTA:	Lipoteichoic acid
NMSC:	Non-Melanoma Skin Cancers
$NO^{\bullet}$:	Nitric Oxide
NOS:	Nitric Oxide Synthase
NSAIDs:	Non-Steroidal Anti-Inflammatory Drugs
$O_2^{\bullet-}$:	Superoxide Anion
PGE_2:	Prostaglandin E_2
SAPs:	Systemic Autoimmune Pathologies
SCC/BCC:	Squamous/Basal Cell Carcinoma
SLE:	Systemic Lupus Erythematosus
Th:	T Helper Lymphocyte
TNF-α:	Tumor Necrosis Factor Alpha
Treg:	T Regulatory Lymphocyte
UVR:	Ultraviolet Radiation

7.0. Introduction

Sunlight is, in many ways, the source of life of our planet. It is impossible to imagine life on earth without it, since the oxygen production by photosynthetic plants (elemental for our atmosphere), is dependent on the absorption of sun-derived photons by chlorophyll and related molecules. However, sunlight is composed by different types of radiations and some of them can be extremely harmful for human health. Particularly, sunlight contains ultraviolet radiation (UVR), with wavelengths below 400 nm, whose harmful effects on human health are well defined. UVR is composed by UVA (320-400 nm), UVB (280-320 nm) and UVC (below 280 nm). The latter is completely blocked by the stratospheric ozone, while both UVA and UVB can reach the earth surface in different amounts [1].

Solar-derived UVA and UVB radiations are known to cause diverse effects on exposed individuals. Some of these effects include acute inflammation with edema and erythema, photoaging, immunosuppression and, carcinogenesis. UVR targets specifically skin cells, since it has low penetration capability; however, the exposed cells can migrate or secrete soluble mediators that can produce systemic effects. In this way, exposure to UVR has both local and systemic effects, which could be studied in different models, from isolated cells (*in vitro* simulations) to experimental animals or even human subjects (specific *in vivo* studies).

The extent of the damage produced by UVR depends on several issues:

1) Skin type: according to the amount of pigmentation, human subjects are divided into six phototypes. Melanin is the natural pigment of the skin and it has the capacity to absorb UVR. The less the content of melanin in the skin, the more susceptible to UVR damage the subject is [2].
2) Exposure time: the radiation dose is directly related to the damage. Moreover, repetitive exposures have different effects than a single one.
3) Latitude: the intensity of the radiation that reaches the earth depends on the geographical site. Near the Equator, the intensity is constant all year long, while it changes during different seasons in north as south latitudes.

UVR impacts on specific molecules of the exposed cells, generating cellular alterations as well as tissue modifications, like epidermis hyperplasia and dermal fibers degradation [3]. In this chapter, we will discuss UVR effects on skin, describing the bibliographic evidence as well as our own experience.

7.1. Skin Tissue and Cellular Components

The skin is a specialized tissue that is composed of a stratified epithelium, the epidermis, the underlying connective tissue, the dermis, and an adipose tissue named hypodermis. Each one of these layers is composed by specific cell types, which have particular roles in the different physiological processes where the tissue is involved. Skin functions were originally related to just a barrier function, including control of body temperature, hydric regulation and, as a physical barrier to microorganisms. However, skin cells have specific roles in other physiological processes like wound healing and active immune responses against pathogens. Moreover, skin cells can specifically respond to external injuries, like UVR. Epidermal and dermal cellular components are briefly described below. Figure 1 shows the structure of the skin, particularly mouse dorsal skin.

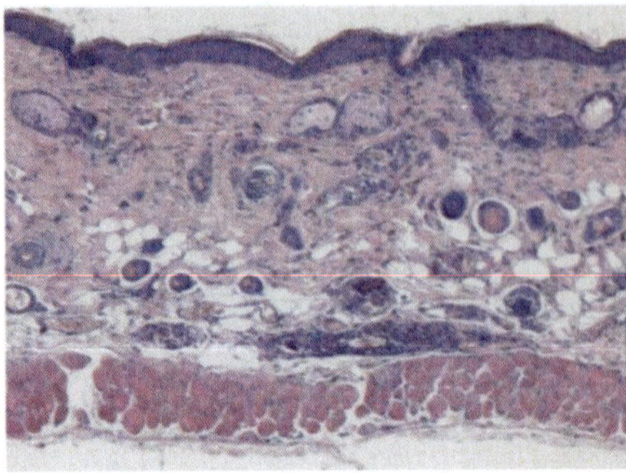

Figure 1. Structure of a SKH:1 hairless mice dorsal skin. Photomicrograph showing epidermis, dermis and hypodermis. H&E staining.

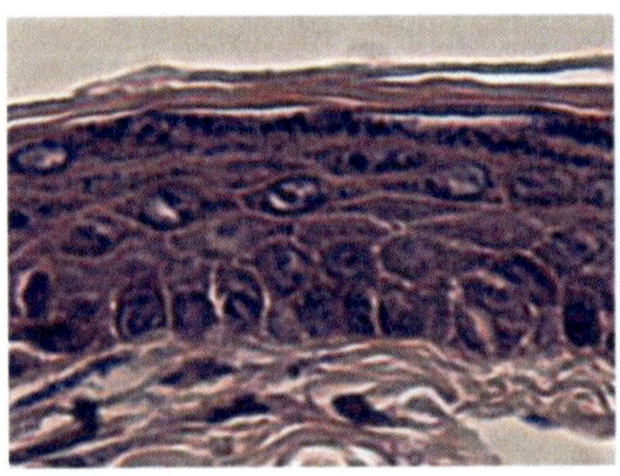

Figure 2. Different strata from a SKH:1 hairless mouse skin sample. Photomicrograph showing stratum spinosum, stratum granulosum and stratum corneum. H&E staining.

7.1.1. Epidermis

The epidermis, the outermost body tissue, is a stratified epithelium mainly composed by keratinocytes. These cells have a unique differentiation process that leads to the production of certain proteins and lipids that ultimately form a corneal keratinized surface all over the body. This corneal, highly hydrophobic layer, represents one of the particular characteristics of this epithelium. It avoids water loss through the surface of our body as well as it impairs the entrance of external microorganisms and molecules. Besides keratinocytes, other cell types are present in the epidermis, like melanocytes, Langerhans cells (LCs) and T lymphocytes (or T cells), each one with their specific role in skin physiological or pathophysiological responses.

7.1.1.1. Keratinocytes

These cells are the main cell type of the epidermis. They have a specific differentiation process from basal keratinocytes to corneal keratinocytes or corneocytes. This differentiation involves the capacity to synthesize specific keratins as well as lipid molecules like ceramides. The diverse differentiation steps leads to four particular layers or strata, called: 1) stratum basale, 2) stratum spinosum, 3) stratum granulosum and 4) stratum corneum. In some specific body location with thicker skin, like the sole, a fifth stratum can be observed: the stratum lucidum. There is a differential keratin production between keratinocytes from different strata: while basal keratinocytes synthesize K5 and K14 keratins, in the stratum spinosum K1 and K10 keratins are produced. Keratinocytes interact with each other by means of tight junctions, which lose strength throughout cellular differentiation. An important characteristic of this cell type is that they are prepared to die, although normal apoptosis and the keratinization process differ in some mechanisms. Keratinocytes are constantly proliferating and dying. Proliferation is a specific characteristic of basal keratinocytes, where other differentiated keratinocytes lack proliferative capacity. Basal keratinocytes, moreover, differentially express an integrin (integrin α2 or CD49b) that promotes the interaction of these cells with the basement membrane, composed of collagen aggregates [4]. Figure 2 shows the different strata from a mouse skin sample.

7.1.1.2. Melanocytes

This cell type is the responsible for melanin synthesis. Melanin is a natural pigment, which absorbs radiation from 300 to 700 nm wavelengths, avoiding part of the damage that UVR can cause. Melanin production is exclusively due to melanocytes. These cells are distributed all along the body, depending on the skin type of each individual. Melanin synthesis (melanogenesis) is induced by UVR and other stimuli. Once synthesized, melanin accumulates in granules forming the melanosome in the cytoplasm of melanocytes. These melanosomes are transported through cytoplasmic prolongations and delivered into keratinocytes cytoplasms. Melanocytes virtually form a net delivering melanin to surrounding keratinocytes, pigmenting the whole skin of the individual [5]. According to the pigmentation level of the skin, human beings are more or less susceptible to sunburn. Consequently, there are six skin phototypes into which people can be classified:

- Phototype I - Always burns, never tans
- Phototype II - Burns easily, tans minimally
- Phototype III - Burns moderately, tans gradually to light brown
- Phototype IV - Burns minimally, always tans well to moderately brown
- Phototype V - Rarely burns, tans profusely to dark
- Phototype VI - Never burns, deeply pigmented

Photodamage in each skin phototype depends on the exposure time, but even the darker ones suffer skin malignancies, highly related to UVR exposure [6].

7.1.1.3. Langerhans Cells

In 1868 a medical student named Paul Langerhans described dendritically shaped cells within the skin, particularly in the epidermis. These Langerhans cells (LCs) could be stained with gold histological techniques, used for the study of nerve cells, and due to their shape and the type of staining used, they were described as nervous cells within the epidermis [7]. Almost a century after that and, by means of electron microscopy, Birbeck described the presence of particular organelles in these cells (that were named after him) that differentiated LCs from melanocytes; besides the lack of melanin synthesis. But it was not until the late 1970's that it was described that LCs do express immunoglobulin Fc and complement receptors as well as major histocompatibility complex type II (MHC II); molecules present in particular immune cells (macrophages and dendritic cells) [8, 9].

LCs are now a very well known type of dendritic cells (DCs). They differentiate, expressing particular surface markers (CD207 or Langerin), as well as common dendritic cells markers (CD11c, MCH II). These cells are distributed all along the body forming a complex net within the epidermis, and they have the capacity to capture foreign molecules and particles and migrate towards the lymph nodes. In this way, LCs work as sentinels of the immune system, activating specific cellular responses in order to protect the individual [10].

7.1.1.4. T Cell Lymphocytes

Different T cell subsets populate the skin. Specifically in mice epidermis, a subset of T cells called "dendritic epidermal T cells" (DETC) have been described. These cells exhibit γδ TCR (T cell receptor), are in intimate contact with keratinocytes and LCs and, control tissue responses [11, 12]. DETCs regulate keratinocyte proliferation and differentiation during wound healing through the recognition of injured or stressed keratinocytes [13]. In human skin, however, αβ TCR-expressing T cells were isolated. These cells were phenotipically characterized as memory/effector cells that express $\alpha^E\beta^7$ integrin, like T cell subsets from other epithelia [14].

7.2. DERMIS

As other connective tissues, the dermis is composed of matrix polymers and cells. Within the polymers, the collagen is the most important. It is mainly synthesized by fibroblasts, which also have the capacity to degrade it through the synthesis and release of matrix metalloproteases (MMPs). The final aspect of the skin depends on the ratio between synthesis of procollagen and MMPs. As an example, the process called photoaging is characterized by wrinkle formation and lose of elasticity of the skin. In this process the balance between synthesis and degradation of collagen is biased towards the latter.

The dermis is highly irrigated with blood and lymphatic vessels. Immune cells trafficking through these vessels in the skin (from the blood to the secondary lymphoid organs) are particularly important. As an example, dendritic cells precursors reach the tissue from blood circulation. Once in the skin, they can differentiate and after their activation by pathogens, for example, they migrate through lymphatic vessels in order to activate specific T cells in the draining lymph node to face the infection [15]. The dermis is also innerved. Depending on the zone of the body there are nervous terminals like Vater-Paccini and Meissner's corpuscles, thermo receptors and nociceptors. In the dermis, numerous hair follicles can be found, according to the zone of the body studied. Other important components of the dermis are the glands: sebaceous and sweat glands reside in this tissue, the former releasing their secretion to the hair follicle channel and the latter directly to the surface of the skin.

As it was mentioned before, there are other populations of dendritic cells in the skin, besides LCs. Dermal DCs are well described and the knowledge of their role in immune response is continuously increasing [16]. Mast cells are also dermis resident cells with immune involvement. Although their role in inflammation is well established, specifically in allergy, novel functions for these cells have been recently described. These functions are related to the induction of tolerance and interaction with regulatory T cells (Tregs) [17].

7.3. SKIN ULTRAVIOLET EXPOSURE: CELLULAR AND TISSULAR EFFECTS

After skin exposure to UVR, a wide variety of molecules are modified in the exposed cells. These modified molecules trigger cellular mechanisms leading to specific responses in the cells (like UVR-induced cell death) and/or the tissue (like dermal inflammation and epidermal hyperplasia). There are many mediators of these cellular and tissular effects, including: 1) extremely small molecules (nitric oxide and superoxide anion are good examples, just composed of two atoms), 2) other molecules that are also small but more complex (like urocanic acid and arachidonic acid-derived molecules) and 3) proteins which act as mediators (enzymes, cytokines, growth factors, matrix metalloproteases and, many others). Every mediator is specifically produced or synthesized by a particular cell type. Besides, each one has specific targets, activating particular cell types by autocrine, paracrine or even, endocrine ways. The complex cellular and tissular response to UVR affects even distant organs, like immune system ones, leading to alterations of the immune response.

Effects produced by a single exposure to UVR might differ from those produced by repetitive or chronic exposures. It is well known, for example, that chronic UVR exposure leads to skin cancer development. Many alterations observed in chronic irradiation models can also be ascertained in single-exposure models. However, there are differences mainly due to a habituation that takes place after repetitive exposures to UVR. This process is limited to a small number of reactions and is known as photoadaptation [18, 19]. Overall, chronic UV radiation is still the main environmental carcinogen on the earth.

7.3.1. Direct Effects on Exposed Cells

UVR has low penetration capacity. It only impacts on epidermal and, in a less extent, on dermal cells. For this reason, one of the most studied cell types, in terms of UVR exposure, is the keratinocyte. Experimental models to evaluate the effects of UVR on keratinocytes include: 1) cell culture (both primary keratinocytes and cell line culture; the most popular among the cell line cultures are HaCaT cells), 2) animal models (including hairy and hairless mice) and 3) humans. However, there are studies for the analysis of UVR effects on other cell types that have also been performed on cell cultures, animal models or, directly on human subjects. These studies include evaluation of dendritic cells' (LCs) functions and fibroblast responses after UVR exposure.

7.3.1.1. Cellular Alterations

One of the best defined UVR effects on cell molecules is the alteration of DNA strands. UVR has the capacity of cross-linking two adjacent thymines, resulting in covalent cross-linked products: 6-4 photoproducts and cyclobutane pyrimidine dimers [20]. These photodimers might produce DNA duplication failures, leading to punctual mutations randomly distributed in cellular DNA, recognizable by CC→TT (UV-specific) and C→T (non UV-specific) transitions. Thymidine dimers are observed in the skin of UVR single exposed animals as soon as 3 minutes after irradiation, with the highest level between half and one hour post-UVR; decreasing towards 36 hours after exposure, to finally being not detectable at 48 hours [21]. The increase in thymine dimers can also be observed in repetitive irradiated healthy subjects, that have a lower response to an acute dose (photoadaptation) [22]. It is worth noting that cells have their own auto-repair mechanisms, which include certain enzymes to detect and eliminate these dimers. The most important mechanisms are the excision of the damaged bases (leaving the DNA backbone intact) and nucleotide excision, which generates nick ends at a certain distances 5' and 3' from the damaged sites [23]. However, these repair mechanisms could be overwhelmed if there is too much damage due to a great exposure or repetitive ones. Repair mechanisms are also less efficient in aged human skin [24].

The presence of DNA alterations triggers the activation of one extremely important protein: p53, whose levels are very low under normal conditions, but increase considerably after keratinocyte UVR exposure [25]. p53 acts as a transcription factor, leading to the expression of several genes involved both in cell cycle arrest and apoptotic cell death activation [26], not only in keratinocytes, but also in all cell types. This is the reason this protein has been designated as the "guardian of the genome" [27]. Cell cycle arrest is performed through p21 up-regulation, which modulates the levels of cyclin D/cyclin dependent kinase 4 (CDK4), in order to let specific enzymes to repair the damage inflicted to the DNA. If genotoxic damage is too large to be repaired, apoptotic cell death is proceeded by regulation of Bax/Bcl-2 levels, mitochondrial release of cytochrome C and, activation of caspases [28, 29]. Cell death of DNA-damaged keratinocytes is intended to avoid the generation and propagation of transformed keratinocytes that may ultimately lead to a non-melanoma skin cancer.

Another important pathway activated by UVR in keratinocytes is the Mitogen-Activated Protein Kinases (MAPK) pathway, which includes several key kinases, like p38, Erk and JNK [30]. This pathway, as the one mediated by NF-κB, is highly important in UVB-mediated inflammatory response and will be discussed later. However, p38 MAPK also has a role in mitochondria-mediated cell death. It is a regulator of cytochrome C released from mitochondria and is independent of caspase cleavage of Bid (a pro-apoptotic member of Bcl-2 family) [31].

Mitochondrial outer membrane permeabilization (produced by pro-apoptotic members of Bcl-2 family) produces, besides the abovementioned release of cytochrome C, loss of mitochondrial membrane potential leading to mitochondrial depolarization, oxidative phosphorylation uncoupling and, superoxide anion production [32–34]. Besides Bid and Bax, there is another important molecule that mediates mitochondrial damage: nitric oxide ($NO^•$). $NO^•$ is produced by NO Synthase (NOS), which has three isoforms: NOS-1 and NOS-3 that are constitutively expressed by many cell types and, NOS-2 that is an inducible isoform which produces large quantities of this important free radical. $NO^•$ mediates lots of physiological functions and it was discovered by its effects as vascular dilator. However, it is also largely produced by macrophages during innate immune responses against bacteria and, by many cell types under different stress conditions [35]. Within the skin, $NO^•$ has different roles and is produced by both constitutive and inducible forms of NOS in many cell types [36]. This free radical mediates mitochondrial damage through the inactivation of specific components of the electron transport chain [37]. The uncoupling of the electron transport chain leads to the incomplete reduction of molecular oxygen (O_2) to superoxide anion ($O_2^{•-}$), such that it may dismutate to hydrogen peroxide (H_2O_2), react in the presence of Fe^{3+} to finally produce hydroxyl radical ($^{\bullet}OH$) or react with NO to produce peroxynitrite ($ONOO^{\bullet}$); all of them important and highly damaging reactive species [38, 39].

Cellular oxidative stress, nitric oxide production, and mitochondrial alterations have been largely studied in UVR skin exposure models. It has been reported that there is a rapid production of $O_2^{\bullet -}$ and $NO^{\bullet}$ within the first hour after keratinocytes UVB irradiation [40]. Besides this almost immediate production, sustained detection of molecules related to $NO^{\bullet}$ were reported. In this way, the detection of both NOS-2 and NOS-1 by Western blot in keratinocytes after UV irradiation [41], and NOS-2 mRNA detection along with increased nitrite levels (ultimate metabolite of $NO^{\bullet}$) in the same model [42], are just a few examples of these $NO^{\bullet}$-related molecules.

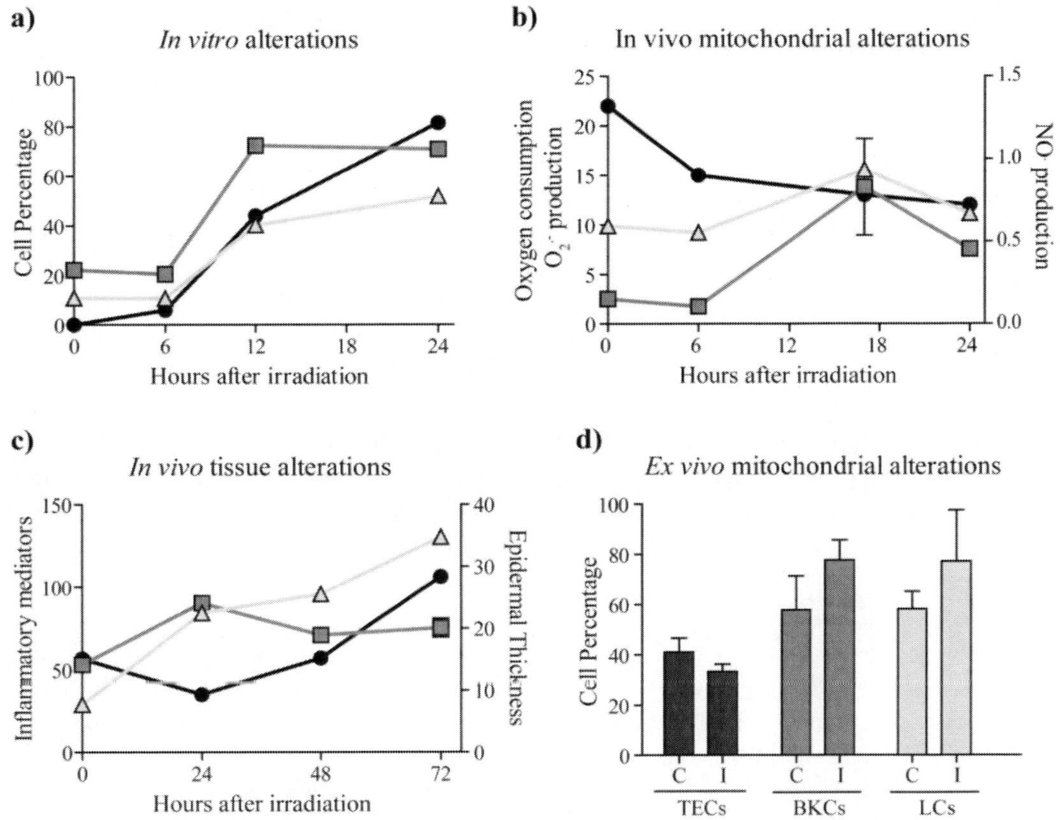

Figure 3. UVR direct effects and cellular alterations evaluated in our laboratory. a) In vitro alterations. Mitochondrial depolarization: ■ DIOC$_6^-$ cells. Cellular apoptosis by microscopy with acridine orange and ethidium bromide staining: ● apoptotic cells. ▲ Superoxide anion production. b) In vivo mitochondrial alterations. ■ Superoxide anion production. Skin mitochondrial respiratory function: ● Oxygen uptake. ▲ Mitochondrial-associated Nitric Oxide production. c) In vivo tissue alterations. ● Epidermal Thickness ■ PGE_2 epidermal production. ▲ TNF-α epidermal production. d) Ex vivo mitochondrial alterations. Percentage of DIOC$_6^+$ cells in TECs: total epidermal cells. BKCs: basal keratinocytes. LCs: Langerhans cells. C: unirradiated control. I: irradiated animals.

In our laboratory, we studied the kinetics of cell death, mitochondrial depolarization and, reactive oxygen/ nitrogen species production using both *in vitro* (human keratinocytes HaCaT cell line) and *in vivo* (SKH:1 hairless mice strain) models. Using fluorescent probes detected by flow cytometry in the *in vitro* model, we observed low levels of apoptotic cell death prior to the loss of mitochondrial polarization, but upon mitochondrial depolarization apoptotic cell death increased with high levels of $O_2^{\bullet-}$, oxidative species production, and low levels of $NO^{\bullet}$ [43]. Loss of mitochondrial function and mitochondrial production of $O_2^{\bullet-}$ and $NO^{\bullet}$ were also detected analyzing epidermis-isolated mitochondria. These mitochondrial dysfunctions were correlated with tissue alterations and inflammation using the *in vivo* model [44]. However, the main difference between the two models is that, *in vivo*, the epidermis can recover from the UVR-induced damage by hyperproliferation of basal keratinocytes, although after chronic irradiation this may lead to carcinogenesis. Moreover, the interaction of keratinocytes with the underlying cells and molecules from the dermis *in vivo* (or that may arrive through blood vessels) as well as the natural differentiation and cornification of the keratinocytes, are other important differences with *in vitro* models. Having these issues in mind, we developed a method to determine mitochondrial damage and $O_2^{\bullet-}$ production in epidermal cells isolated from UV-irradiated animals immediately after their sacrifice. In this way, we performed the challenge *in vivo*, where all the cellular and molecular interactions can take place, and the analysis of the damage afterwards, using fluorescent probes. At first, we observed that mitochondrial polarization as well as $O_2^{\bullet-}$ production in freshly isolated keratinocytes is completely different from those in cultured ones. Secondly, we noticed that basal keratinocytes and Langerhans cells (identified by flow cytometry using specific antibodies) respond differentially to UVR exposure, not only with no detectable loss of mitochondrial polarization but also with a slight, but significant, increase in this functional parameter 24 hours after irradiation [45]. Figure 3 shows a summary of the results obtained [44-46].

This increment in mitochondrial functionality reflects the higher metabolic activity of basal keratinocytes, necessary for cell proliferation and epidermal recovery. In this sense, both cellular and mitochondrial damage are counterbalanced by basal keratinocytes' hyperactivity and proliferation. However, this response may ultimately be also harmful due to cancer development related to keratinocytes' hyperproliferation.

7.3.1.2. Inflammatory Response

It is well known that an excessive solar exposure of the skin produces erythema afterwards, according to the particular skin phototype. This well-described local inflammatory response and the many molecules involved, as well as different treatments to confront inflammation, have been published during the last decades.

As it was mentioned before, $O_2^{\bullet-}$ and $NO^{\bullet}$ are mediators commonly produced after skin (or keratinocytes) UVR exposure. These two reactive species have important roles in the generation of the post-irradiation inflammatory response of the skin. Just to correlate these free radicals with a well known inflammatory response, $O_2^{\bullet-}$ is largely produced by macrophages after internalization of whole bacteria or after recognition of certain molecules, like lipopolysaccharide (LPS), in order to kill the infecting pathogen. It is also of common knowledge that extracellular release of this free radical, as it may happen after irradiation, impacts on blood vessels by increasing vascular permeability [47]. $NO^{\bullet}$ also produces an increase in vascular endothelium permeability as well as vasodilation of blood vessels, favoring the arrival of blood cells (specially neutrophils and lymphocytes) to the exposed site [48]. Indeed, NOS-2 expression is a parameter which is commonly evaluated after UVR exposure and it was used to study differences in the response of normal as well as photosensitive skin after irradiation [49]. NOS-2, through its product $NO^{\bullet}$, is capable of inducing the expression of another important key enzyme of the inflammatory response: cyclooxygenase (COX) [50]. COX catalyzes the conversion of arachidonic acid (released from the cellular membrane) into prostaglandin H_2, which in turns into other prostaglandins (PGs) and thromboxanes (Tx) (Figure 4). One of the most relevant prostaglandins in the skin response to UVR is PGE_2. Indeed, both PGE_2 and $NO^{\bullet}$ combined, mediate UVB-induced erythema. This was demonstrated using different specific enzymatic inhibitors in human healthy subjects [51]. Non-steroidal anti-inflammatory drugs (NSAIDs) have been widely used in the treatment of UVR effects on the skin. Some of their effects are decrease of epidermal proliferation, skin erythema, p53 activation (with consequent decrease in PCNA and cyclin D1) and their direct effect: decrease in PGE_2 levels [52–54]. NSAIDs have been applied both orally and topically on animal models and human subjects, showing that pharmacological COX inhibition could be achieved using different pharmaceutical forms.

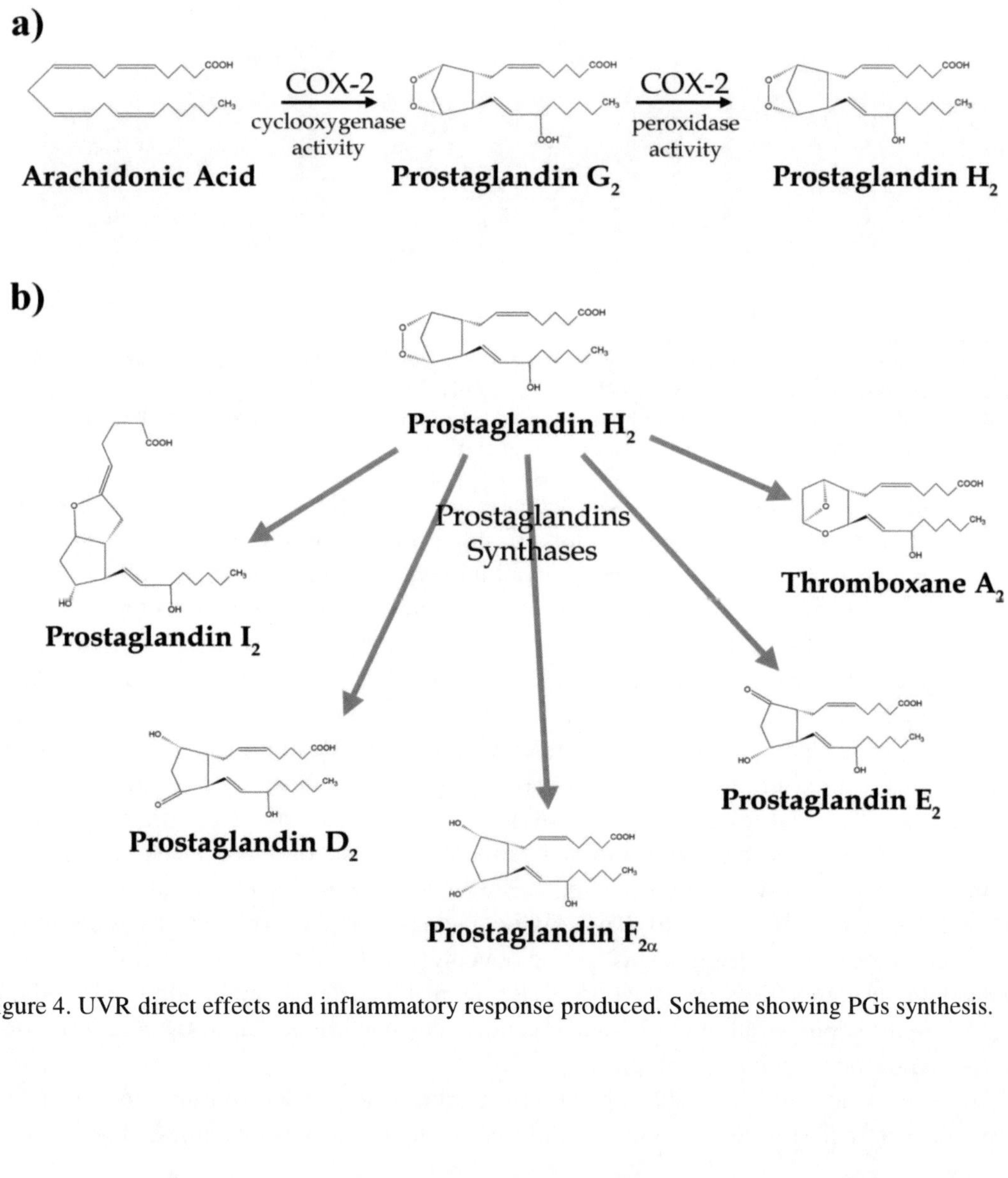

Figure 4. UVR direct effects and inflammatory response produced. Scheme showing PGs synthesis.

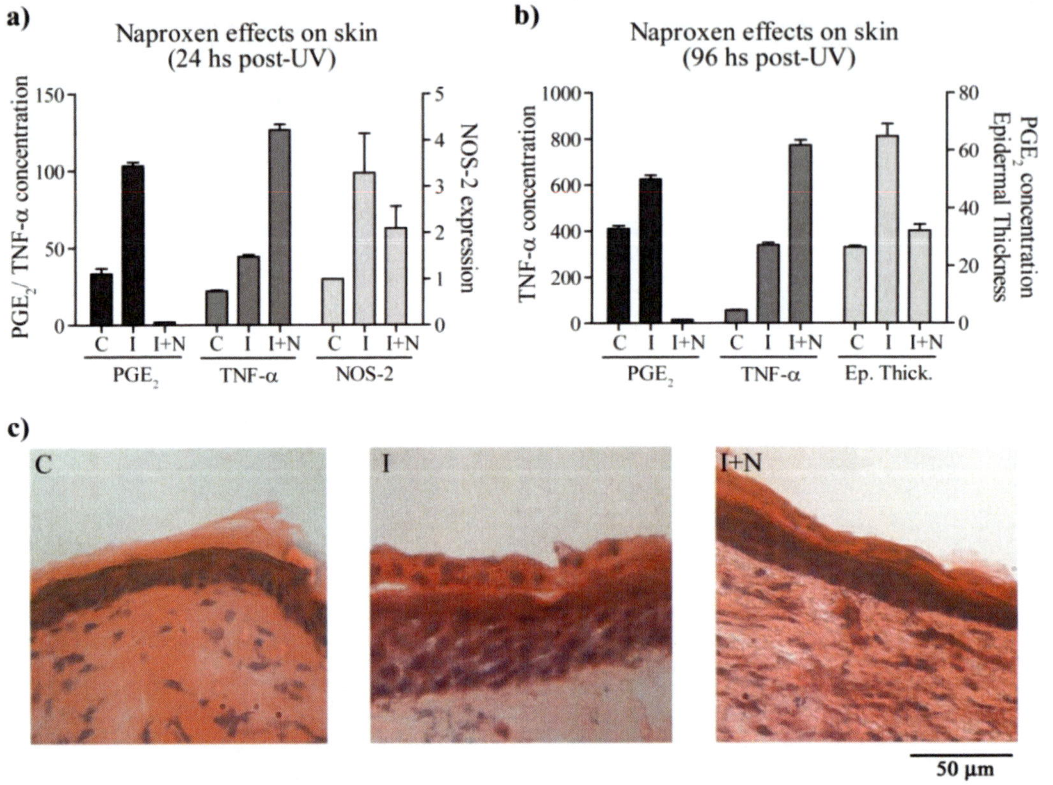

Figure 5. UV direct effects and inflammatory response and treatment evaluated in our laboratory. a) and b) Naproxen effects on skin 24 or 96 hs post-UV respectively: PGE_2 and TNF-α production, NOS-2 expression, Epidermal Thickness. c) Skin histology: H&E staining. C: unirradiated control. I: irradiated animals. I+N: irradiated animals treated with three applications of Naproxen.

COX inhibition not only produces a decrease in the levels of its products, especially PGE_2, but it also modifies the production of other important inflammatory mediators: cytokines. Between the proinflammatory cytokines and chemokines, the most highly produced by keratinocytes after UV irradiation are: 1) TNF-α (tumor necrosis factor-α), 2) IL-6 (interleukin-6), 3) IL-1β, 4) IL-8, 5) IFN-γ and, 6) MIP-1β [55]. We have reported that naproxen topical treatment of UV irradiated hairless mice produced almost an absolute decrease of PGE_2 production and a concomitant decrease in the transcription levels of NOS-2. Unexpectedly, it also led to an increase in epidermal levels of TNF-α measured 24 hours after irradiation [46]. Moreover, by applying topical naproxen three times (24, 48 and 72 hours post-irradiation) and analyzing the effects 96 hours post-irradiation, TNF-α increment was even more evident. This three application treatment was also efficient in decreasing the epidermal hyperplasia observed in irradiated animals [56]. Figure 5 shows a summary of the results previously reported [46, 56]. This TNF-α increase can be explained by the role of PGE_2 as an activator of adenylate cyclase, producing cAMP, through its interaction with prostaglandin receptors EP2 and EP4 [57]. Being that cAMP is an inhibitor of TNF-α transcription, the suppression in PGE_2 signal on adenylate cyclase produced by naproxen also suppress TNF-α transcription inhibition [58].

Keratinocytes are the main cell population suffering from UVR exposure, because they are the most external cells. They produce plenty of mediators such as cytokines, chemokines, lipid mediators and, small mediators (like $NO^{\bullet}$), many of them mentioned in previous paragraphs. These mediators impact on the underlying connective tissue, the dermis. This tissue responds with many modifications after skin UVR exposure, due to either a direct effect of the radiation on dermal cells (those that are capable of penetrating the epidermis) and/or to an indirect effect mediated by the abovementioned inflammatory molecules. Within dermis alterations there are vasodilation, *de novo* vascularization, tissue infiltration (by mononuclear and polymorphonuclear cells) and connective fibers alterations. After a single irradiation, fibroblasts are able to produce VEGF (vascular endothelial growth factor), which is the most important mediator of the neo-vascularization in the dermis. This effect on dermal fibroblasts is also produced by PGE_2 stimulation [59], demonstrating the interaction between the connective tissue and the epithelium (source of PGE_2). Photoaging is one of the harmful effects of chronic UV irradiation and it is related to extracellular matrix degradation with elastic fibers alterations [60]. These effects are also direct and indirectly produced by UVR. Fibroblast exposure to UVR leads to the production of MMPs (matrix metalloproteases, responsible for extracellular matrix degradation), however, keratinocyte-produced mediators after UVR exposure can also have the same effect [61]. Photoaging is also explained by a decreased capacity of collagen synthesis by dermal fibroblasts. Indeed, UVR also promotes a decrease in the synthesis of type I procollagen, mediated by a UVR-induced down-regulation of TGF-β receptor II, which in this cell type activates Smad 2/3 after TGF-β binding leading to procollagen synthesis [62].

7.4. Effects on Distant Organs

Since UVR exposed cells are able to secrete soluble mediators, such as cytokines and prostaglandins, it is not surprising to find effects of radiation on distant, non-exposed, organs. Indeed, these effects have been described and studied using specific blocking antibodies and knockout animals. However, soluble mediators' distribution along the body through lymphatic and blood circulation is not the unique explanation for UVR-induced effects on distant organs. Specific skin-homing cells are also capable of leaving the tissue after UV irradiation and afterwards interacting with cells in distant organs, activating or modifying their activity. The most significant systemic effect of UVR is, beyond any doubt, immune response modification. Hereafter, we briefly describe the mechanisms underlying UV-induced systemic immunosuppression.

7.4.1. Effects on Immune System: Systemic Immunosuppression

Skin UVR exposure has dramatic effects on distant non-irradiated organs. The best described systemic effect is a down regulation of immune response or immunosuppression. This suppression of specific immune responses was studied in several animal models. It was more than thirty years ago when Michael Fisher and Margaret Kripke first described that UVR-induced skin tumors were rejected when they were transplanted into control animals, but not into irradiated mice [63, 64]. This lack of response against those highly antigenic tumors was further investigated, showing that the specific suppressive response was particularly located within the immune cells of secondary lymphoid organs, and it could be transferred to naïve animals. These cells were characterized as Lyt-1^+2^- Ia^- and it was shown that removing them from transferred T lymphocytes, abolish suppressive activity not only on recipient T cells, but also in co-transferred cytotoxic cells [65].

Systemic immunosuppression was also extensively demonstrated using specific immune responses against protein antigens (DTH, delayed type hypersensitivity) and haptens (CHS, contact hypersensitivity). The most commonly used challengers are Ovalbumin (Ova) for DTH reactions and oxazolone (Oxa) or 2,4-dinitro-1-fluorobenzene (DNFB) for CHS ones. These reactions require two phases: a first contact with the antigen or hapten (sensitization phase) and a second contact that leads to a quantifiable inflammatory response (challenge phase). Using these reactions it was possible to assess the cellular mechanisms underlying the immunessuppression. UVR exposure is responsible for the induction of regulatory T cells (Treg) that suppress specific immune responses against Oxa and DNFB. CHS reaction was abolished when irradiated mice were sensitized with haptens after irradiation. Moreover, lack of response was successfully transferred to naïve animals through immune cells transference, particularly Treg. However, if Treg were transferred after sensitization in non-irradiated mice, CHS reaction was not abolished, showing that once activated, specific cells were not able to be stopped in lymph nodes by Treg. Finally, authors proved that specific Treg were capable of blocking CHS reaction in sensitized mice if they were injected directly in the challenged tissue (the ear), showing that Treg were responsible of UV-induced suppression of CHS reaction, not only during the sensitization phase but also during the elicitation one [66].

Besides cellular mediators of UV-induced immunosuppression, there are many molecules involved in this phenomenon. Some of these mediators are synthesized by keratinocytes after UV exposure, such as Interleukin 10 (IL-10) [67]. This cytokine is crucial in the development of the innmunosuppressed state, since it has been shown that its blockade, using specific antibodies, abolished the UV-induced lack of response to CHS reaction [68]. However, IL-10 is not the only soluble mediator involved in this immunosuppression, a cascade that also includes PGE_2 and IL-4 is additionally related to the suppression [69]. Moreover, knockout mice for IL-10 and platelet-activating factor receptor (PAFR) also failed to set an efficient CHS reaction, indicating their importance in the process [68, 70]. Another relevant molecule in skin photobiology is vitamin D, which is synthesized in the skin as a precursor in a specific photoreaction. This vitamin is believed to have some suppressive effects on immune cells, however, since UV-derived immunosuppression has been proven without the involvement of the vitamin D pathway, its role in immunosuppression is not fully understood yet [71, 72].

Finally, antigen presenting cells (APCs) activity can also be modified by UV irradiation. LCs absolute number have been found to be decreased in epidermis after single or chronic UV exposure [73]. Moreover, DCs that leave the skin after irradiation (isolated from efferent lymphatic vessels) have shown a decreased capacity to induce T cells proliferation [74]. Additionally, systemic innate immune response is also affected. It has been reported that circulating cytokines produced after an LPS challenge (an innate immune challenge) are altered by UV exposure. UV irradiation normally promotes a decrease in IL-12p70 serum levels *in vivo*, while splenic cells isolated from UV-irradiated mice and challenged *in vitro* with LPS produced higher levels of TNF-α, IL-1, IL-6 and PGE_2 compared to cells isolated from non-irradiated mice. These changes in APCs activity are also reflected in T cell activation *in vitro*, where APCs isolated from UV-irradiated animals were able to induce a decrease in T cell-derived IFN-γ production [75]. A recent publication established that LCs are not responsible for the observed immunosuppression, while this modulation of the immune response is more related to dermal DCs activity [76].

In our laboratory, we have analyzed the effects of two different UV irradiation schedules: an acute exposure schedule consisting of 4 doses of 100 mJ/cm^2, and a chronic exposure one consisting of 20 doses of 50 mJ/cm^2. Using these models, we have analyzed alterations in epidermal tissue and in the T cell populations' quantity and functionality in skin, regional lymph nodes, and spleen. We have found that both acute and chronic UVR led to epidermal hyperplasia and also to a decrease in epidermal T-cell density. But chronic UVR shifted the cytokine production towards a Th2/Treg profile, while acute UVR produced the opposite effect, increasing the T-cell proliferative response. Our results indicate that, even though acute irradiation has a direct effect on skin, it does not modify the overall T-cell response considerably. In contrast, chronic UVR has marked effects in the immune cell response of the irradiated animals [77]. Figure 6 shows a summary of the results previously reported [77].

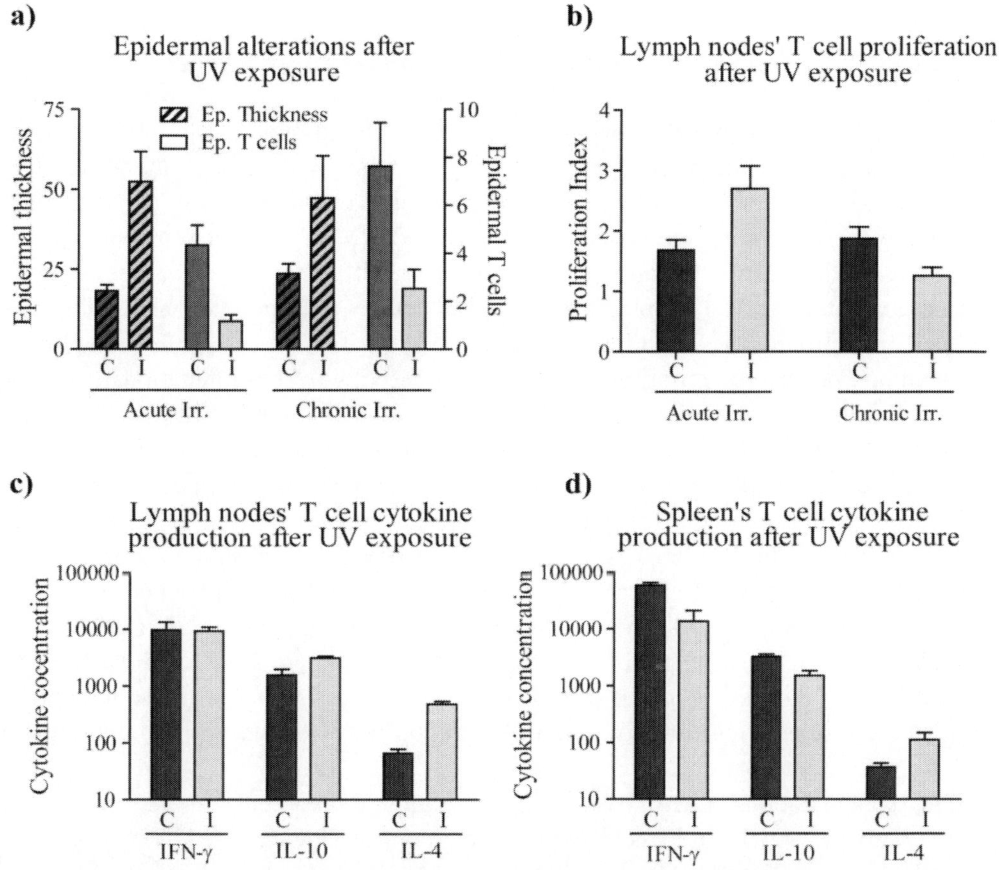

Figure 6. UVR effects on distant organs and immunosupression evaluated in our laboratory. a) Epidermal alterations after an acute and also a chronic UVR exposure: epidermal thickness and epidermal T cells' percentage. b) Lymph nodes' T cell ConA proliferation index after an acute and also a chronic UV exposure. c) Lymph nodes' T cell ConA-induced cytokine production after a chronic UV exposure d) Spleen's T cell ConA-induced cytokine production after a chronic UV exposure. C: unirradiated control, I: irradiated animals.

7.5. SKIN CANCER

Skin cancer is the proliferation and expansion of uncontrolled skin cells. This unregulated growth is highly related to sun or tanning bed exposure behavior of the individuals. Virtually every cell type of the skin is able to mutate and generate cancer. Within skin cancer, the most important in incidence according to the Skin Cancer Foundation are: 1) cutaneous malignant melanoma (CMM), 2) squamous cell carcinoma (SCC), 3) basal cell carcinoma (BCC), 4) actinic keratosis and 5) dysplastic nevi. These malignancies include the proliferation of abnormal melanocytes (CMM) and keratinocytes (SCC and BCC). The former is commonly known as melanoma and the latter are globally known as non-melanoma skin cancers (NMSCs). There are mechanisms involved in NMSC development which are very well documented [78, 79]. Hereafter, we will discuss some of them, highlighting the importance of key molecules that may be targeted in experimental procedures.

7.5.1. Malignant Transformation: Non-Melanoma Skin Cancer

UVR is capable of generating direct damage to DNA of exposed cells, as it was described before. Particularly, the skin is the tissue which is the most exposed to sunlight, and to the UV radiation that it contains. As the epidermis is composed mainly by keratinocytes, this cell type originates the most common types of skin cancers: NMSC. As abovementioned, UVR exposure may produce DNA mutations, randomly distributed along the genome. However, when genes encoding for key proteins needed for cell cycle regulation are mutated, the affected cell may progress to a malignant phenotype. Specifically, a high frequency of p53 mutations among patients suffering of SCC has been described. These mutations exhibited a pattern of UV-induced mutations (CC→TT and C→T) and generated changes in the protein sequence [80]. When p53 mutations were investigated in aggressive and non-aggressive SCC and BCC, a great incidence of p53 mutations in both NMSC was found, however, these sequence modifications were not related to the aggressiveness of the cancer [81]. Moreover, p53 mutations (CC→TT) at codons 245 and 247/248 were identified in a high proportion of 17/23 of normal skin from UV-exposed sites in Australian skin cancer patients, compared to 1/20 of mutated codons in non-exposed skin from the same patients [82]. Finally, it was reported that a high percentage of keratinocytes surrounding BCC and SCC, which supposedly suffered the same sun exposure as the malignant cells, had missense mutations of p53 [83]. These results demonstrate the role of sunlight in UVR-induced p53 mutations that may lead to NMSC development.

Besides p53 mutations, there is an important pathway commonly altered in BCC: the Hedgehog pathway, involved in cell proliferation control. Patched (PTCH) protein is a receptor which acts negatively in this pathway, and it has been found mutated in BCC. Particularly, the mutation rate was found increased in patients suffering xeroderma pigmentosum (XP, a pathology where a defect of the DNA repair enzyme NER is always present) [84]. The localization of the mutations found in the PTCH gene in BCC patients was studied in people with a XP history as well as in others with no previous disease records. This comparison showed the random localization of the mutations but did not demonstrate a pattern associated with the presence of NER deficiency. Moreover, it revealed that although mutations were produced in different domains of the protein, they always lead to the lack of functionality of the receptor [85].

Apoptosis (programmed cell death) and cell cycle control are important cellular mechanisms involved in the control of transformed cells. As it was mentioned before, missense mutations of key molecules are necessary events for tumor development. However, there are other important events mediated by soluble molecules secreted by tumor cells that trigger cell proliferation and neo-vascularization, among others. Between these soluble molecules there is a crucial molecule that seems to be highly important in NMSC development: PGE_2.

As abovementioned, PGE_2 is a product of the COX pathway. COX-2, the inducible isoform of the enzyme, is over-expressed in mouse and human NMSCs [86]. Indeed, knocking down any of the isoforms (COX-1 or COX-2) in mice results in a decrease in NMSC formation [87]. The same effect is produced by blocking the enzymes with non steroidal anti-inflammatory drugs (NSAIDs), both COX-2 specific or not [56, 88]. PGE_2 induces keratinocyte proliferation, by increasing cAMP through EGF receptor (EGFR) activation; besides, it also activates the ERK1/2 and PI3K pathways [89]. Activation of multiple nuclear factors, including NF-κB, AP-1 and, CREB, leads to synthesis of many carcinogenesis-related molecules. One important molecule is the vascular endothelial growth factor (VEGF), which mediates neo-vascularization and is produced by both keratinocytes and fibroblasts after PGE_2 stimulation [59, 89]. Cellular effects of PGE_2 depend on the expression of the E prostanoids receptors (EP1 to 4). The role of the different receptors in the development of NMSC was evaluated using knockout mice and a specific antagonist. One group described, using ONO-8713 - a specific EP1 receptor antagonist-, that the interaction of PGE_2 with this receptor mediates UV-induced carcinogenesis. The ONO-8713 tumor inhibitory effect was similar to the one obtained with celecoxib treatment, which produces the abrogation of PGE_2 synthesis [90]. Another report, demonstrated that the importance of EP2 receptor, but not EP3, as mediator of PGE_2 effects on skin cancer development [91]. It can be concluded that both EP1 and EP2 receptors are significant in mediating PGE_2 skin carcinogenesis.

Another main pathway involved in NMSC development is the one initiated by EGFR activation. UVR, as it was already mentioned, is able to induce epidermal hyperplasia due to an increase in keratinocyte proliferation. This hyperplasia is mediated by EGFR activation, since knockout mice for the receptor failed to generate UVR-induced keratinocyte proliferation. Moreover, the EGFR knockout mice respond to UVR not only failing to induce proliferation, but also increasing apoptotic cell death (via increased expression of p53 and p21, compared to wild type mice) [92]. An additional relevant effect of UVR-induced EGFR activation in keratinocytes is the induction of COX-2 gene expression, with the abovementioned effects of its product, PGE_2 [93]. An interesting fact about EGFR activation is the rapid cellular response, which takes place as fast as a few minutes after irradiation. This almost immediate response is mediated by ROS, which are capable of inhibiting cellular phosphatases due to oxidation. These phosphatases act constitutively inhibiting EGFR phosphorilation, and consequently, activation. Once phosphatases are oxidatively inhibited, EGFR auto-phosphorylates and activates its own pathway [94]. Finally, it is possible to correlate the role of activated EGFR (and other extracellular receptors) as essential mediators of NMSC development due to UVR exposure [95, 96].

7.5.2. Non-Melanoma Skin Cancer Prevention and Treatment

The management of squamous cell carcinomas has long been influenced by the work of Lund et al, where he reported that metastasis occurred in only 0.1% of a series of these tumors. The same happened with the probability of metastasis from basal cell carcinomas; there are works which report very low proportions such as 0.1-0.54% or 0.0028%. These reports have led to a widespread belief that NMSCs developing in a site exposed to sun are very unlikely to metastasize, conditioning in this way the clinical management of this kind of tumors [97].

Several later studies, however, have suggested higher rates of metastasis ranging from 2.6% to 3.3%. This evidence has motivated physicians and researchers to give greater importance to the clinical management of NMSCs, focusing on prevention strategies and new better alternatives for the treatment of these skin cancers.

7.5.2.1. Clinical Management

Early skin biopsy and treatment of premalignant and malignant lesions are essential for treating these patients successfully. There are two kinds of treatments available, surgical and pharmacological ones, using topical as well as systemic drugs.

Treatment should begin with surgical resection of the affected areas in an attempt to obtain clear margins. Among surgical methods, destructive ones such as electrocautery, cryotherapy, curettage, and excision can all be used for low risk NMSC. The later has the additional benefit of the histological examination of the excision margins, in order to check for tumor clearance, which is impossible to perform in the first three methods mentioned. Mohs' micrographic surgery is a specialized technique for NMSC at high risk sites (central face), with maximal preservation of normal tissues. A five year cure rates for all these treatments are in the order of 95% or better. Following surgery, adjuvant radiation could be considered.

Radiotherapy is an alternative for patients who are poor surgical candidates, such as elderly patients with extensive lesions. It is not recommended for young patients, as the late cosmetic results are inferior to those of surgery. The five year cure rate has been estimated at about 90%. For patients with metastatic NMSC, adjuvant radiation, chemotherapy, and staging by sentinel lymph node dissection may be employed [98].

One of the available pharmacological topical therapies consists of a 5-fluorouracil (5-FU) cream, which is frequently used alone or in combination with topical tretinoin or alpha/beta hydroxyl acids to reduce the numbers of NMSCs and their precursors, especially multiple superficial tumors on the trunk and limbs. 5-FU inhibits thymidylate synthetase and subsequently halts DNA synthesis. Pretreatment with topical tretinoin may help with reducing the thickness of the stratum corneum, thereby aiding in topical drug delivery. Another topical treatment includes Imiquimod as an immunomodulator; its mechanism of action is through activation of the toll-like receptor 7 (TLR-7), generating various cytokines (including IFN-γ) and other mediators that are the products of an activated innate immune response. Other treatments comprise 3% topical diclofenac or retinoids alone. Complete clearance was appreciated in approximately 40-60% of the patients using topical therapies.

Topical photodynamic treatment is effective for superficial NMSCs. This technique uses aminolaevulinic acid, which is applied topically to the lesion, as a precursor of protoporphyrin IX. Tumor tissue absorbing this porphyrin metabolite becomes photosensitive with its conversion to protoporphyrin IX and subject to photodestruction when exposed to light (620-670 nm) and oxygen by reactive oxygen species that causes cell death in rapidly dividing cells. The average clearance rate for superficial basal cell carcinoma from a review on photodynamic therapy was around 87% [99].

Pharmacological systemic therapies are not the most used alternative, but systemic retinoids are vitamin A derivatives that have been shown to be effective in the treatment of NMSCs and their precursors.

It has been shown that organ transplantation increases the risk of developing SCC and BCC by 65- and 10-fold, respectively. Therefore, in the general population and especially in organ transplantation recipients, the greatest impact should be made in the prevention of NMSCs through simple, but rigorous, patient education on the benefits of UV protection, periodic self-skin examinations, and regular follow-ups. Accordingly, vitamin D and calcium supplementation should also be incorporated in transplant recipients [100].

7.5.2.2. Experimental Treatments

Although the clinical management of NMSCs is nowadays well established and there are different surgical options as well as pharmacological ones, new alternative treatments are always being developed and tested. They range from well-known drugs but delivered in different ways, to natural compounds or newly-developed drugs.

Intra-lesional therapy is not widely used, however, various well-known agents have been used successfully in the treatment of primary cutaneous malignancies when delivered intra-lesionally. This therapy can be used to treat many lesions simultaneously, provides good cosmesis, and can be done quickly with very little downtime. However, the effectiveness of this treatment modality is unknown, injections can be painful, medications are costly. The patients may require multiple treatment visits. Intra-lesional interferon-α (IFN-α) is an experimental treatment which has been under study. The postulated mechanism is unclear, but it is thought to induce apoptosis of tumor cells via CD95 receptor-ligand interaction, inducing infiltration of cytotoxic lymphocytes. In a series of 140 patients with basal cell carcinoma treated with intra-lesional and peri-lesional interferon-α 2b, 67% achieved cure, sustained over a mean follow up period of three years [98]. Although treatment with intra-lesional 5-FU shows promise, further study is needed. Bleomycin is a group of sulfurcontaining glycopeptide cytotoxic antibiotics derived from *Streptomyces verticillus* and has been shown to have anti-viral, anti-tumoral, and anti-bacterial effects. Bleomycin affects the G2 and S phases of rapidly dividing cells, leading to breakage of single-stranded DNA, inhibiting DNA repair by the inhibition of DNA ligase, causing apoptosis and, subsequent epidermal necrosis. Although bleomycin is most commonly used to treat warts, treatment with intra-lesional bleomycin has also been described as an effective treatment of multiple BCCs [99].

There is a wide variety of newly-developed drugs, which comprises: 1) monoclonal antibodies, 2) synthetic drugs, 3) improved versions of well-known ones and also 4) natural compounds. Afamelanotide is a synthetic analog of α-MSH (melanocyte stimulating hormone) that may protect skin from UVB damage through the induction of melanogenesis. It is currently being evaluated in a multicenter, randomized, double-blind, placebo controlled trial to help prevent AKs and SCCs (ClinicaTrials.gov ID: NCT00829192). As it was mentioned before, EGFR activation is involved in NMSCs development and growth. Cetuximab is a monoclonal antibody which inhibits the EGFR receptor, and it has been used primarily in metastatic SCC of the head and neck, though it is being experimentally tested for other forms of NMSC [101]. Resiquimod is a TLR 7/8 agonist that has stimulatory effects on monocytic cells, although it is 10- to 100-times more potent than Imiquimod. One phase II study consisting of 132 patients showed an efficacy of 40% to 74% in the treatment of AKs when treating 4 to 8 lesions on the scalp with various concentrations of Resiquimod 3 times weekly. There was no difference in efficacy among the concentrations tested, but lower concentrations produced better tolerability [99]. It has been shown that blocking the hedgehog pathway stops growth of BCC. As it was described, this pathway is active in most BCCs and only in a few normal tissues, such as hair follicles; it becomes inactive in most adult tissues, with the exception of roles in tissue maintenance and repair. Results of a phase II trial indicate that the new hedgehog inhibitor, Vismodegib, is effective in this setting. The drug was approved by the FDA in 2012 for the use in patients with locally advanced BCC who are not candidates for surgery or radiation therapy and for patients whose cancer has metastasized [102].

Plenty of natural compounds are nowadays being tested as alternative treatments for NMSCs, some of them renowned and also several novelties. Ingenol mebutate is a new topical medication that has recently been approved by the FDA. It is a biologically active macrocyclic diterpene ester extracted from *Euphorbia peplus* plant. There are two proposed mechanisms of action of this novel drug. The first involves direct damage to the mitochondrial membrane, leading to rapid cellular necrosis and the release of pro-inflammatory mediators. The second may involve the maturation of B cells that produce antibodies that lead to antibody-dependent cellular cytotoxicity. Early results are promising and this treatment has the additional benefit of requiring fewer applications (2–3 days) in contrast to imiquimod, 5-FU, and diclofenac that can require treatment regimens that last from weeks to months [100]. Many natural compounds have been proven to down regulate COX-2 over-expression and/or PGE_2 over-production, aberrations which are commonly seen in NMSCs. Examples of some of them are curcumin (*Curcuma longa*), apigenin (fruits and vegetables), cyanidin (fruits and vegetables) and resveratrol (*Vitis vinífera*). A number of these molecules also have the additional abilities to be anti-oxidants (curcumin), to have anti-inflammatory and anti-proliferative properties (resveratrol), to have radical-scavenging effects (cyanidin) or to induce caspase-dependent p53-mediated apoptosis (apigenin) [103].

Regarding the relationship between NMSCs, COX-2 and PGE_2, NSAIDs have also been tested in order to treat this kind of tumor [104]. There has been plenty of research publications involving orally administered NSAIDs, not only non specific NSAIDs like indomethacin, but also COX-2 specific ones like celecoxib [88, 105]. In spite of this fact, not much has been researched about the topical application of these drugs. In this way we have tested topically applied Naproxen - a non selective NSAID - in a mouse model of SCC induced by UVB chronic irradiation. SKH:1 hairless mice were irradiated with 50 mJ/cm^2 of UVB, three times a week for 4 months. A 10% naproxen aqueous solution was applied immediately after each irradiation. We have observed a significant reduction in the incidence of pretumoral or tumoral lesions after the 4 months, in the group treated with Naproxen with respects to the group irradiated but not treated. [56]. We have also registered some epidermal alterations which occurred during SCC development, such as the hyperproliferation and the increment in apoptosis, with a concomitant decrease in the epidermal T cell percentage. Figure 7 shows a summary of these results [56].

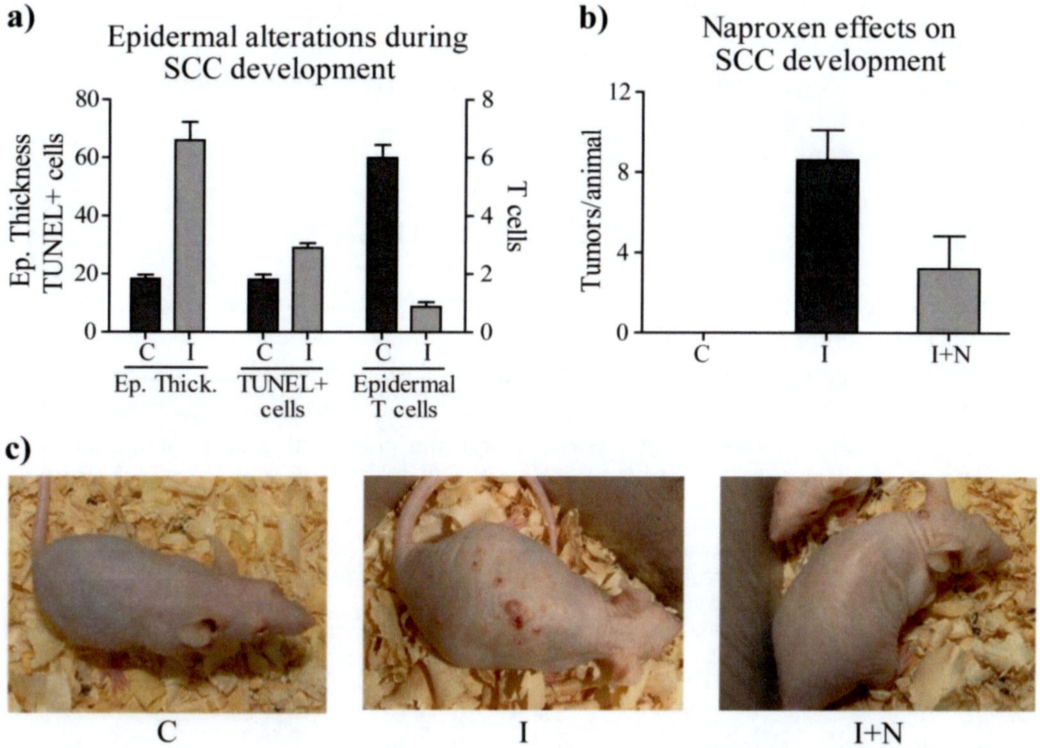

Figure 7. Skin cancer and experimental treatments evaluated in our laboratory. a) Epidermal alterations during SCC development: proliferation (epidermal thickness), apoptosis (TUNEL^{+} cells) and T cell percentage (epidermal T cells). b) Naproxen effects on SCC development: number of tumors per animal after 4 months of chronic irradiation. c) Photoghraphs of the skin aspect of the SKH:1 hairless mice after the 4 months. C: unirradiated control. I: irradiated animals. I+N: irradiated animals treated with Naproxen.

There is also another novel group of therapies being tested for cancer treatment which includes probiotics, not only as entire living microorganisms, but also some of their isolated molecules. These treatments were proved to be useful in animal models of colon and breast cancer, both kind of tumors related to the common mucosal immune system [106]. In relationship with UVR skin exposure, it was shown that lactobacilli oral administration reverted some detrimental effects of UV-single exposure, such as the decrease in contact hypersensitivity reaction, also in epidermal Langerhans cells density, and the increase in IL-10 serum concentration [107]. Having in mind these two facts (anti-tumoral and anti-UV effects of probiotics) and also that one possible mechanism of interaction of immune cells with probiotics is mediated by bacterial surface molecules (such as Lipoteichoic Acid or LTA) and innate receptors (such as toll-like receptors TLRs), we decided to test them. We evaluated in our laboratory the anti-SCC effects of orally administered LTA purified from the probiotic bacteria *Lactobacillus rhamnosus*, in a chronic irradiation mouse model. SKH:1 hairless mice were irradiated with 50 mJ/cm^2 of UVB, 3 times a week for 6 months. One hundred microliters of a 1 mg/ml of LTA solution was administered immediately after each irradiation. We demonstrated a clear activation of the immune system (gut-associated as well as systemic immunity) in LTA-treated mice, with increases in IgA+ cells in the small intestine, in IFN-γ production by inguinal lymph node cells and, in the number of total helper and cytotoxic T-cells. This modulation of the immune system during the chronic irradiation schedule led to a delay in the appearance of SCC tumors in our mice, comparing irradiated LTA-treated mice vs control irradiated vehicle-treated ones [108]. Figure 8 shows a summary of the results obtained [108].

7.6. Skin Photosensitivity

Cutaneous photosensitivity (CP) is a condition where a skin abnormal reaction to UVR takes place. This reaction includes a greater erythema formation which also lasts longer than usual, as well as induction or exacerbation of cutaneous lesions in particular patients such as lupus erythematosus (LE) ones. All of these occurring even at low doses of radiation or after short exposures. UVR involved in this phenomenon can come either from solar radiation (specifically UVBr) or from artificial sources as well [109, 110].

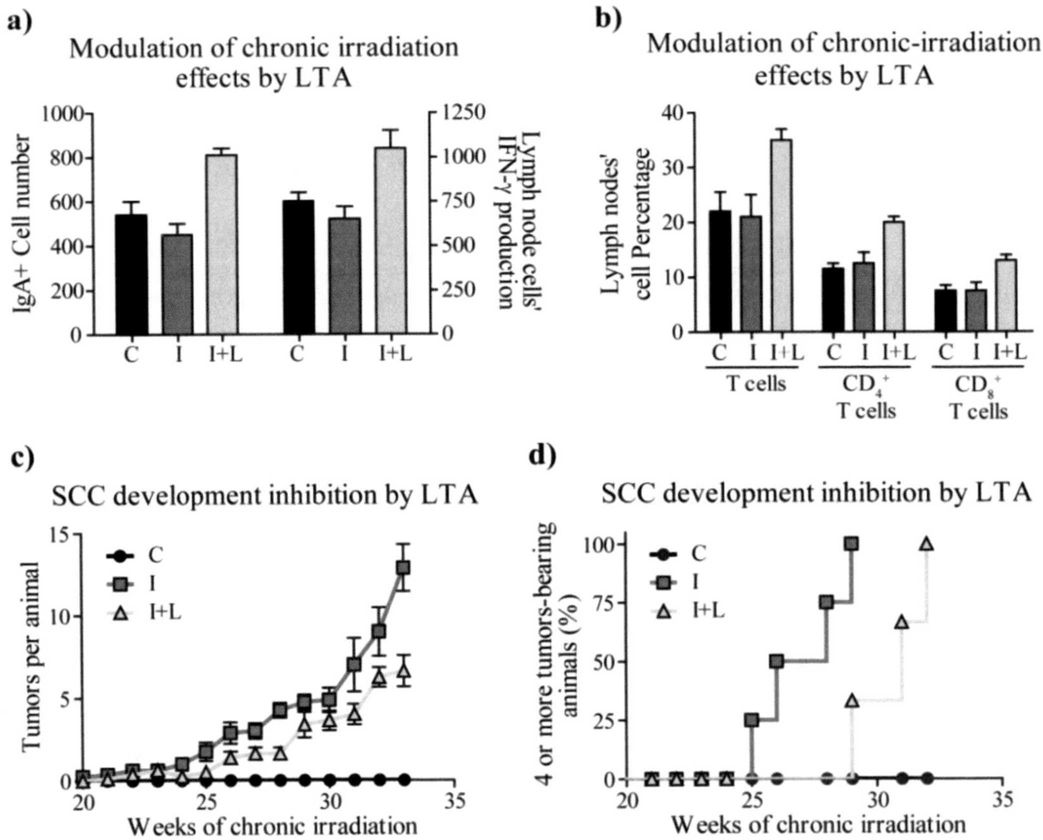

Figure 8. Skin cancer and experimental treatments evaluated in our laboratory. Modulation of chronic irradiation effects by LTA: a) IgA+ cell number and lymph node cells' IFN-γ production. b) lymph nodes' percentage of T cells, LT CD4$^+$ and LT CD8$^+$. SCC development inhibition by LTA: c) number of tumors per animal d) number of animals with 4 or more tumors. ●/C: unirradiated control. ■/I: irradiated animals. ▲/I+L: irradiated animals treated with LTA.

7.6.1. General Remarks

According to the American College of Rheumatology, CP consists of a skin rash as a result of an unusual reaction to sunlight, by patient history or physician observation. CP phenomena as already described, conforms one of the eleven diagnostic criteria for Systemic LE (SLE) [111].

The science of photosensitivity began to be defined in the 1960s and early 1970s, and its presence in lupus patients was studied initially by Baer and Harber and, Everett and Olson. The mechanisms underlying photosensitivity in LE have been investigated ever since Epstein et al., and knowledge has been built upon in the last decades, although many issues remain unresolved [112–114].

7.6.2. Mechanisms Involved and Relationship with Autoimmunity

The mechanisms involved in the development of CP have been extensively studied focusing on different approaches; these include the presence of certain auto-antibodies (autoAbs) [115], the deposition of immunoglobulins in the dermo-epidermal junctions [116], the expression of auto-antigens (autoAgs) in the skin [117], and also the relation with apoptotic cell death induced by UVR [118, 119]. However, the exact mechanism that triggers CP has not been fully elucidated yet (Figure 9).

CP is a common feature between patients with a certain group of systemic auto-immune pathologies (SAPs) with rheumatic component, such as SLE. The incidence of CP is variable among different SAPs, and it can reach up to 90% in SLE patients [120]. These patients also share the presence of some autoAbs among them and also with other non-CP related SAPs, like Sjögren's syndrome (SS) and rheumatoid arthritis (RA) [121, 122]. Most of these autoAbs are raised against nuclear components, for example ribonucleoproteins (RNPs) like La/SS-B 47 kDa, Ro/SS-A 52 kDa and Ro/SS-A 60 kDa [123, 124].

Apoptosis is an active process that leads to the ordered removal of living cells avoiding the generation of an inflammatory environment. This is desirable in most of the cases, though in SAPs this might not be so favorable. It has been demonstrated that UVR driven apoptotic keratinocytes relocate RNPs towards the cell membrane [125, 126] and this event could, on a first instance, induce the production of autoAbs [127]. Subsequently these circulating autoAbs could bind to the apoptotic keratinocytes' membrane, producing autoAbs mediated cytotoxicity, finally leading to an unwanted inflammatory response, which under normal conditions would not take place [128]. These events are proposed to be relevant in the generation of cutaneous damage in SAPs patients.

There has been a long-standing issue as to whether or how anti-nuclear antibodies interact with epidermal cells to produce photosensitive skin lesions, making special emphasis in anti-Ro/SS-A autoAbs. There are many theories that relate specifically the presence of circulating anti-Ro/SS-A autoAbs with the CP phenomenon [109, 115, 119]. This idea of positive correlation is supported by studies where the expression of Ro/SS-A proteins was measured in skin biopsy specimens of patients with photosensitive forms of LE.

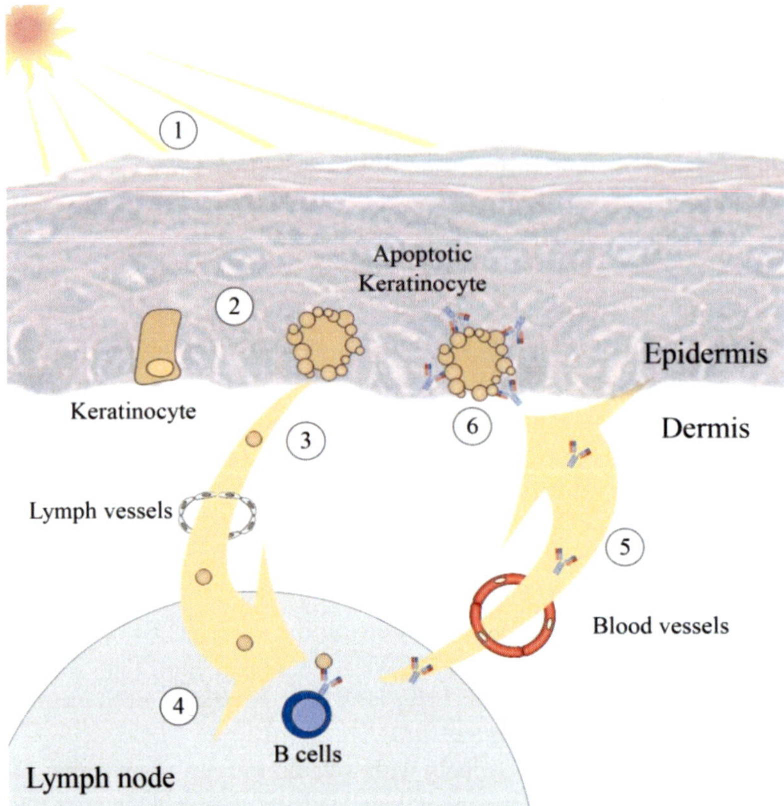

Figure 9. Cutaneous photosensitivity and theories about its etiology. Scheme showing 1: UVR impacts on the skin 2: autoantigens relocate on the surface of apoptotic keratinocytes 3: autoantigens are carried to the lymph nodes through lymph vessels 4: autoantibodies' generation is triggered 5: autoantibodies reach the skin through blood vessels 6: autoantobodies bind to the surface of newly apoptotic keratinocytes and induce citotoxicity.

These authors found that the expression of the antigens was greater in LE patients with CP than in LE patients without CP, and also that there was a strong correlation between the skin expression of these autoAgs and the presence and titer of circulating autoAbs against them [117, 129]. Nevertheless, there are several studies which are in disagreement with this positive association; thus, there has been great controversy about this subject over many years, and it is still a topic of discussion between scientists. To begin with, there is epidemiological data which indicate that there is not a significant direct association between anti-Ro/SS-A autoAbs and CP [130, 131]. Moreover there is the fact that 75% of SS's patients do have anti-Ro/SS-A autoAbs, many with high autoAbs titers, but not CP [132]. Furthermore there is the occurrence of CP in individuals that do not show any kind of disease, nor any circulating autoAbs [133].

As it can be observed, anti-Ro/SS-A autoAbs are not a *sine qua non* of photosensitivity. In our laboratory, we decided to conduct and analyze this matter from different aspects. First, we evaluated the autoAbs profile and titer (anti-Ro/SS-A 52 kDa or Ro52, anti-Ro/SS-A 60 kDa or Ro60, anti-La/SS-B, and anti-Sm, by enzyme-linked immunosorbent assay [ELISA]) in a local cohort of 169 SAP patients with/out CP. We found that there was no association between the presence of anti-Ro/SS-A autoAbs and the occurrence of CP. This was true for anti-Ro52 as well as for anti-Ro60, both for presence and titer. Half of the anti-Ro52 positive patients were photosensitive and half were not, the same happened for anti-Ro60 positive patients. On the other hand, a statistically significant positive association was found between CP and the presence as well as high reactivity anti-Sm autoAbs, although this fact could be biased by the incidence of both events in SLE patients [134].

In a second approach, we studied the binding capacity of purified autoAbs to apoptotic keratinocytes cell surfaces by flow cytometry and also by fluorescent microscopy in an *in vitro* model. We selected 10 patients from the previous study, according to their clinical characteristics, laboratory findings and affinity-purified anti-Ro52 and anti-Ro60 autoAbs from their serum samples. We analyzed the existence of a relationship between the binding profile to apoptotic keratinocytes of the serum and pure autoAbs of each patient, and the presence or absence of CP.

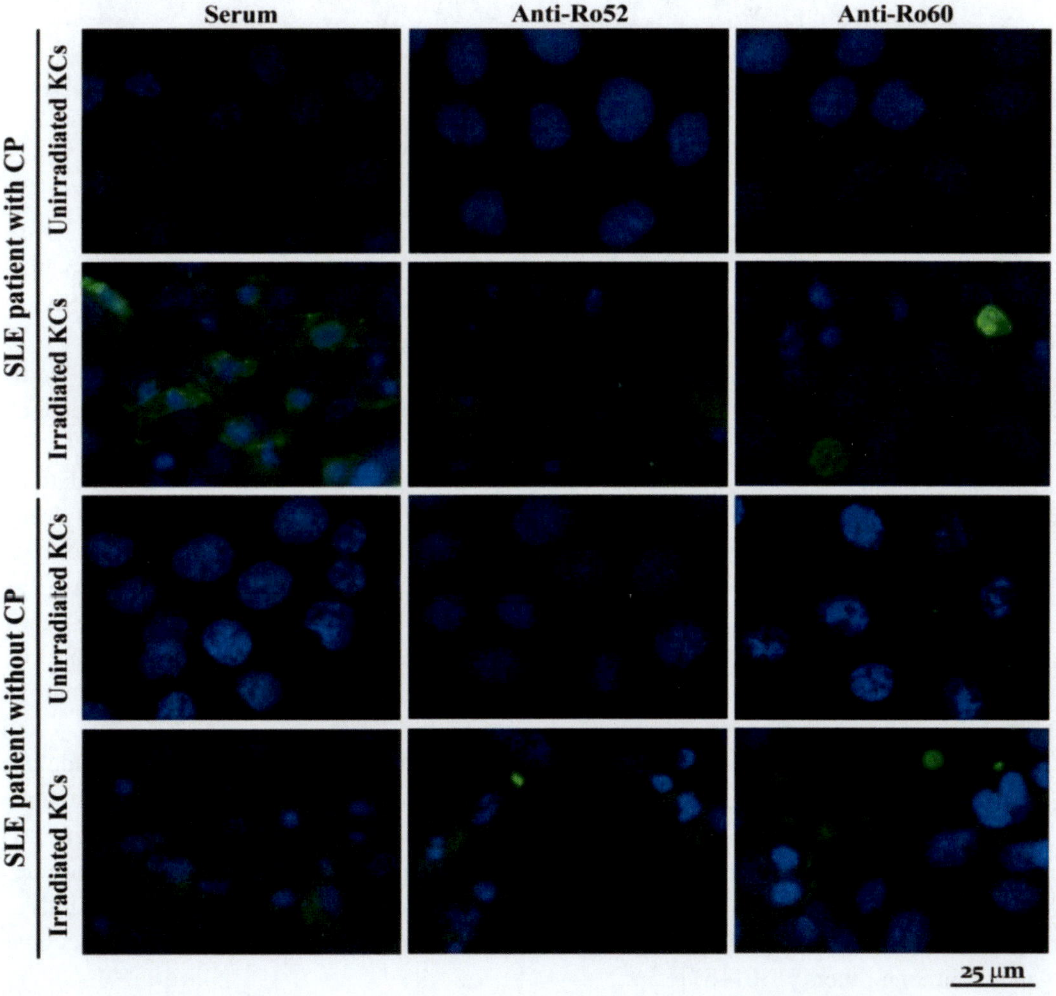

Figure 10. Cutaneous photosensitivity and anti-Ro autoantibodies evaluated in our laboratory. Photograph showing an immunoflourescence of HaCaT cells (irradiated and unirradiated) interacting with purified anti-Ro autoantibodies (anti-Ro 52 or anti-Ro 60) as well as with total serum from SLE patients with and without CP.

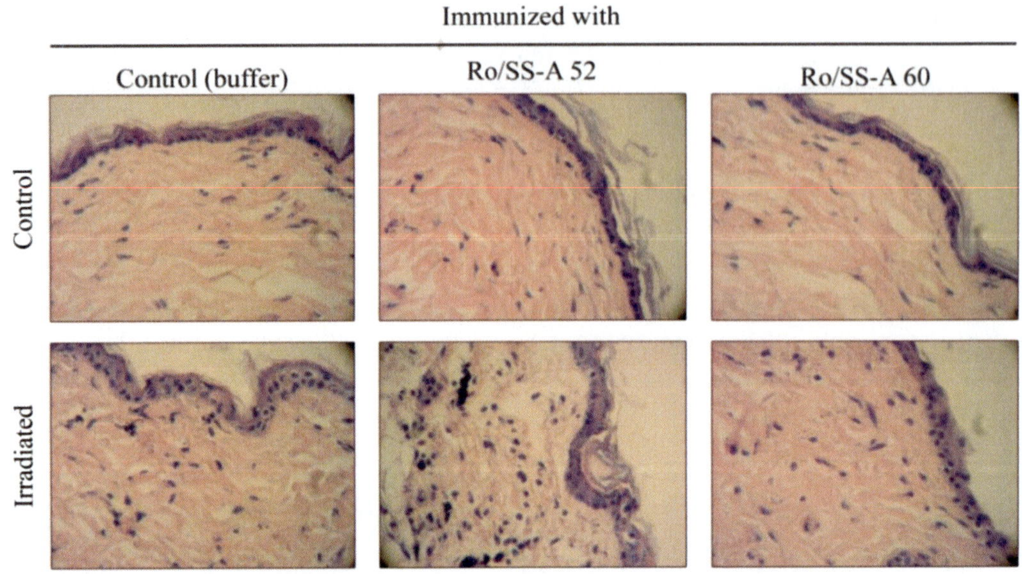

Figure 11. Cutaneous photosensitivity and anti-Ro autoantibodies evaluated in our laboratory. Photomicrograph showing skin histology of irradiated mice with circulating autoantibodies.

We concluded that serum of the 10 patients did bind to apoptotic keratinocytes, as well as their corresponding purified anti-Ro52 and anti-Ro60 autoAbs, but there was no relationship between the sera binding capacity and the presence or absence of CP, nor with the SAPs [135] (Figure 10).

Afterwards, we used an *in vivo* model that consisted of hairless SKH:1 mice, which were induced to produce anti-murine Ro52 and/or Ro60 autoAbs and were subsequently irradiated with UVB light. We evaluated the skin histology and also the epidermal production of TNF-α. Mice with circulating anti-Ro/SS-A autoAbs were divided in 2 groups, one was irradiated and the other mocked-irradiated. TNF-α production showed no statistical differences between the 3 immunized-irradiated groups and the non immunized-irradiated control group; neither did the evaluated parameters by histological analysis, except for the slight inflammatory infiltrate found in the irradiated group immunized with Ro52. This fact suggests that the single presence of anti-Ro/SS-A autoAbs is not the only requirement for the development of CP, and many other mediators might be involved in the pathogenesis of this manifestation [135] (Figure 11).

In light of our own results and in view of the fact that a high percentage of individuals can show CP without having circulating anti-Ro/SS-A autoAbs, it is logical to consider the possibility of the existence of other humoral factors not yet identified but, different to autoAbs which might be responsible for the CP pathogenesis. These factors could include molecules with apoptotic-cell-surface binding capacity (such as lactadherine), C reactive protein, vitamin K-dependent S protein, C4 binding protein, seric amiloid protein or, lectins in general, among others [136, 137].

It may also be highly plausible that CP is related to non-sera related factors, like for example, the intracellular functions of some proteins, like Ro/SS-A. It has been shown that Ro/SS-A 52 *knock out* (-/-) mice developed not only systemic autoimmunity lupus-like, but also an exaggerated skin inflammatory response. This is thought to be related with one of

Ro/SS-A 52 functions which involves the negative regulation of pro-inflammatory cytokines [138] and the loss of its functions when it is induced to relocate towards the cell surface in apoptotic cells.

In genetically susceptible individuals, a conjunction of factors are then likely to conform to the CP scenario: keratinocytes with higher susceptibility to UVR could generate a higher amount of apoptotic cells in which Ro/SS-A 52 protein is relocalized from the nucleus to the plasma membrane, therefore not being able to perform its physiological functions. This might cause a great inflammatory environment that together with: 1) the possible deficiencies in apoptotic cell elimination, 2) the opsonization of these cells by specific autoAbs, 3) the phagocytosis of opsonized keratinocytes by macrophages and, 4) the subsequent release of more proinflammatory cytokines, could engender the great inflammation which finally causes CP lesions.

7.6.3. Clinical Management

CP is a manifestation with no specific treatment available. All actions that can be undertaken are preventive, and named "photoprotection". Photoprotection can be classified as conventional or not conventional.

Conventional photoprotection includes avoiding sun exposure and, when it is inevitable, the usage of appropriate special clothes and UVA/UVB broad spectrum sun-screens. Artificial UVR sources should also be avoided, as well as photosensitizing drugs intake, such as phenothiazines or sulphonamides, among many others.

Non-conventional photoprotection comprises the usage of special sun-screens which have been supplemented with antioxidants (such as vitamin C and E) and topical or intravenous administration of melanocyte stimulating hormone as well as thymidine dinucleotides. All of these strategies have been proved in experimental studies as effective, in the protection against UVR.

Photoprotection is a fundamental aspect in LE patients, because it has been reported to be effective avoiding the formation of cutaneous lesions, as well as associated with a better clinical outcome and also with less renal complications or even the need of an immunosupresive treatment [139].

As a final thought it is worth to notice that CP phenomenon has consequences in quality of life of suffering individuals, as it has been reported in a recent study performed in the UK. As stated before, conventional photoprotection includes avoidance of sun exposure or the usage of integral protection (clothes and sun-screens). These actions could limit affected people's lifestyle. In fact, the study reported that photosensitive individuals spent less time outdoors than healthy people, being the difference statistically significant during weekends and summer days, making an impact in the individual's social life. The authors also found a significant difference in the unemployment rate for CP patients, due to their inability to work in open fields or under strong artificial light sources. Likewise, photosensitive persons show vitamin D deficiencies. All of these events cause the high prevalence of anxiety and depression that is seen in CP people. Additionally, in the mentioned study, they also found an occurrence of 40% of psychiatric disorders in CP patients [140].

Conclusion

Marvelous sunlight, source of light and warmth, can be as healthy as harmful. The final effect of UVR on the skin and immune system depends exclusively on the intensity of the radiation as well as on the length and frequency of the exposure. UVR modifies skin cell functions, from genetic material (direct damage) to alterations in cellular metabolism, leading to mitochondrial dysfunction, reactive oxygen and nitrogen species production, cell cycle arrest and apoptotic cell death. These changes affect mostly the epidermis, as it is the tissue directly exposed to the radiation. Injured keratinocytes produce soluble mediators which can spread through, and impact on, the dermis as well as on lymph nodes resulting in local inflammation and systemic immunosupression. Moreover, dermal and epidermal DCs can also migrate to lymph nodes and promote T cell differentiation to a regulatory phenotype. The persistence of these effects can lead to long-term alterations, ending up in serious diseases like CP or NMSC. Although none of these illnesses has a high incidence of mortality, they affect people's lifestyle in many ways, limiting their everyday life in significant aspects.

References

[1] Diffey BL (1991) Solar ultraviolet radiation effects on biological systems. *Physics in Medicine and Biology* 36:299–328.

[2] Magin P, Pond D, Smith W, et al. (2012) Reliability of skin-type self-assessment: agreement of adolescents' repeated Fitzpatrick skin phototype classification ratings during a cohort study. *Journal of the European Academy of Dermatology and Venereology* 26:1396–9. doi: 10.1111/j.1468-3083.2011.04298.x.

[3] Osmola-Mańkowska A, Silny W, Dańczak-Pazdrowska A, et al. (2012) The sun – our friend or foe ? *Annals of Agricultural and Environmental Medicine* 19:805–809.

[4] Grenache DG, Zhang Z, Wells LE, et al. (2007) Wound Healing in the a2b1 Integrin-Deficient Mouse : Altered Keratinocyte Biology and Dysregulated Matrix Metalloproteinase Expression. *Journal of Investigative Dermatology* 127:455–466. doi: 10.1038/sj.jid.5700611

[5] Slominski A, Tobin DJ, Shibahara S, Wortsman J (2004) Melanin pigmentation in mammalian skin and its hormonal regulation. *Physiological reviews* 84:1155–228. doi: 10.1152/physrev.00044.2003

[6] Pennello G, Devesa S, Gail M (2000) Association of Surface Ultraviolet B Radiation Levels with Melanoma and Nonmelanoma Skin Cancer in United States Blacks Association of Surface Ultraviolet B Radiation Levels with Melanoma and Nonmelanoma Skin Cancer in United States Blacks. 291–297.

[7] Jolles S (2002) Paul Langerhans. *Journal of clinical pathology* 55:243.

[8] Romani N, Holzmann S, Tripp CH, et al. (2003) Langerhans cells – dendritic cells of the epidermis. *Apmis* 111:725–740.

[9] Becker Y (2003) Milestones in the research on skin epidermal Langerhans/dendritic cells (LCs/DCs) from the discovery of Paul Langerhans 1868-1989. *Virus genes* 26:131–4.

[10] Merad M, Ginhoux F, Collin M (2008) Origin, homeostasis and function of Langerhans cells and other langerin-expressing dendritic cells. *Nature Reviews Immunology* 8:935–47. doi: 10.1038/nri2455.

[11] Chodaczek G, Papanna V, Zal MA, Zal T (2012) Body-barrier surveillance by epidermal γδ TCRs. *Nature immunology* 13:272–82. doi: 10.1038/ni.2240.

[12] Jameson JM, Cauvi G, Witherden D a, Havran WL (2004) A keratinocyte-responsive gamma delta TCR is necessary for dendritic epidermal T cell activation by damaged keratinocytes and maintenance in the epidermis. *Journal of immunology* (Baltimore, Md : 1950) 172:3573–9.

[13] Jameson J, Ugarte K, Chen N, et al. (2002) A Role for Skin gd T Cells in Wound Repair. *Science* 296:747–749.

[14] Spetz a L, Strominger J, Groh-Spies V (1996) T cell subsets in normal human epidermis. *The American journal of pathology* 149:665–74.

[15] Randolph GJ, Angeli V, Swartz M a (2005) Dendritic-cell trafficking to lymph nodes through lymphatic vessels. *Nature Reviews Immunology* 5:617–28. doi: 10.1038/nri1670.

[16] Nestle FO, Di Meglio P, Qin JZ, Nickoloff BJ (2009) Skin immune sentinels in health and disease. *Nature Reviews Immunology* 9:679–691. doi: 10.1038/nri2622

[17] Harvima IT, Nilsson G (2011) Mast cells as regulators of skin inflammation and immunity. *Acta Dermato-Venereologica* 91:644–50. doi: 10.2340/00015555-1197

[18] Narbutt J, Lesiak A, Boncela J, et al. (2007) Repeated exposures of humans to low doses of solar simulated radiation lead to limited photoadaptation and photoprotection against UVB-induced erythema and cytokine mRNA up-regulation. *Journal of Dermatological Science* 45:210–212. doi: 10.1016/j.jdermsci.2006.11.001

[19] Norval M, McLoone P, Lesiak A, Narbutt J (2008) The Effect of Chronic Ultraviolet Radiation on the Human Immune System. *Photochemistry and Photobiology* 84:19–28.

[20] Brash DE (1988) UV mutagenic photoproducts in Escherichia coli and human cells: a molecular genetics perspective on human skin cancer. *Photochemistry and photobiology* 48:59–66.

[21] Lu YP, Lou YR, Yen P, et al. (1999) Time course for early adaptive responses to ultraviolet B light in the epidermis of SKH-1 mice. *Cancer research* 59:4591–602.

[22] Narbutt J, Lesiak A, Jochymski C, et al. (2007) Increased cyclooxygenase expression and thymine dimer formation after repeated exposures of humans to low doses of solar simulated radiation. *Experimental Dermatology* 16:837–843. doi: 10.1111/j.1600-0625.2007.00598.x.

[23] Britt AB (1995) Repair of DNA Damage lnduced by Ultraviolet Radiation. *Plant Physiology* 108:891–896.

[24] Yamada M, Udono MU, Hori M, et al. (2006) Aged human skin removes UVB-induced pyrimidine dimers from the epidermis more slowly than younger adult skin in vivo. *Archives of dermatological research* 297:294–302. doi: 10.1007/s00403-005-0618-0

[25] Decraene D, Agostinis P, Pupe A, et al. (2001) Acute response of human skin to solar radiation : regulation and function of the p53 protein. *Journal of Photochemistry and Photobiology B: Biology* 63:78–83.

[26] Yoshida K, Miki Y (2010) The cell death machinery governed by the p53 tumor suppressor in response to DNA damage. *Cancer science* 101:831–5. doi: 10.1111/j.1349-7006.2010.01488.x

[27] Lane DP (1992) p53, guardian of the genome. *Nature* 358:15–16.

[28] Bowen AR, Hanks AN, Allen SM, et al. (2003) Apoptosis Regulators and Responses in Human Melanocytic and Keratinocytic Cells. *Journal of Investigative Dermatology* 120:48–55.

[29] Israels LG, Israels ED (1999) Apoptosis. *Stem Cells* 17:306–313.

[30] Chang L, Karin M (2001) Mammalian MAP kinase signalling cascades. *Nature* 410:37–40. doi: 10.1038/35065000.

[31] Assefa Z, Vantieghem a, Garmyn M, et al. (2000) p38 mitogen-activated protein kinase regulates a novel, caspase-independent pathway for the mitochondrial cytochrome c release in ultraviolet B radiation-induced apoptosis. *The Journal of biological chemistry* 275:21416–21. doi: 10.1074/jbc.M002634200.

[32] Madesh M, Antonsson B, Srinivasula SM, et al. (2002) Rapid kinetics of tBid-induced cytochrome c and Smac/DIABLO release and mitochondrial depolarization. *The Journal of biological chemistry* 277:5651–9. doi: 10.1074/jbc.M108171200.

[33] Kluck RM, Degli Esposti M, Perkins G, et al. (1999) The Pro-apoptotic Proteins, Bid and Bax, Cause a Limited Permeabilization of the Mitochondrial Outer Membrane that Is Enhanced by Cytosol. *The Journal of cell biology* 147:809–822.

[34] Düssmann H, Kögel D, Rehm M, Prehn JHM (2003) Mitochondrial Membrane Permeabilization and Superoxide Production during Apoptosis. *The Journal of biological chemistry* 278:12645–12649. doi: 10.1074/jbc.M210826200.

[35] Jones KJ, Perris AD, Vernallis AB, et al. (2005) Induction of inflammatory cytokines and nitric oxide in J774.2 cells and murine macrophages by lipoteichoic acid and related cell wall antigens from Staphylococcus epidermidis. *Journal of medical microbiology* 54:315–21. doi: 10.1099/jmm.0.45872-0.

[36] Cals-Grierson M-M, Ormerod AD (2004) Nitric oxide function in the skin. *Nitric Oxide* 10:179–193. doi: 10.1016/j.niox.2004.04.005.

[37] Poderoso JJ, Carreras MC, Lisdero C, et al. (1996) Nitric oxide inhibits electron transfer and increases superoxide radical production in rat heart mitochondria and submitochondrial particles. *Archives of biochemistry and biophysics* 328:85–92.

[38] Squadrito GL, Pryor WA (1998) Oxidative Chemistry of Nitric Oxide: The Roles of Superoxide, Peroxynitrite, and Carbon Dioxide. *Free Radical Biology and Medicine* 25:392– 403.

[39] Bauer G (2002) Signaling and proapoptotic functions of transformed cell-derived reactive oxygen species. *Prostaglandins, Leukotrienes and Essential Fatty Acids* 66:41–56. doi: 10.1054/plef.2001.0332

[40] Aitken GR, Henderson JR, Chang S-C, et al. (2007) Direct monitoring of UV-induced free radical generation in HaCaT keratinocytes. *Clinical and experimental dermatology* 32:722–7. doi: 10.1111/j.1365-2230.2007.02474.x.

[41] Chang H-R, Tsao D-A, Wang S-R, Yu H-S (2003) Expression of nitric oxide synthases in keratinocytes after UVB irradiation. *Archives of dermatological research* 295:293–6. doi: 10.1007/s00403-003-0433-4.

[42] Seo SJ (2002) Time course of expression of mRNA of inducible nitric oxide synthase and generation of nitric oxide by ultraviolet B in keratinocyte cell lines. *British Journal of Dermatology* 28:45–662. doi: 10.1046/j.1365-2133.2002.04849.x.

[43] Paz ML, González Maglio DH, Weill FS, et al. (2008) Mitochondrial dysfunction and cellular stress progression after ultraviolet B irradiation in human keratinocytes. *Photodermatology, Photoimmunology & Photomedicine* 24:115–122.

[44] Gonzalez Maglio DH, Paz ML, Ferrari A, et al. (2005) Skin damage and mitochondrial dysfunction after acute ultraviolet B irradiation: relationship with nitric oxide production. *Photodermatology, Photoimmunology and Photomedicine* 21:311–317.

[45] Gonzalez Maglio DH, Cela EM, Ferrari A, Leoni J (2011) Mitochondrial function evaluation in epidermal cells ex vivo after ultraviolet irradiation. *Experimental dermatology* 20:947–950. doi: 10.1111/j.1600-0625.2011.01342.x.

[46] Paz ML, Ferrari A, Weill FS, et al. (2008) Time-course evaluation and treatment of skin inflammatory immune response after ultraviolet B irradiation. *Cytokine* 44:70–7. doi: 10.1016/j.cyto.2008.06.012.

[47] Guzik TJ, Korbut R, Adamek-Guzik T (2003) Nitric Oxide and Superoxide in Inflammation and Immune Regulation. *Journal of Physiology and Pharmacology* 54:469–487.

[48] Holan V, Krulová M, Zajícová A, Pindjákova J (2001) Nitric oxide as a regulatory and effector molecule in the immune system. *Molecular Immunology* 38:989–995.

[49] Kuhn a, Fehsel K, Lehmann P, et al. (1998) Aberrant timing in epidermal expression of inducible nitric oxide synthase after UV irradiation in cutaneous lupus erythematosus. *The Journal of investigative dermatology* 111:149–53. doi: 10.1046/j.1523-1747.1998.00253.x.

[50] Chun K-S, Cha H-H, Shin J-W, et al. (2004) Nitric oxide induces expression of cyclooxygenase-2 in mouse skin through activation of NF-kappaB. *Carcinogenesis* 25:445–54. doi: 10.1093/carcin/bgh021.

[51] Rhodes L, Belgi G, Parslew R, Friedmann P (2001) Ultraviolet-B-Induced Erythema is Mediated by Nitric Oxide and Prostaglandin E2 in Combination. *Journal of Investigative Dermatology* 117:880–885.

[52] Athar M, An KP, Tang X, et al. (2004) Photoprotective effects of sulindac against ultraviolet B-induced phototoxicity in the skin of SKH-1 hairless mice. *Toxicology and Applied Pharmacology* 195:370 – 378. doi: 10.1016/j.taap.2003.09.030.

[53] Tripp CS, Blomme EAG, Chinn KS, et al. (2003) Epidermal COX-2 Induction Following Ultraviolet Irradiation: Suggested Mechanism for the Role of COX-2 Inhibition in Photoprotection. *Journal of Investigative Dermatology* 121:853–861.

[54] Wilgus TA, Ross MS, Parrett ML, Oberyszyn TM (2000) Topical application of a selective cyclooxygenase inhibitor suppresses UVB mediated cutaneous inflammation. *Prostaglandins & other Lipid Mediators* 62:367–384.

[55] Yoshizumi M, Nakamura T, Kato M, et al. (2008) Release of cytokines/chemokines and cell death in UVB-irradiated human keratinocytes, HaCaT. *Cell Biology International* 32:1405–1411. doi: 10.1016/j.cellbi.2008.08.011.

[56] González Maglio DH, Paz ML, Ferrari A, et al. (2010) Alterations in Skin Immune Response Throughout Chronic UVB Irradiation—Skin Cancer Development and Prevention by Naproxen. *Photochemistry and Photobiology* 86:146–152.

[57] Negishi M, Sugimoto Y, Ichikawa a (1995) Prostaglandin E receptors. *Journal of lipid mediators and cell signalling* 12:379–91.

[58] Grandjean-Laquerriere a (2003) Differential regulation of TNF-α, IL-6 and IL-10 in UVB-irradiated human keratinocytes via cyclic AMP/protein kinase A pathway. *Cytokine* 23:138–149. doi: 10.1016/S1043-4666(03)00224-2.
[59] Trompezinski S, Pernet I, Schmitt D, Viac J (2001) UV radiation and prostaglandin E2 up-regulate vascular endothelial growth factor (VEGF) in cultured human fibroblasts. *Inflammation research: official journal of the European Histamine Research Society.* (et al) 50:422–7.
[60] Sharma MR, Werth B, Werth VP (2011) Animal models of acute photodamage: comparisons of anatomic, cellular and molecular responses in C57BL/6J, SKH1 and Balb/c mice. *Photochemistry and photobiology* 87:690–8. doi: 10.1111/j.1751-1097.2011.00911.x.
[61] Fagot D, Asselineau D, Bernerd F (2002) Direct role of human dermal fibroblasts and indirect participation of epidermal keratinocytes in MMP-1 production after UV-B irradiation. *Archives of dermatological research* 293:576–583. doi: 10.1007/s00403-001-0271-1.
[62] Quan T, He T, Kang S, et al. (2004) Solar ultraviolet irradiation reduces collagen in photoaged human skin by blocking transforming growth factor-beta type II receptor/Smad signaling. *The American journal of pathology* 165:741–51.
[63] Fisher MS, Kripke ML (1977) Systemic alteration induced in mice by ultraviolet light irradiation and its relationship to ultraviolet carcinogenesis. *Proceedings of the National Academy of Sciences* 74:1688–1692.
[64] Fisher MS, Kripke ML (1978) Further studies on the tumor-specific suppressor cells induced by ultraviolet radiation. *Journal of immunology* (Baltimore, Md : 1950) 121:1139–44.
[65] Ullrich SE, Kripke ML (1984) Mechanisms in the suppression of tumor rejection produced in mice by repeated UV irradiation. *Journal of immunology* (Baltimore, Md : 1950) 133:2786–90.
[66] Schwarz A, Maeda A, Wild MK, et al. (2004) Ultraviolet radiation-induced regulatory T cells not only inhibit the induction but can suppress the effector phase of contact hypersensitivity. *Journal of immunology* (Baltimore, Md : 1950) 172:1036–43.
[67] Enk CD, Sredni D, Blauvelt A, Katz SI (1995) Induction of IL-10 Gene Expression in Human Keratinocytes by UVB Exposure In Vivo and In Vitro. *The Journal of Immunology* 154:4851–4856.
[68] Zhang Q, Yao Y, Konger RL, et al. (2008) UVB radiation-mediated inhibition of contact hypersensitivity reactions is dependent on the platelet-activating factor system. *The Journal of investigative dermatology* 128:1780–7. doi: 10.1038/sj.jid.5701251
[69] Shreedhar V, Giese T, Sung VW, Ullrich SE (1998) A cytokine cascade including prostaglandin E2, IL-4, and IL-10 is responsible for UV-induced systemic immune suppression. *Journal of immunology* (Baltimore, Md : 1950) 160:3783–9.
[70] Loser K, Apelt J, Voskort M, et al. (2007) IL-10 controls ultraviolet-induced carcinogenesis in mice. *Journal of immunology* (Baltimore, Md : 1950) 179:365–71.
[71] Ghoreishi M, Bach P, Obst J, et al. (2009) Expansion of Antigen-Specific Regulatory T Cells with the Topical Vitamin D Analog Calcipotriol. *The Journal of Immunology* 182:6071–6078. doi: 10.4049/jimmunol.0804064.

[72] Xu H, Elmets C a (2012) UVB immunosuppression: vitamin D or not vitamin D? That is the question. *The Journal of investigative dermatology* 132:2676–8. doi: 10.1038/jid.2012.327.

[73] McLoone P, Woods GM, Norval M (2005) Decrease in Langerhans Cells and Increase in Lymph Node Dendritic Cells Following Chronic Exposure of Mice to Suberythemal Doses of Solar Simulated Radiation Wavelength nm. *Photochemistry and Photobiology* 81:1168–1173.

[74] Dandie GW, Clydesdale GJ, Jacobs I, Muller HK (1998) Effects of UV on the migration and function of epidermal antigen presenting cells. *Mutation research* 422:147–54.

[75] Boonstra a, van Oudenaren a, Barendregt B, et al. (2000) UVB irradiation modulates systemic immune responses by affecting cytokine production of antigen-presenting cells. *International immunology* 12:1531–8.

[76] Wang L, Jameson SC, Hogquist K a (2009) Epidermal Langerhans cells are not required for UV-induced immunosuppression. *Journal of immunology* (Baltimore, Md : 1950) 183:5548–53. doi: 10.4049/jimmunol.0900235.

[77] Weill FS, Cela EM, Ferrari A, et al. (2011) Skin Exposure to Chronic But Not Acute UV Radiation Affects Peripheral T-Cell Function. *Journal of toxicology and environmental health Part A* 74:838–47. doi: 10.1080/15287394.2011.570228.

[78] Chen AC, Halliday GM, Damian DL (2013) Non-melanoma skin cancer: carcinogenesis and chemoprevention. *Pathology* 45:331–41. doi: 10.1097/PAT.0b013e32835f515c.

[79] De Gruijl FR, van Kranen HJ, Mullenders LHF (2001) UV-induced DNA damage, repair, mutations and oncogenic pathways in skin cancer. *Journal of Photochemistry and Photobiology B: Biology* 63:19–27.

[80] Brash DE, Rudolph JA, Simon JA, et al. (1991) A role for sunlight in skin cancer: UV-induced p53 mutations in squamous cell carcinoma. *Proceedings of the National Academy of Sciences* 88:10124–10128. doi: 10.1073/pnas.88.22.10124

[81] Bolshakov S, Walker CM, Strom SS, et al. (2003) P53 Mutations in Human Aggressive and Nonaggressive Basal and Squamous Cell Carcinomas. *Clinical cancer research: an official journal of the American Association for Cancer Research* 9:228–34.

[82] Nakazawa H, English D, Randell PL, et al. (1994) UV and skin cancer: specific p53 gene mutation in normal skin as a biologically relevant exposure measurement. *Proceedings of the National Academy of Sciences* 91:360–364.

[83] Bäckvall H, Strömberg S, Gustafsson A, et al. (2004) Mutation spectra of epidermal p53 clones adjacent to basal cell carcinoma and squamous cell carcinoma. *Experimental Dermatology* 13:643–50. doi: 10.1111/j.0906-6705.2004.00211.x.

[84] Errico MD, Calcagnile A, Canzona F, et al. (2000) UV mutation signature in tumor suppressor genes involved in skin carcinogenesis in xeroderma pigmentosum patients. *Oncogene* 19:463–467.

[85] Daya-Grosjean L, Sarasin A (2000) UV-specific mutations of the human patched gene in basal cell carcinomas from normal individuals and xeroderma pigmentosum patients. *Mutation Research* 450:193–9.

[86] An KP, Athar M, Tang X, et al. (2002) Cyclooxygenase-2 expression in murine and human nonmelanoma skin cancers: implications for therapeutic approaches. *Photochemistry and photobiology* 76:73–80.

[87] Tiano HF, Loftin CD, Akunda J, et al. (2002) Deficiency of either cyclooxygenase (COX)-1 or COX-2 alters epidermal differentiation and reduces mouse skin tumorigenesis. *Cancer research* 62:3395–401.

[88] Pentland a P, Schoggins JW, Scott G a, et al. (1999) Reduction of UV-induced skin tumors in hairless mice by selective COX-2 inhibition. *Carcinogenesis* 20:1939–44.

[89] Ansari KM, Rundhaug JE, Fischer SM (2008) Multiple signaling pathways are responsible for prostaglandin E2-induced murine keratinocyte proliferation. *Molecular cancer research : MCR* 6:1003–16. doi: 10.1158/1541-7786.MCR-07-2144.

[90] Tober KL, Wilgus TA, Kusewitt DF, et al. (2006) Importance of the EP1 Receptor in Cutaneous UVB-Induced Inflammation and Tumor Development. *Journal of Investigative Dermatology* 126:205–211. doi: 10.1038/sj.jid.5700014.

[91] Sung YM, He G, Fischer SM (2005) Lack of expression of the EP2 but not EP3 receptor for prostaglandin E2 results in suppression of skin tumor development. *Cancer research* 65:9304–11. doi: 10.1158/0008-5472.CAN-05-1015.

[92] El-Abaseri TB, Putta S, Hansen LA (2006) Ultraviolet irradiation induces keratinocyte proliferation and epidermal hyperplasia through the activation of the epidermal growth factor receptor. *Carcinogenesis* 27:225–231. doi: 10.1093/carcin/bgi220.

[93] Ashida M, Bito T, Budiyanto a, et al. (2003) Involvement of EGF receptor activation in the induction of cyclooxygenase-2 in HaCaT keratinocytes after UVB. *Experimental dermatology* 12:445–52.

[94] Xu Y, Shao Y, Voorhees JJ, Fisher GJ (2006) Oxidative inhibition of receptor-type protein-tyrosine phosphatase kappa by ultraviolet irradiation activates epidermal growth factor receptor in human keratinocytes. *The Journal of biological chemistry* 281:27389–97. doi: 10.1074/jbc.M602355200.

[95] El-Abaseri TB, Hansen LA (2007) EGFR Activation and Ultraviolet Light-Induced Skin Carcinogenesis. *Journal of biomedicine & biotechnology*. doi: 10.1155/2007/97939.

[96] Martínez-Carpio P a, Trelles M a (2010) Cutaneous epidermal growth factor receptor system following ultraviolet irradiation: exploring the role of molecular mechanisms. *Photodermatology, photoimmunology & photomedicine* 26:250–6. doi: 10.1111/j.1600-0781.2010.00534.x.

[97] Harvey I, Shalom D, Marks RM, Frankel RM (1989) Non-melanoma skin cancer. *British Medical Journal* 299:4–6.

[98] Wong CSM, Strange RC, Lear JT (2003) Basal Cell Carcinoma. *British Medical Journal* 327:794–798.

[99] Galiczynski EM, Vidimos AT (2011) Nonsurgical Treatment of Nonmelanoma Skin Cancer. *Dermatologic Clinics* 29:297–309. doi: 10.1016/j.det.2011.01.011.

[100] Bangash HK, Colegio OR (2012) Management of Non- Melanoma Skin Cancer in Immunocompromised Solid Organ Transplant Recipients. *Current Treatment Options in Oncology* 13:354–376. doi: 10.1007/s11864-012-0195-3.

[101] Wollina U (2012) Cetuximab in non-melanoma skin cancer. *Expert Opinion on Biological Therapy* 12:949–956.

[102] Roth BJ, Krilov L, Adams S, et al. (2013) Clinical Cancer Advances 2012: Annual Report on Progress Against Cancer From the American Society of Clinical Oncology. *Journal of Clinical Oncology* 31:131–161. doi: 10.1200/JCO.2012.47.1938.

[103] Filip GA, Clichici S (2011) Chemoprevention of Skin Cancer by Natural Compounds. Skin Cancers. Risk factors, prevention and therapy. pp 197–220.

[104] González Maglio DH, Paz ML, Cela EM, Leoni J (2011) Cyclooxigenase-2 Overexpression in Non-Melanoma Skin Cancer: Molecular Pathways Involved as Targets for Prevention and Treatment. In: La Porta CA (ed) Skin Cancers. Risk factors, prevention and therapy. *In Tech*, pp 175–196.

[105] Reeve VE, Matheson MJ, Bosnie M, Boehm-Wilcox C (1995) The protective effect of indomethacin on photocarcinogenesis in hairless mice. *Cancer Letters* 95:213–219. doi: 10.1016/0304-3835(95)03886-2.

[106] De Moreno de LeBlanc A, Matar C, Perdigón G (2007) The application of probiotics in cancer. *The British journal of nutrition* 98 Suppl 1:S105–10. doi: 10.1017/S0007114507839602.

[107] Guéniche A, Benyacoub J, Buetler TM, et al. (2006) Supplementation with oral probiotic bacteria maintains cutaneous immune homeostasis after UV exposure. *European journal of dermatology : EJD* 16:511–517.

[108] Weill FS, Cela EM, Paz ML, et al. (2013) Lipoteichoic acid from Lactobacillus rhamnosus GG as an oral photoprotective agent against UV-induced carcinogenesis. *British journal of nutrition* 109:457–466. doi: 10.1017/S0007114512001225

[109] Scheinfeld N, Deleo VA (2004) Photosensitivity in lupus erythematosus. *Photodermatology, Photoimmunology & Photomedicine* 20:272–279.

[110] Morison WL (2004) *Photosensitivity*. 1111–1117.

[111] Hochberg M (1997) Updating the American College of Rheumatology revised criteria for the classification of systemic lupus erythematosus. *Arthritis and Rheumatism* 40:1725.

[112] Baer RL, Harber LC (1965) Photobiology of lupus erythematosus. *Archives of Dermatology* 92:124–128.

[113] Everett MA, Olson RL (1964) Response of Cutaneous Lupus Erythematosus to Ultraviolet Light. *Journal of Investigative Dermatology* 44:133–134.

[114] Epstein JH, Tuffanelli D, Dubois EL (1965) Light Sensitivity and Lupus Erythematosus. *Archives of Dermatology* 91:483–485.

[115] Furukawa F (1999) Antinuclear antibody-keratinocyte interactions in photosensitive cutaneous lupus erythematosus. *Histology and histopathology* 14:627–33.

[116] Lee L a, Gaither KK, Coulter SN, et al. (1989) Pattern of cutaneous immunoglobulin G deposition in subacute cutaneous lupus erythematosus is reproduced by infusing purified anti-Ro (SSA) autoantibodies into human skin-grafted mice. *The Journal of clinical investigation* 83:1556–62. doi: 10.1172/JCI114052.

[117] Ioannides D, Golden BD, Buyon JP, Bystryn JC (2000) Expression of SS-A/Ro and SS-B/La antigens in skin biopsy specimens of patients with photosensitive forms of lupus erythematosus. *Archives of dermatology* 136:340–6.

[118] Furukawa F, Kanauchi H, Imamura S (1994) Susceptibility to UVB Light in Cultured Keratinocytes of Cutaneous Lupus erythematosus. *Dermatology* 189:18–23.

[119] Furukawa F, Kashihara-Sawami M, Lyons MB, Norris DA (1990) Binding of antibodies to the extractable nuclear antigens SS-A/Ro and SS-B/La is induced on the surface of human keratinocytes by ultraviolet light (UVL): implications for the pathogenesis of photosensitive cutaneous Lupus. *Journal of Investigative Dermatology* 94:77–85.

[120] Wysenbeek a J, Block D a, Fries JF (1989) Prevalence and expression of photosensitivity in systemic lupus erythematosus. *Annals of the rheumatic diseases* 48:461–463.

[121] Von Mühlen C a, Tan EM (1995) Autoantibodies in the diagnosis of systemic rheumatic diseases. *Seminars in Arthritis and Rheumatism* 24:323–358.

[122] Schneeberger E, Citera G, Heredia M, Maldonado Cocco J (2008) Clinical significance of anti-Ro antibodies in rheumatoid arthritis. *Clinical Rheumatology* 27:517–519. doi: 10.1007/s10067-007-0812-x.

[123] Fabini G, Rutjes S a, Zimmermann C, et al. (2000) Analysis of the molecular composition of Ro ribonucleoprotein complexes. Identification of novel Y RNA-binding proteins. *European journal of biochemistry / FEBS* 267:2778–89.

[124] Reichlin M (1996) SS-A (Ro) autoantibodies. In: Peter JP, Shoenfeld Y (eds) *Autoantibodies*. pp 783–788.

[125] Casciola-Rosen L a, Anhalt G, Rosen a (1994) Autoantigens targeted in systemic lupus erythematosus are clustered in two populations of surface structures on apoptotic keratinocytes. *The Journal of experimental medicine* 179:1317–30.

[126] Ayukawa K, Taniguchi S, Masumoto J, et al. (2000) La autoantigen is cleaved in the COOH terminus and loses the nuclear localization signal during apoptosis. *The Journal of biological chemistry* 275:34465–70. doi: 10.1074/jbc.M003673200.

[127] Mevorach D, Zhou JL, Song X, Elkon KB (1998) Systemic exposure to irradiated apoptotic cells induces autoantibody production. *The Journal of experimental medicine* 188:387–92.

[128] Furukawa F, Itoh T, Wakita H, et al. (1999) Keratinocytes from patients with lupus erythematosus show enhanced cytotoxicity to ultraviolet radiation and to antibody-mediated cytotoxicity. *Clinical and experimental immunology* 118:164–70.

[129] Oke V, Vassilaki I, Espinosa A, et al. (2009) High Ro52 expression in spontaneous and UV-induced cutaneous inflammation. *The Journal of investigative dermatology* 129:2000–10. doi: 10.1038/jid.2008.453.

[130] Sanders CJG, Van Weelden H, Kazzaz G a a, et al. (2003) Photosensitivity in patients with lupus erythematosus: a clinical and photobiological study of 100 patients using a prolonged phototest protocol. *The British journal of dermatology* 149:131–7.

[131] Sutej PG, Gear AJ, Morrison RCA, et al. (1989) Photosensitivity and Anti-Ro (SS-A) Antibodies in Black Patients with Systemic Lupus Erythematosus (SLE). *Rheumatology* 28:321–324.

[132] Nardi N, Brito-Zerón P, Ramos-Casals M, et al. (2006) Circulating auto-antibodies against nuclear and non-nuclear antigens in primary Sjögren's syndrome: prevalence and clinical significance in 335 patients. *Clinical rheumatology* 25:341–346. doi: 10.1007/s10067-005-0059-3

[133] Doria A (1996) Photosensitivity in systemic lupus erythematosus: laboratory testing of ARA/ACR definition. *Lupus* 5:263–268.

[134] Paz ML, González Maglio DH, Pino M, et al. (2011) Anti-ribonucleoproteins autoantibodies in patients with systemic autoimmune diseases. Relation with cutaneous photosensitivity. *Clinical rheumatology* 30:209–16. doi: 10.1007/s10067-010-1481-8

[135] Paz ML, Cela EM, Ferrari A, et al. (2014) Two Approaches to the Study of a Controversial Relationship: Cutaneous Photosensitivity and Anti-Ro/SS-A

Autoantibodies. *Open Journal of Rheumatology and Autoimmune Diseases* 4:1-12. doi:10.4236/ojra.2014.41001.

[136] Webb JH, Blom AM, Dahlbäck B (2003) The binding of protein S and the protein S – C4BP complex to neutrophils is apoptosis dependent. *Blood Coagulation and Fibrinolysis* 14:355–359. doi: 10.1097/01.mbc.0000061318.06975.a4.

[137] Blaschke F, Bruemmer D, Yin F, et al. (2004) C-reactive protein induces apoptosis in human coronary vascular smooth muscle cells. *Circulation* 110:579–87. doi: 10.1161/01.CIR.0000136999.77584.A2.

[138] Espinosa A, Dardalhon V, Brauner S, et al. (2009) Loss of the lupus autoantigen Ro52/Trim21 induces tissue inflammation and systemic autoimmunity by disregulating the IL-23-Th17 pathway. *The Journal of experimental medicine* 206:1661–71. doi: 10.1084/jem.20090585.

[139] Kuhn a, Ruland V, Bonsmann G (2010) Photosensitivity, phototesting, and photoprotection in cutaneous lupus erythematosus. *Lupus* 19:1036–46. doi: 10.1177/0961203310370344.

[140] 139. Stafford R, Farrar MD, Kift R, et al. (2010) The impact of photosensitivity disorders on aspects of lifestyle. *The British journal of dermatology* 163:817–22. doi: 10.1111/j.1365-2133.2010.09905.x.

In: UV Radiation: Properties, Effects, and Applications
Editor: James A. Radosevich
ISBN: 978-1-63321-090-5

Chapter 8

Mechanisms of Nitric Oxide Mediated UV Resistance in Solid Tumors

***Madeeha Aqil*[1] *and James A. Radosevich*[1,2*]**
[1]Oral Medicine and Diagnostic Sciences, College of Dentistry,
University of Illinois at Chicago, Chicago, IL, US
[2]Jesse Brown VAMC, Chicago, IL, US

Abstract

Ultraviolet radiation is known to induce DNA damage via the production of thymine-thymine dimers. This insult to DNA integrity leads to the development of tumors primarily due to mutations in oncogenes. One of the natural agents playing a dual role in tumor biology is nitric oxide. It is a diffusible, ubiquitous secondary messenger, produced biologically by nitric oxide synthase (NOS) family of enzymes. Over-expression of NOS is responsible for maintaining tumor vasculature and facilitating neoplastic transformation. In the appropriate microenvironment and at higher concentration for extended duration, NO exposure initiates inflammation leading to mutations and clinical presentation of cancers.

In previous clinical studies, we and others have reported that patients presenting with or converting to, tumors with high levels of nitric oxide, tended to do more poorly than those which had low NO expressing tumors. We also reported for the first time that human head and neck cancer cells over-express nitric oxide and can produce high levels of nitric oxide *in vivo*. Nitrogen oxides have been shown to interact with DNA leading to carcinogenic adducts. Thus to better understand the role nitric oxide plays in tumor progression and to simulate the conditions observed in clinic, we produced cell lines that grow in comparatively high nitric oxide environments. This was done by gradually exposing parent cells to increasing NO levels, resulting in a new set of 'HNO' cell lines. Interestingly, the HNO cell lines show much more aggressive growth behavior and express characteristics present in tumor stem cells. They show better UV resistance compared to parent cell lines. Further, the cellular protective mechanisms and antiapoptic

* Corresponding author: James A. Radosevich, Ph.D. Oral Medicine and Diagnostic Sciences, College of Dentistry, University of Illinois at Chicago, 801 S. Paulina St., Chicago, IL, 60612, USA; Phone: (312) 996-9538; Fax: (312) 996-9592; E-mail address: jrados@uic.edu.

pathways are induced in HNO cells along with deregulation of cell cycle. In this review, we will discuss various mechanisms that lead to development of UV resistance in high NO cells.

Keywords: High Nitric Oxide, solid tumors, Upper aerodigestive tumor, Ultraviolet radiation resistance, GST-pi, cell cycle, apoptosis

ABBREVIATIONS

ATM	ataxia telangiectasia mutated
ATR	ataxia telangiectasia and *Rad3*-related
Cdk	cyclin dependent kinase
CHK	checkpoint kinase
GST	Glutathione-S-Transferase
HNO	High nitric oxide
HNSCC	Head and Neck squamous cell carcinoma.

8.0. INTRODUCTION

The portion of electromagnetic spectrum between x rays and visible light, i.e., between 40 and 400 nm (30–3 eV) is defined as Ultraviolet (UV) radiation. The UV spectrum is divided into Vacuum UV (40-190 nm), Far UV (190-220 nm), UVC (220-290 nm), UVB (290-320), and UVA (320-400 nm). Vacuum UV, Far UV and UVC do not reach the earth's surface and are completely absorbed in the atmosphere because of the ozone layer.

Most commonly encountered UV radiation is UV-A which leads to skin tanning. It is essential for synthesis of vitamin D; however, overexposure can lead to skin toughening, suppression of immune system, and cataract formation. The most destructive of UV radiation is UV-B since it can lead to photochemical DNA damage [1]. Although in small amounts, it is important for vitamin D synthesis but excess exposure can lead to skin cancer. Some part of UV-B radiation is absorbed by ozone layer whereas the UV-A passes through ozone unfiltered [2].

The most detrimental effect of UV radiation is initiation of oncogenic processes through direct or indirect DNA damage [3]. A critical compound determining the behavior of tumor and neighboring normal cells is Nitric Oxide (NO). NO is known to play a dual role in tumor development and progression based on the microenvironment and concentration. We developed High Nitric Oxide adapted cancer cell lines (HNO) to study the role of NO in tumors. Interestingly, we found HNO cells to show aggressive growth properties and resistance to UV induced DNA damage. In the present chapter, we discuss the cellular responses to DNA damage, the different molecular mechanisms observed in HNO cells that impart UV resistance, and hence a more aggressive growth profile.

8.1. UV Induced DNA Damage

UV radiation is one of the most detrimental exogenous agents for cellular DNA. UV-B can be directly absorbed by DNA whereas UV-A generates free radicals which indirectly affects DNA. These interactions affect the genome integrity and eventually perturb normal physiological processes [4]. Different types of DNA damage are discussed below in brief.

8.1.1. Purine and Pyrimidine Photoproducts

Absorption of UV-B radiation by DNA leads to formation of two major types of DNA lesions - cyclobutane pyrimidine dimers (CPDs) and pyrimidine 6-4 pyrimidone photoproducts (6-4PPs) [5]. The CPD is a four-member ring structure involving C5 and C6 of both neighboring bases. The 6-4PPs are formed by a non-cyclic bond between C6 (of the 5 -end) and C4 (of the 3 -end) of the involved pyrimidines [6]. It has been observed that T-T and T-C sequences are more photoreactive than C-T and C-C sequences. In single stranded DNA and at the flexible ends of poly (dA)-(dT) tracts, the CPD lesions are formed at higher quantity. However, bending of DNA towards the minor groove reduces CPDs formation [7]. Apart from these dimeric lesions, monomeric photochemical products are also known for cytosine. Recently, one group characterized 5-methylcytosine by mass spectrometry and high performance liquid chromatography [8]. Another cytosine photochemical product has been reported called "cytosine photohydrate" (6-hydroxy-5,6-dihydrocytosine) as a result of photohydration reaction. The hydration reactions also lead to formation of pyrimidine glycols [9].

Many oxidation products of purine bases such as 8-oxo-Ade, 8-oxo-7,8-dihydroguanyl (8-oxoGua), 2,6-diamino- 4-hydroxy-5-formamidoadenine (FapyAde), FapyGua, and oxazolone have been reported to form upon UV exposure of DNA. This oxidative DNA damage is caused mainly by UV induced reactive oxygen species [10, 11].

8.2. DNA Double Strand Breaks

The DNA lesions and other photoproducts lead to stalled replication forks and/or transcriptional blockage, thus forming double strand breaks (DSBs). When the DNA replication machinery encounters any lesion, the DNA polymerase is stalled at the blockage point leading to formation of a Y shaped structure. This is recognized by specific endonucleases that makes a nick in the template strand thus inducing a DSB close to the lesion [12]. In addition, blocking of topoisomerase I complexes due to DNA lesions also lead to DSB formation. Free radicals have been shown to prevent topoisomerase II mediated DNA re-ligation which can also cause DSB formation [13].

8.3. DNA Damage Repair

UV induced DNA lesions inhibit replication and transcription processes thus causing genomic instability. To counteract these effects and maintain genetic homeostasis eukaryotes have evolved detailed DNA repair mechanisms [14]. Nucleotide Excision Repair (NER), Base Excision (BER) Repair, and mismatch repair pathways mainly fix single strand breaks while the Non homologous end joining and Homologous recombination pathways repair double strand breaks [15]. These are discussed in details below.

8.3.1. Nucleotide Excision Repair

NER is a highly conserved and a critically important pathway of DNA repair in eukaryotes. This pathway is mainly involved in removing helix distorting lesions like CPDs and 6-4PPs [16]. In eukaryotic cells, it is further subdivided into two pathways: Global Genome Repair (GGR) and Transcription Coupled Repair (TCR). GGR repairs lesions anywhere in the genome whereas TCR repairs genes in the transcribed strand of active genes. The sequential events in these pathways involve damage recognition, opening of DNA double strand at damage site, double nicks on both sides of lesion, re-synthesis and finally ligation [17]. The initial damage recognition proteins are different; however, a common set of repair factors is involved in both. The most critical DNA damage recognition protein in GGR is XPC-HR23B centrin 2 complex and/or the heterodimer complex of DDB1 and DDB2 (DNA damage binding protein 1 and 2) [18]. In the TCR pathway, the ATP dependent chromatin remodeling protein Cockayne syndrome protein B (CSB) and the WD40 domain containing protein Cockayne syndrome protein A (CSA) are recruited to the site of damage. Then the RNA polymerase 2 complex is abrogated and the damage is repaired. The following steps and proteins are common to both GGR and TCR. These are the unwinding of DNA at the site of lesion by Xeroderma Pigmentosum group B and D helicases (XPB and XPD), which are subunits of the transcription factor TFIIH complex [19]. Then the repair complex is stabilised by XPA. Two structure specific endonucleases, XPG and ERCC1-XPF, then orient the incision in the damaged area eventually leading to the removal of a part of single stranded DNA, with a gap of 25-30 nucleotides. The DNA polymerase complex then fills up this gap and the DNA ligase reseals the strand [20].

8.3.2. Base Excision Repair (BER)

This pathway primarily removes non helix distorting, small single base oxidative DNA lesions which are generally induced by UV-A. Highly reactive oxygen radicals cause these oxidative lesions. The most common types of base lesions repaired by BER include oxidized bases (e.g. 8 oxoguanine), deaminated bases (hypoxanthine and xanthine) and alkylated bases (e.g. 7 methyl guanine) [21]. The first step involves removal of the damaged base by DNA glycosylase, thus creating apurinic/apyrimidinic (AP) site. This AP site is then removed by AP endonuclease 1 (APE1) along with phosphodiesterase that creates nicks in the DNA strand 5' or 3' to the AP site [22]. The resulting single strand break is then synthesized wherein either a single base is added (short patch BER) or 2-10 nucleotides are synthesized (long patch BER) [23]. DNA polymerase β catalyzes strand synthesis in short patch BER and DNA ligase1 along with its cofactor XRCC1 reseals the strand. The long patch BER is carried out by actions of FEN1 endonuclease and DNA polymerases pol δ and pol ε along with proliferating cell nuclear antigen (PCNA) [24].

8.3.3. Homologous Recombination

The homologous recombination pathway mainly repairs DNA double strand breaks. It occurs during S and G2 phases of the cell cycle after the DNA replication has occurred and before the cell enters mitosis, because sister chromatids are easily available at these stages. They act as templates for accurate synthesis of damaged DNA [25]. A trimeric complex consisting of Mre11, Rad50 and Nbs1 proteins, also known as MRN complex, initiate the repair process by binding to DSBs. MRN complex recruits helicase and nuclease complexes including the Exo1 and Dna2 nuclease that carry out DNA resection generating 3' overhangs. These overhangs are bound by RPA protein that have high affinity for ssDNA. Then the RAD51 protein along with other mediators binds to RPA coated ssDNA, forming a nucleoprotien filament which then searches for homologous sequence to the 3' overhang [26]. Generally, this homologous sequence is provided by the sister chromatid and a strand invasion process occurs. After strand invasion, the DNA polymerase extends the end of the invading strand, synthesizing new DNA that anneals to the other complementary strand followed by ligation [27].

8.3.4. Non Homologous end Joining (NHEJ)

The NHEJ pathway is initiated by binding of Ku heterodimer complex to the DSB breaks. This complex consists of Ku70 and Ku80 proteins, together forming a basket shaped structure that binds to the DNA end and then slides down to allow binding of more Ku molecules. Ku heterodimer forms a complex with the DNA-dependent protein kinase catalytic subunit (DNA-PKcs) to form active DNA-dependent protein kinase, DNA-PK [28]. It is believed that Ku acts as a docking site for other NHEJ proteins. Once bound, the end processing starts which involves removal of damaged DNA by Artemis nuclease, among others, and re-synthesis by DNA polymerases (mainly the X family DNA polymerases). The final step of ligation is performed by DNA ligase IV complex that also contains XRCC4 and XLF as cofactors [29].

8.4. Cellular Responses to DNA Damage

The assault on genomic integrity by diverse genotoxic agents including ultraviolet radiation leads to activation of a global signaling network termed DNA Damage response (DDR) [30]. The DDR signaling cascade involves complex and tightly regulated molecular events that eventually culminate in 1) cell cycle arrest to allow time for DNA damage repair or, 2) apoptosis, if the damage is excessive and lethal or, 3) tumorigenesis if not properly repaired and bypassed [31]. The cell cycle can be subdivided into four distinct phases: the mitotic division phase or M phase, the first gap phase or G1 phase, the DNA replication phase or S phase, the second gap phase or G2 phase. The first and foremost event in DDR is activation of cell cycle checkpoint proteins [32]. Discovered in 1980s, these surveillance mechanisms can delay the cell cycle in presence of DNA damage. Mammalian cells have 3 main checkpoints: G1/S, intra-S, and G2/M. These are mainly controlled by Cyclins and Cyclin-dependent kinases (Cdks), which sequentially act in G_1 to initiate the S phase and in G_2 to initiate mitosis [33]. Figure 1 shows cell cycle regulation in different phases.

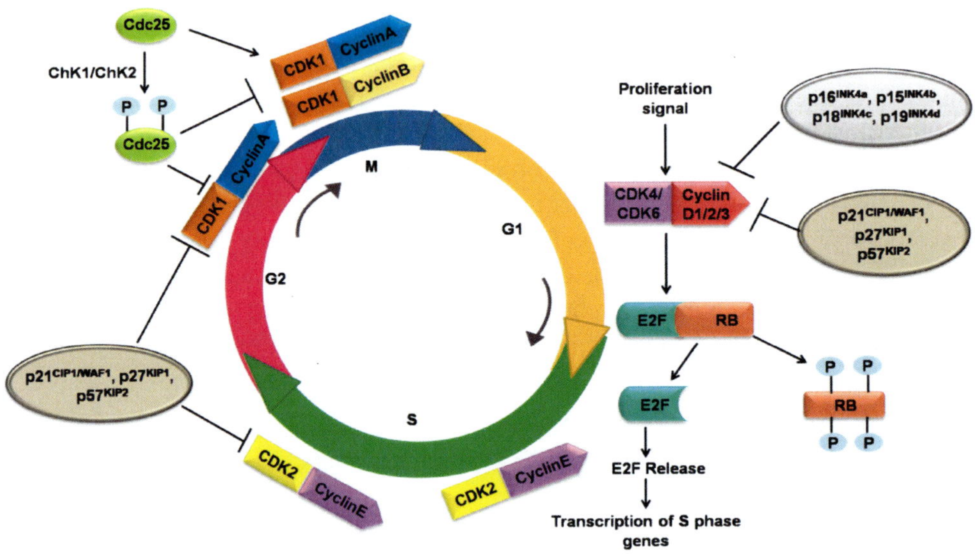

Figure 1. Passage of cells through different phases of cell cycle is regulated by specific cyclin/ck complexes as shown above. The negative regulators of INK4 and Cip/Kip family inhibit the cyclin/cdk complexes.

Cdks are Ser/Thr kinases that form heterodimer complex with Cyclins. The Cyclin-Cdk complex is generally present in inactive state due to inhibitory phosphorylation by Wee1 and Myt1. The expression of cyclins is tightly regulated during different phases of the cell cycle, thus regulating Cdk activity in a timely manner. Specific Cdks form complex with specific cyclins to allow progression of cell cycle [34].

Regulation of checkpoint signaling proceeds through a network of proteins which can be broadly divided into four groups: damage sensors, signal mediators, signal transducers and effectors [31]. Figure 2 shows main proteins involved in cellular DNA Damage Response (DDR). The main sensor protein is MRN complex, which binds to DSBs during all stages of cell cycle. In addition to MRN complex, the Ku70/80 heterodimer also recognizes DSBs. The RPA protein complex recognizes the ssDNA damage. These sensor complexes then recruit the transducer kinases, namely the ATM (ataxia telangiectasia mutated), ATR (ATM and RAD3 related) kinase and DNA-PKcs. They belong to the phhosphoinositide-3 kinase related protein kinase (PIKK) group and play a central role in coordinating DDR. The ATM is recruited by MRN complex, whereas ATR and DNA-PKcs are recruited by RPA and Ku80 respectively. With the help of other signal mediators such as TopBP1(topoisomerse binding protein 1) and Claspin1 for ATR, MDC1(mediator of DNA damage checkpoint), 53BP1 (p53 binding protein1) and BRCA1 for ATM, the transducer kinases activate the effector kinases Chk1 and Chk2, which then spread the damage signal [35].

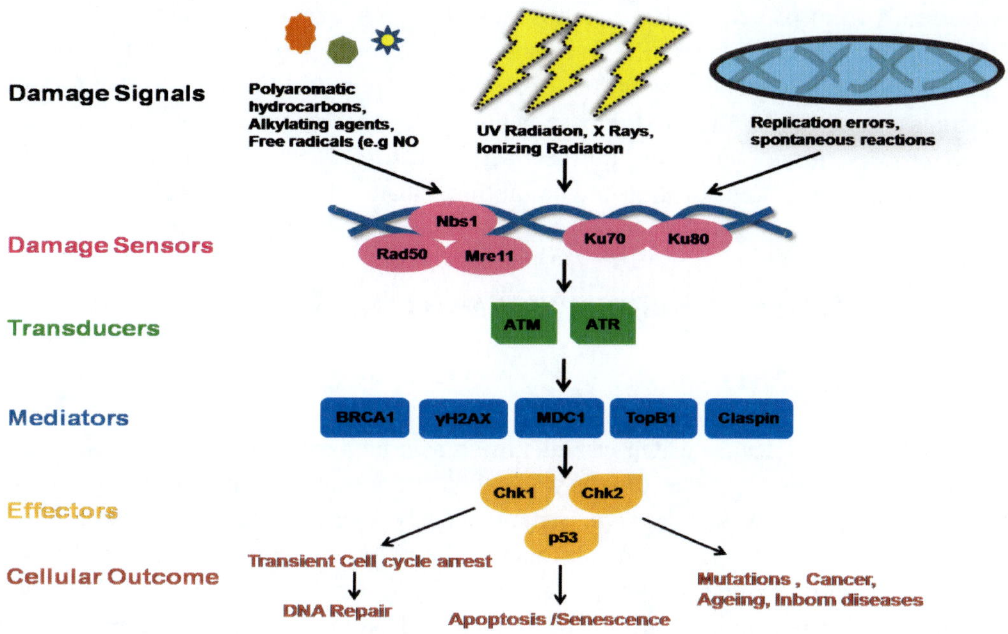

Figure 2. Schematic of cellular DNA damage Response showing key proteins at different steps and the possible cellular fate.

The Chk2 activation via ATM is mainly restricted to DSBs (e.g. by inonizing radiation) whereas Chk1 is activated in response to ssDNA breaks. The resection at DSB sites results in formation of ssDNA intermediate structure which promotes slower activation of Chk1 via RPA-ATR/ATRIP pathway [36]. ATM predominantly phosphorylates Chk2 at Thr68. This activates intramolecular trans-autophosphorylation at Thr 383/387, thus leading to homodimerization and activation of Chk2. However, Chk1 activation does not require dimerization or trans-autophosphorylation. Predominantly, ATR and to a lesser extent ATM, phosphorylates Chk1 at Ser317/345, directly leading to its activation. The activated signal mediators remain at the site of damage while Chk1/Chk2 are released and further activate the effector proteins. These include the cell cycle regulators, transcription factors, the DNA repair factors, and the apoptotic machinery [37]. The two main effector targets that control cell cycle and apoptosis are the Cdc25 phosphatases and the transcription factor p53. The Cdc25 phosphatases regulate activation of cyclin dependent kinases (Cdks) through inhibitory site dephosphorylation [38]. In humans, there are three isoforms of Cdc25 phosphatases: Cdc25A, Cdc25B, and Cdc25C. Cdc25A activates both Cdk2–cyclin E (an S-phase kinase) and Cdk1–cyclin B, a key regulator of G2-M transition, thus functioning throughout the cell cycle [39]. In response to DNA damage, Chk1 and Chk2 phosphorylate Cdc25A, which is rapidly degraded through ubiquitin pathway. This prevents activation of S phase and M phase cyclin-Cdk complexes, thus leading to cell cycle arrest. Cdc25B mainly functions at the G2/M transition of the cell cycle and Cdc25C acts during M phase to sustain the activity of Cdk1–Cyclin B [38]. The cell cycle is negatively regulated by Cdk inhibitors that include: 1) the

INK4 family including p16INK4A, p15INK4B, p18INK4C and, 2) the Cip and Kip (kinase inhibitor protein) family of inhibitors including p21CIP1, p27KIP1 and P57KIP2. Whereas the INK4 family, specifically inhibit CDK4 and CDK6, the Cip/Kip family inhibit all cyclins D, E, and A [40]. The transcription of p21CIP1 is regulated by p53. High levels of p21CIP1 can protect cells from p53 dependent apoptosis during DNA damage and induce cell cycle arrest [41]. High levels of p21CIP1 can lead to both pro-apoptotic and anti-apoptotic signaling depending upon cellular context. It can bind to apoptotic signaling kinase (ASK1) in the cytoplasm and inhibit the MAP kinase pathway. It can also bind to pro-caspase 3 and prevent its processing and activation. Further, activated ERK2 can phosphorylate p21CIP1 leading to its cytoplasmic translocation and ubiquitin mediated degradation [42].

8.5. Nitric Oxide (NO) and DNA Damage

8.5.1. Biosynthesis of NO

NO is a diatomic molecule which was identified as a pleiotropic signaling messenger in mammalian cells in 1990s. At that time it was called "endothlium derived relaxing factor," which Ignaro et al and Moncada *et.al.* simultaneously identified as Nitric oxide. It is lipophillic and easily diffuses across the plasma membrane [43]. The term "nitric oxide" may refer to any of the nitric oxide reactive species (NO●, NO-, and NO+) however, in biological systems, nitric oxide generally refers to the free radical NO●. The Nitric Oxide Synthase (NOS) family of enzymes catalyze the biosynthesis of NO. Three isoforms of NOS exists namely: 1) NOS1/neuronal NOS (nNOS) 2) NOS2/inducible NOS (iNOS); and 3) NOS3/endothelial NOS (eNOS) (44). While neuronal and endothelial NOS are calcium dependent and constitutively expressed, iNOS is a calcium independent enzyme and is induced by inflammatory stimuli. Almost every cell throughout the body can express iNOS [45]. The full active form of NOS requires two NOS monomers associated with two molecules of the Calcium binding protein- Calmodulin. The NOS2 is maximally activated even at a basal calcium concentration; however the NOS1 and NOS3 are highly dependent on intracellular calcium concentrations [45]. The expression of NOS isoforms is differentially regulated at transcriptional, translational and post-translational levels. The biosynthetic reaction catalyzed by NOSs involves L-argininie, NADPH and oxygen as reactants, producing L-citrulline, NADP and NO. Thus, classically NO is produced from L-argnine in the presence of oxygen. However, a parallel pathway for NO generation also exists in mammals. This pathway involves reduction of nitrate and nitrite to form bioactive NO in blood and tissues during physiological hypoxic conditions. To a certain extent, these pathways work in parallel, but when oxygen availability is reduced and NOS activity is decreased, nitrite reduction to NO becomes more prominent [46].

8.5.2. Role of NO in Normal Physiology

NO is a critical signaling molecule in several physiological processes including smooth muscle relaxation, inhibition of platelet aggregation and adhesion, neurotransmission, cellular proliferation, migration, anti-tumor and anti-microbial immune responses [43]. There are three main biomolecular reactions that NO can undergo to mediate these numerous biological effects. These are: 1) redox interaction with thiols, 2) coordination reaction with metal functional centers, and 3) through protein nitrosylation [45]. Generally, the NO signaling can be classified into cGMP dependent and cGMP independent pathways [47]. The classical NO signaling is through the cGMP dependent pathway which is triggered by activation of soluble guanylyl cyclase enzyme [48]. This occurs through binding of NO to the sixth coordinating position of iron in haem moiety of sGC resulting in a conformational change and hence almost 200 fold activation. The high levels of cGMP thus produced activate Ras and cGMP dependent protein kinase-PKG. PKG can interact with Raf which can further activate MAPK or ERK. The Ras-Raf-MEK-ERK pathway then enhances DNA binding of AP1 (activator protein1), eventually resulting in increased cell proliferation and migration [49]. The cGMP independent activity of NO involves covalent modifications of proteins through S-nitrosylation and other redox modifications that do not involve NO addition. S-Nitrosylation is the most important reaction of NO with the thiol group of cysteine residues leading to formation of more stable nitrosothiols. Apart from nitrosothiol, other modifications such as disulfide formation with reduced glutathione (GSNO) or oxidation towards sulfenic acid are also important since they are reversible. Different proteins such as ryanodine receptors, ras, glyceraldehyede-3-phosphate dehydrogenases, caspases, and DNA repair proteins are usually post-transationally modified by nitrosylation [50]. Through these mechanisms, NO regulates a number of physiological processes including vasodilation, smooth muscle relaxation, wound healing, angiogenesis, cell proliferation, inflammation and cytotoxic effects by innate immune cells. NO and NO derived reactive nitrogen species are also involved in many redox reactions which may induce oxidative and nitrosative stress potentially leading to DNA damage [51]. NO can induce mutations via several mechanisms including DNA oxidation and deamination leading to DNA strand breaks and possible oncogenic transformation [52].

8.5.3. NO in Tumor Biology

NO plays a dual role in tumor biology wherein it can both promote and inhibit tumor development and metastasis [53]. The dichotomous effect of NO is dependent on its concentration, interactions with other chemical species, the NO producer cell type, and the target cell type [54-56]. Persistent inflammation and continuous exposure to moderate to high concentrations of NO produced by iNOS are thought to initiate oncogenic transformation [55, 57]. NO and NO derived reactive nitrogen species lead to nitrosative and oxidative stress resulting in DNA damage and also inhibition of DNA repair enzymes through direct or indirect mechanisms [58, 59]. Studies show that iNOS knockout mice show reduced incidence of gastric carcinomas and significant reduction in urethane induced lung carcinomas [60, 61]. Elevated NOS activity and NO levels have been found in lung cancer patients [62, 63], and malignant neoplasms of the central nervous system [56, 64].

Sustained exposure to higher NO concentration in appropriate microenvironment initiates inflammation and can facilitate tumor growth and/or metastasis [65]. Several studies show positive correlation between iNOS expression and poor clinical outcome of patients. Histological investigations show a positive correlation between high VEGF expression or angiogenic activity and iNOS expression in human brain, lung, head and neck, breast, stomach and colon tumors [65-69]. In malignant and benign lesions of the breast, both eNOS and iNOS are expressed [70]. Further, it has been shown that iNOS activity is higher in less-differentiated breast carcinomas and is mainly present in tumor infiltrating macrophages [71]. iNOS is also expressed in human ovarian cancer cells [72, 73] and 25% of colon carcinomas [74, 75]. We have shown increased expression of endothelial constitutive NOS (NOS3) in human head neck squamous cell carcinoma (HNSCCa). We immunostained eleven paraffin-embedded samples of normal oral mucosa, 3 reactive oral lesions, 13 samples of squamous dysplasia, and 120 specimens of HNSCCa with an anti-NOS3 monoclonal antibody and graded the intensity pattern on a scale of 0 to +4. On comparison to normal squamous mucosa, we found significantly higher expression of NOS3 ($p<0.001$) in squamous hyperplasia, dysplasia and HNSCCa [76]. Furthermore variation in NOS3 expression in different regions of mucosa is positively associated with inflammation [76]. In accordance with these observations, we have also found higher protein nitrosylation through increased nitrotyrosine staining levels in HNSCCa and other reactive lesions. Interestingly the neighboring inflammatory cells also showed intense staining for nitrotyrosine indicating role of NO in immune suppression [77, 78]. This is shown in Figure 3 below.

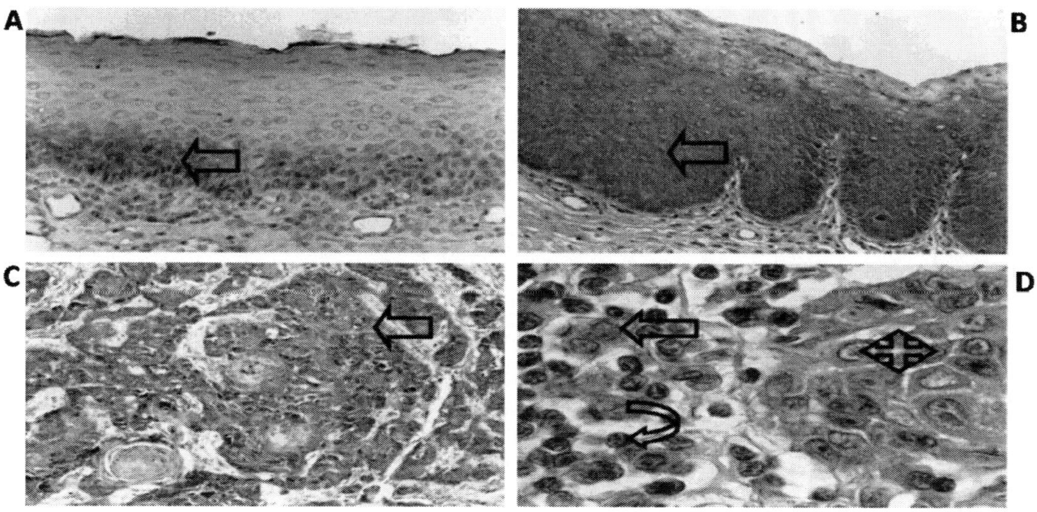

Figure 3. (A) Normal oral mucosa (original magnification, 400×) demonstrating localization of nitrotyrosine to the basal and parabasal levels (⇦). (B) Squamous dysplasia (original magnification, 400×) demonstrating intense global staining for nitrotyrosine (⇦). (C) Squamous cell carcinoma (original magnification, 400×) demonstrating intense staining for nitrotyrosine (⇦). (D) Perilesional inflammatory infiltrate (original magnification, 1000×) demonstrating staining intense staining of lymphocytes (⤺) and macrophages (⇦), and tumor cells (✥) with nitrotyrosine.

We have also found limited expression of eNOS in salivary gland tumors whereas in esophageal lesions NOS3 expression was significantly higher in squamous mucosa and adenocarcinomas, compared to glandular metaplasia [79, 80]. In cases of cervical squamous cell carcinomas, we have demonstrated increased expression of iNOS that correlated with advanced tumor stage [45]. This shows a common functional characteristic of NO in promoting tumor growth in squamous mucosa at different sites in the body. We have also reported cytoplasmic and nuclear expression of eNOS in endometrial carcinomas, where increased staining frequency correlated with increased myometrial invasion and decreased disease free survival [81]. In cases of lung cancer, we have demonstrated increased nitrotyrosine expression and enhanced DNA strand breaks with increasing concentrations of nitric oxide donors in both the lung adenocarcinoma cell line, A549 and the control lung fibroblast, WI38 [82]. We found that genotoxic effect is highly dependent on the concentration and kinetics of delivery of NO. The nitrosative stress did not cause increased apoptosis in the tumor cell line compared to the non-tumor immortalized cell. This is partly explained by nitrosylation and hence inhibition of caspase 9 and 3 by thiol based, but not NONOate based NO donors [58, 82]. Thus, it is believed that the expression level, kinetics of NO delivery, the cellular microenvironment, the cell type, and the genetic background are critical factors in determining NO sensitivity and overall effect of NO [83]. Furthermore, continuous exposure to low levels of NO can produce NO resistance, which might allow tumor progression even at cytotoxic concentrations, thus leading to NO adaptation.

Apart from the findings discussed above, several conflicting observations have been reported in many studies regarding the critical role of NO in tumor development [43, 64, 84]. Transfection with iNOS has been shown to inhibit tumor growth and invasion in mouse tumor models (85) as well as human renal and colon carcinoma cells [86-88]. NO derived from iNOS has also been shown to suppress lymphoma formation by promoting apoptosis and inhibiting tumor proliferation in p53 knockout mice [89, 90]. Thus, it is believed that NO has context and concentration dependent effect for tumor formation. The definition of high and low can vary depending upon the particular system under study and how NO can modulate the cellular mechanisms under that particular biochemical milieu [91]. For example, sustained exposure to 10-30 nM NO results in ERK (Extracellular signal Regulated Kinase) activation in a cGMP dependent manner [92]. As the concentration increases to 60 nM, Akt gets phosphorylated. With further increase to 100nM, Hif-1α is stabilized, whereas at 400nM p53 gets phosphorylated and acetylated [93, 94]. Nitrosation of important proteins like PARP and caspases occurs at further high concentrations of up to 1μM [95, 96]. However, the effect can be toxic or protective depending upon the unique cellular microenvironment. In general, at low concentrations NO is believed to regulate several physiological processes, however at higher concentrations it can be damaging with pathological consequences [97, 98].

8.6. *In Vitro* Model System to Study Role of NO in Tumor Biology

As stated above, we and others have reported a spectrum of NOS expression in patient populations of different types of tumors. In clinical studies we and others have shown that patients presenting with (or converting to), tumors with high levels of NO tended to do more poorly than those patients which had low expressing tumors [45, 99, 100]. Studies show that high free radical NO• encountered in the tumor microenvironment can also kill immune cells [57]. Since it is practically impossible to study the effects of high NO in human patients, we sought to produce a unique, *in vitro* tissue culture model system of NO stressed tumor cells to determine if they could adapt to increasing levels of NO•. The resulting cell culture system would mimic the spectrum of NO• expression found clinically. The model system was developed by "adapting" low NO• expressing cell lines to increasing levels of NO• donor

[101-103]. The adaptation process began by first exposing the cells to fresh media containing media supplemented with 50 μM DETA-NONOate. Cells were then cultured until they reached ~90% confluency. At this time, the cells were again passaged, and subsequently grown in media treated with 75 μM DETA-NONOate. In this manner, the NO donor concentration was increased at 25 μM increments with each passage and the cell lines successfully withstood this treatment approach [101, 103]. Figure 4 shows the adaptation curve of A549 cells to NO.

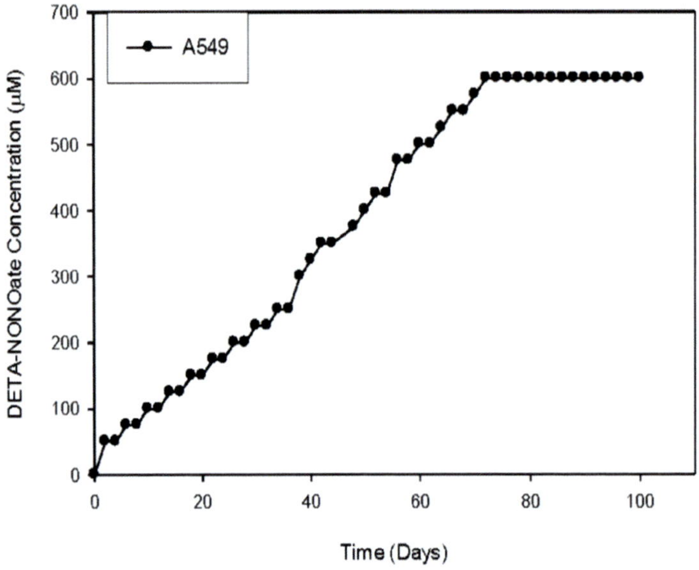

Figure 4.

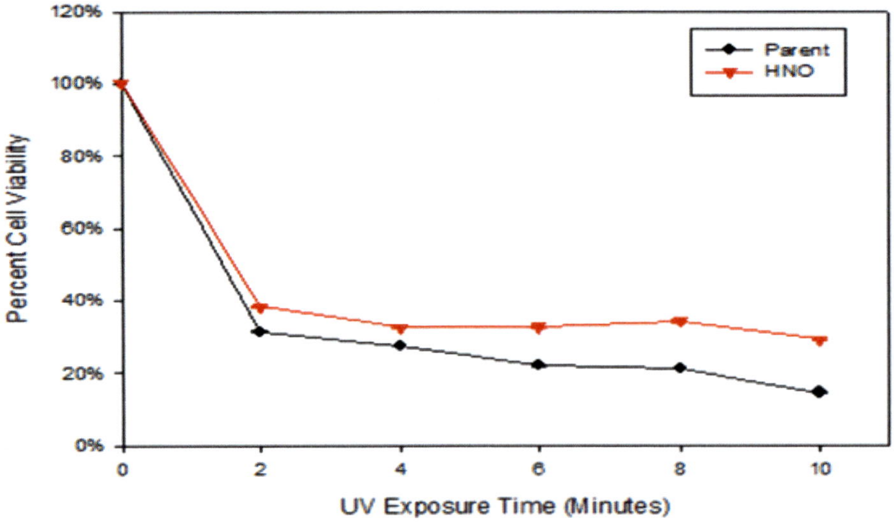

Figure 5. Comparison of growth profiles of parent and HNO cells after UV treatment for A549.

For each cell line, the adaptation endpoint was selected as the concentration in which the exogenous NO● introduced to the cells was lethal to the parent cell lines; this was approximately 600μM. At this endpoint concentration, the HNO cells grow fine and are not morphologically different from the original (untreated) parent cells. Thus, we developed a new set of HNO cell lines. We have successfully adapted 10 human cancer cell lines. These are four breast adenocarcinomas (T-47D, Hs578t, BT-20, and MCF-7), one lung adenocarcinoma cell line (A549), and five head and neck squamous cell carcinoma (HNSCC) cell lines (three originating from the tongue:SCC016, SCC040, and SCC056; one from the floor of mouth: SCC114; and one from the alveolar ridge: SCC116) [101, 103, 104]. These cell line pairs provide a useful model system to study role of NO in modulating tumor cell signaling. Though morphologically the HNO cells are similar to the parent cells, nevertheless they show aggressive growth behavior and are resistant to X-ray and UV radiation [105-107].

Below we discuss the unique characteristics of the HNO cell lines that impart them aggressive growth behavior and radio resistance.

8.6.1. Mechanism of UV Radiation Resistance in HNO Cells

8.6.1.1. Enhanced Expression of Free Radical Scavenging Enzymes

It is known that exposure to UV-A radiation leads to generation of free radicals and reactive oxygen species that indirectly leads to DNA damage, and may result in skin cancer [108]. One of the universal free radical scavenging system is the Glutathione-S-Transferases (GSTs) group of enzymes. GSTs catalyze thioether bond formation between electrophilic chemical moieties and glutathione (GSH). Thus they are critical for cellular detoxification of superoxide radicals, endogenous toxic metabolites, and exogenous toxic chemicals [109]. GSTs are divided into soluble cytosolic {cGST alpha (A, α), mu (M, μ), Pi (pi, π), omega (ω), delta (δ), sigma (σ), theta (T, θ), zeta (ζ), and mitochondrial (mGST α, π, μ and kappa (K)} and structurally distinct membrane bound form (MGST) [110]. GSTs also play an important role in the activation of mitogen activated protein (MAP) kinases and various transcription factors that regulate apoptosis and cell survival. The cytosolic GST-pi is expressed in normal epithelial tissues of the respiratory, digestive, and urinary tract. It is also the most predominant form expressed in many tumors and drug resistant cell lines [111]. In fact, GST-pi has been hypothesized to be a marker of increased exposure to carcinogens [112]. We have shown increased expression of GST-pi in clinical samples from head and neck cancer patients which correlated with higher expression of eNOS and nitrotyrosine [113]. Further, we observed high levels of GST-pi expression in both parent and HNO cells for the three head and neck cell line pairs, SCC016, SCC040, SCC056. While the SCC016 cell line pair showed a significant difference in GST-pi expression between parent and HNO cells, the SCC040 and SCC056 showed comparable expression between parent and HNO [105]. It has been reported that decreases in GST-pi expression is associated with poor prognosis in squamous esophageal carcinoma [114, 115]. In the case of gliomas, high GST-pi expression in tumor cell nuclei is associated with clinically more aggressive tumors and is a strong predictor of poor patient survival [116]. We also observed higher expression of GST-pi in A549-HNO cell line compared to A549 parent cells [45]. In the breast cancer cell line pairs, we have shown elevated expression in Hs578T-HNO and MCF-7-HNO, relative to their parent cell lines. T-47D had little expression in both the parent and HNO cell lines while BT-20 had significant expression in both parent and the HNO cell lines [117]. Importantly for all these cell line pairs, the HNO cells show more resistance to UV exposure when compared to the parent cell lines. High levels of GST-pi in these cells suggest that it might facilitate in better tolerance to free radical stress generated by UV.

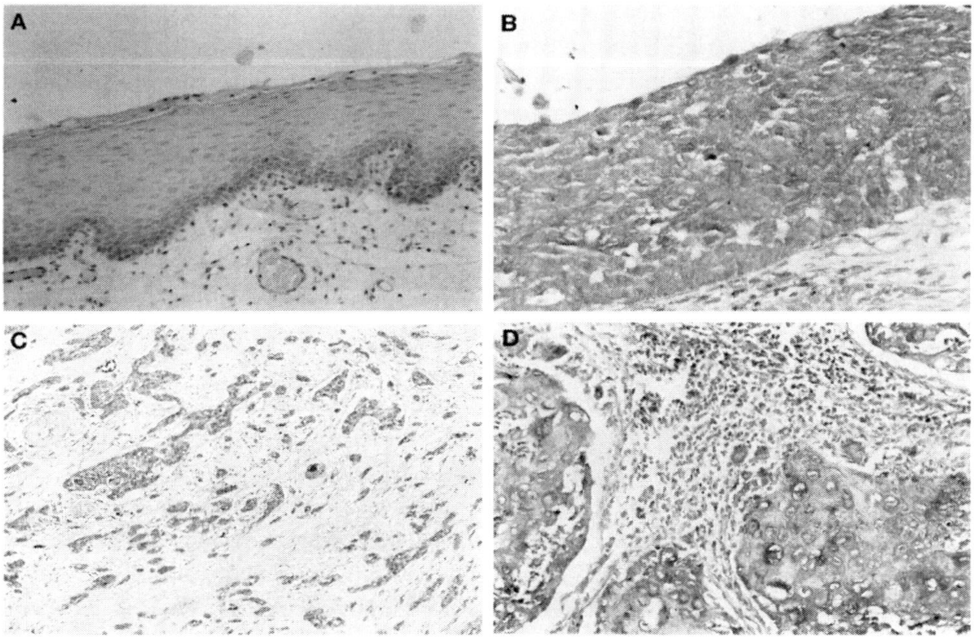

Figure 6. A. Normal squamous mucosa staining very little with diaminobenzidine against glutathione *S*-transferase p (GST-p). B. Staining of squamous dysplasia demonstrating an increased staining intensity against GST-p. C. Squamous cell carcinoma demonstrating global, intense staining for GST-p. D. Perilesional inflammatory cells staining very little for GST-p when compared with adjacent squamous cell carcinoma. Original magnification x200.

8.6.2. High Expression of DDR Pathway Protein

An assault on genomic integrity by UV radiation activates the cellular DNA Damage Response (DDR). The DDR is complex network of signaling cascades including DNA repair, cell cycle checkpoints and damage tolerance [35, 118]. The DDR signaling pathways consist of a) sensor proteins that recognize damaged DNA sites and activate the signaling cascade by interacting with b) transducer proteins that pass on the signal to other downstream proteins and finally to the c) effector proteins that interact with other proteins resulting in a biological response [119]. The key protein complexes involved in sensing the DNA damage are Ku80/Ku70 heterodimer in the NHEJ pathway and the MRN comlpex in the homologous repair pathway. The MRN complex has MRe11, Rad50 and Nbs-1 proteins. Ku80/Ku70 recruits and activates DNA-PKcs (DNA dependent protein kinase, catalytic subunit) which then recruits the XLF-XRCC4 ligase complex, among other cofactors [120]. We assessed the expression levels of these proteins in A549 parent vs A549 HNO cells (Figure 7). Up to 8 h after exposure, we found constitutive expression of Mre11, Rad52 and Rad50 in both parent and HNO cells, however at 16 hr only HNO cells showed expression of these MRN complex proteins. The expression of Ku80 was similar in both the cell lines however, it decreased in parent cells at 16hr. For DNA-PK the expression was similar initially and then increased until 8 hr in both cell lines, however with prolonged NO exposure it was sustained only in HNO cells [121].

A sudden burst of increase in XLF levels in A549 parent cells was observed with NO treatment but it was diminished at 16 hr. Only HNO cells maintained XLF expression. Thus,, we found that the proteins of Non-Homologous End Joining (NHEJ) and Homologous Recombination (HR) were initially expressed in both cell types however at 16 hr only HNO cells showed sustained expression of these proteins. The parent cells showed low expression of only NHEJ pathway proteins. The HR pathway mediates error free DNA repair while the NHEJ pathway is error prone. Thus, it is highly likely that A549 parent cells with only the NHEJ repair pathway accumulate enough mutations that become lethal for these cells. On the other hand, the A549-HNO cells, with both repair pathways, eventually survive under harsh conditions [106, 121]. They would also be primed to better survive UV radiation as both the repair systems are active.

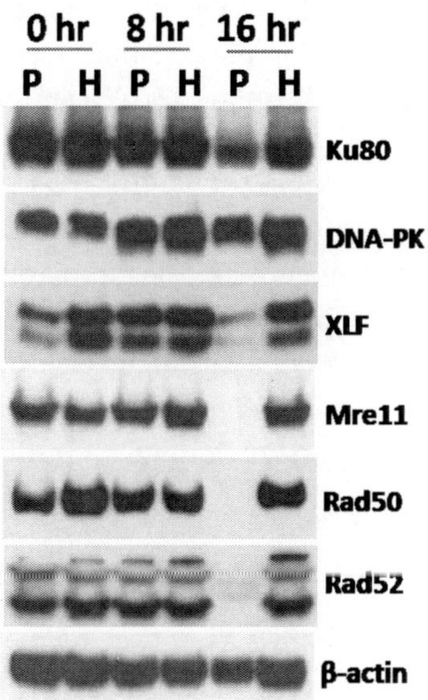

Figure 7. Expression levels of DNA damage sensor proteins at different time points post NO exposure in parent and HNO cell lines.

The main transducers of DNA damage signaling are ATM and ATR kinases belonging to phosphoinositide-3 kinase related protein kinase family (122). They can directly bind to DNA or to proteins in the damage sensor complex through conserved motifs and thus get activated by phosphorylation [36]. ATM is recruited to the DNA damage site through interaction with MRN complex. Upon recruitment to the DSB site, ATM gets activated through auto-phosphorylation at Ser1981and further leads to phosphorylation and activation of ATR [15]. The ATM and ATR kinases transmit the DNA damage signal to checkpoint kinases-Chk1 and Chk2 [118, 119]. These checkpoint kinases regulate the surveillance signaling mechanisms and delay the cell cycle in the presence of damaged DNA. Activation of all these proteins leads to downstream signaling cascades eventually modulating the cellular processes including the cell cycle, senescence, and apoptosis. On comparing expression of these molecules in A549-HNO cells with the parent cells, we observed an initial increase in phospho ATM (Ser1981), which stabilized with prolonged NO exposure. The phospho ATR (Ser428) level remained similar with prolonged exposure in both parent and HNO cells. Thus, we concluded that ATM and ATR are constitutively activated in both parental and A549-HNO cells, however, enhanced and sustained phosphorylation is only found in HNO cells. Usually, Chk1 is activated by ATR while ATM activates Chk2, although there is considerable cross talk between ATM and ATR. In HNO adapted A549 cells, Chk1 levels were similar to parent cells initially, however, after 8 hr the Chk1 expression increased in an NO dependent manner and was sustained with prolonged NO exposure at 16 hr and 72 hr. The Chk2 expression in A549-HNO cells is high compared to parent cells, but with prolonged NO donor treatment the levels remained similar to parent cells. However, expression of Ser19 phosphorylated Chk2 is higher in HNO cells at all time points [121]. The two important targets of Chk1 and Chk2 are BRCA1 and H2AX. BRCA-1 is found in macro-complexes and is recruited to DNA damage sites through interaction with Ku80 sensor protein [123]. Phosphorylation of BRCA1 at Serine 1524 by ATM is important for error-free end-joining process. Phosphorylated γH2AX (Ser 139) directs recruitment of DNA damage response components and chromatin modifying complexes, thus facilitating the DNA repair process [124]. The expression of γH2AX and phospho H2AX (Ser 139) is high in A549-HNO cells and is sustained with extended NO donor treatment. It should be noted that H2AX dephosphorylation attenuates the checkpoint signaling and allows the cell cycle to resume. The phospho BRCA1 levels (Ser 1524) increased at 16 hr in A549 HNO cells. With extended NO treatment it was enhanced in HNO cells but not in parent cells. These mediator proteins remain at the site of damage, while Chk1 and Chk2 are released to activate soluble targets. Thus, we showed that the A549 HNO cells sustained ATR/Chk1 DNA damage checkpoint activity which correlates with enhanced DNA repair capacity to maintain genomic integrity of A549-HNO cells under harmful conditions including UV radiation [121].

8.6.3. Modulation of Cell Cycle Checkpoints

The DDR signaling cascades activate cell cycle checkpoints, thus allowing time for DNA repair before entering a new phase of the cell cycle. We carried out detailed investigation into the molecular mechanisms involved in cell cycle regulation in HNO adapted cells to understand their enhanced growth profile after UV exposure (106). The initial cell cycle analysis shows prolonged S phase in HNO cells (106), wherein the cells can repair the damaged DNA through HR pathway, before proceeding to the G2/M phase. It is well established that the cell cycle is regulated in different phases through coordinated action of special protein kinases termed cyclin dependent protein kinases (Cdks) [125]. Cdks are relatively small proteins with molecular weight ranging from 35 to 40 kDa. Sequential activation and inactivation of Cdks is regulated by periodic expression of its regulatory subunit termed cyclin [33]. Together, the cyclin-cdk forms an active kinase complex phosphorylating the target proteins mainly at serine -threonine residues [34]. Specific activator and inhibitor proteins regulate the activity of Cdks. The Cdc25 phosphatase family of proteins are the activators of Cdks [126]. There are three isoforms in humans that include

Cdc25A, Cdc25B and Cdc25C. All three are involved in regulating G1-S and G2-M checkpoints. Cdk2, which plays a key role in late G1-S transition and S phase is an important target of Cdcc25A. It should be noted that phosphorylation at different sites in Cdc25A regulates its different activities under various biological conditions. Phosphorylation of Cdc25A at Serine123 by Chk1 and Chk2 is a pre-requisite for ubiquitin mediated degradation of Cdc25A. Whereas the Cyclin B-Cdk1 mediated phosphorylation at Ser 17 and Ser 115 during mitosis stabilizes Cdc25A levels. Interestingly, the HNO cells maintain high levels of Cdc25A that increased with time in an NO dependent manner. This may result in accelerated entry of HNO cells in S phase with bypass of the G1-S checkpoint; therefore resulting in increased mitosis. It should be noted that high levels of stabilized Cdc25A in mitosis are important for continued activation of Cdk1, and thus enhanced cell proliferation [127]. This could be one of the mechanisms supporting high proliferation rate of HNO cells. Several other studies have reported a high frequency of Cdc25A over-expression in various human cancers including carcinomas of the head & neck, esophagus, breast and lung [128, 129]. Furthermore, there are reports showing a correlation between Cdc25A over-expression and both clinically aggressive tumor behavior with poor survival of patients [130, 131].

Another important regulator of cell cycle in G1 phase and G1-S transition is the Retinoblastoma (Rb) tumor suppressor protein. Rb is hypophosphorylated and bound to the transcription factor E2F in early G1 phase and in quiescent cells [132]. This inhibits expression of E2F regulated genes including Cdc25A and Cyclin E. As G1 phase progresses phosphorylation of Rb occurs by both Cdk4/Cdk6-Cyclin D and Cdk2-cyclin E. This results in the dissociation of Rb-E2F complexes, releasing E2F, which can then activate transcription of its target genes [133]. The high NO adapted A549-HNO cells show constitutive, and high levels of hyperphosphorylated Rb protein at Ser780, Ser795 and Ser807 (Figure 8) [106]. This may lead to accelerated progression of the cell cycle in A549-HNO cells exposed to NO. Our studies on expression of different Cyclin/Cdk complexes demonstrated increased expression of phase specific Cyclin/Cdk complexes in A549-HNO cells compared to parental cells. This indicates bypass of G1-S and G2-M checkpoints during early time points after treatment with NO donor. Further, with extended NO donor treatment, the intra-S phase checkpoint was activated in HNO cells as evident by reduced Cdk1 levels. Thus the A549-HNO cells show prolonged S phase [106].

We also studied cell cycle inhibitors in the HNO cells versus the parent cells. There are two main protein families of these inhibitors namely, the INK4 family and Cip/Kip family. The Cip/Kip family includes p21CIP1, p27KIP1, and P57KIP2 and the INK4 family includes INK4A, INK4B, INK4C, and INK4D. While the Cip/Kip family inhibits all cyclins D, E, and A the INK4 family inhibits CDK4 and CDK6 [126]. In response to NO treatment, we found the expression levels of p15INK4B and p27Kip1 to be moderately reduced in both parental and HNO cells. Reduced p15INK4B may result in enhanced Cdk4 and Cdk6 activity in G1 phase resulting in hyperphosphorylation and inactivation of Rb thus contributing to an enhanced proliferation rate. On the other hand, the expression of p21CIP increased to very high levels with prolonged NO donor treatment (Figure 8) [106]. This may lead to the inhibition of the cell cycle at G1, G2 or S phase depending on which Cyclin/Cdk complex it is inactivating. Interestingly, the Cyclin D3/Cdk6 complex is not inactivated by p21CIP1. Thus, cells utilizing this complex in G1 phase can still bypass high levels of p21CIP1 and enter S phase. Enhanced expression of p21CIP1 can induce both apoptotic and anti-apoptotic signaling based on the cellular context. When bound to Apoptotic Signaling Kinase (ASK1) in the cytoplasm, it can inhibit the MAP kinase pathway. It can also bind to pro-caspase 3 and prevent its processing and activation [4]. Thus, it is highly likely that the inhibition of apoptosis and slowing of HNO cells in S phase might be due to high levels of p21CIP1. Further, sustained and enhanced ATR/Chk1/p53/p21CIP1 DNA damage checkpoint activation in A549-HNO cells may lead to cell cycle arrest; in this case, in S phase for HNO cells and provide time for DNA repair. This correlates with enhanced DNA repair capacity and thus, maintenance of genome integrity of A549-HNO cells. Hence, we concluded that gradual NO exposure over prolonged time periods and subsequent adaptation of A549 cells to HNO steadily led to an enhanced cell cycle regulation network. This distinguishes HNO cells from parental cells and provides the ability for survival in harsh conditions.

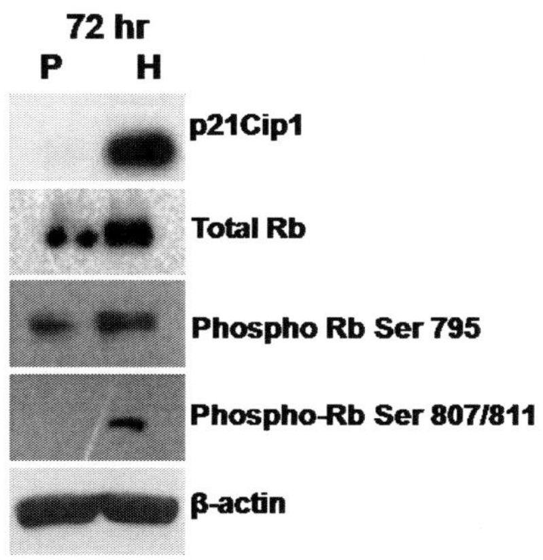

Figure 8. Expression levels of cell cycle regulator at 72hr post NO exposure in parent and HNO cell lines.

8.6.4. Enhanced Anti-Apoptotic Signaling

It is well established that p53 is one of the important effector molecules targeted by Chk1 and Chk2. DNA damage is one of the many signals for activation of cellular p53 pathway which ultimately concludes with cell cycle modulation and/or apoptosis [125]. To better understand p53 functions in HNO cells we addressed the apoptotic regulation by p53 in A549 parent and A549-HNO cells. Morphological analysis revealed initiation of apoptosis in the parent cells at 16 hr time point however, the HNO adapted cells exhibited normal growth at all time points. Higher expression of p53 and phospho p53 (Ser15, 46) was found in HNO cells with prolonged exposure to NO donor and not in parent cells [106].

It should be noted that serine 15 phosphorylation of p53 induces p21CIP1 expression, thus activating G1-S checkpoint. We further analyzed the levels of anti-apoptotic proteins in both parental and HNO cells in response to NO treatment. We found constitutive expression of Bcl-2, Phospho Bcl-2 (Ser70) and Bcl-xl in an NO independent manner in both parental and HNO cells. However, sustained phosphorylation of Bcl-2 at Serine 70 was only observed in HNO cells. We assessed levels of another important anti-apoptotic Bcl-2 family member, Mcl-1. The elimination of Mcl-1 by ubiquitin mediated proteasomal degradation is an early and required step for DNA damage-induced apoptosis. Interestingly, Mcl1 was up-regulated in an NO responsive manner in both parent and HNO cells. In A549-Parent cells the degradation of Mcl-1 after 16 hours of NO treatment was associated with the onset of apoptosis, whereas there was sustained accumulation of Mcl-1 in HNO cells following prolonged NO exposure [106]. This suggests an enhanced anti-apoptotic pathway that may facilitate resistance and continued proliferation of HNO cells under UV radiation. The above mechanisms are summarized in Figure 10.

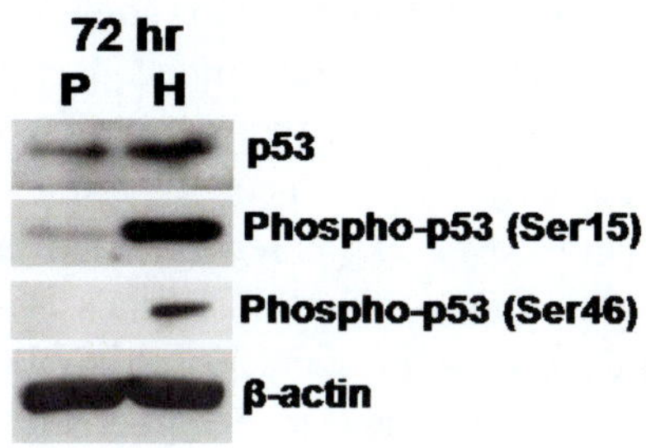

Figure 9. Expression level total and phospho p53 post NO exposure in parent and HNO cell lines.

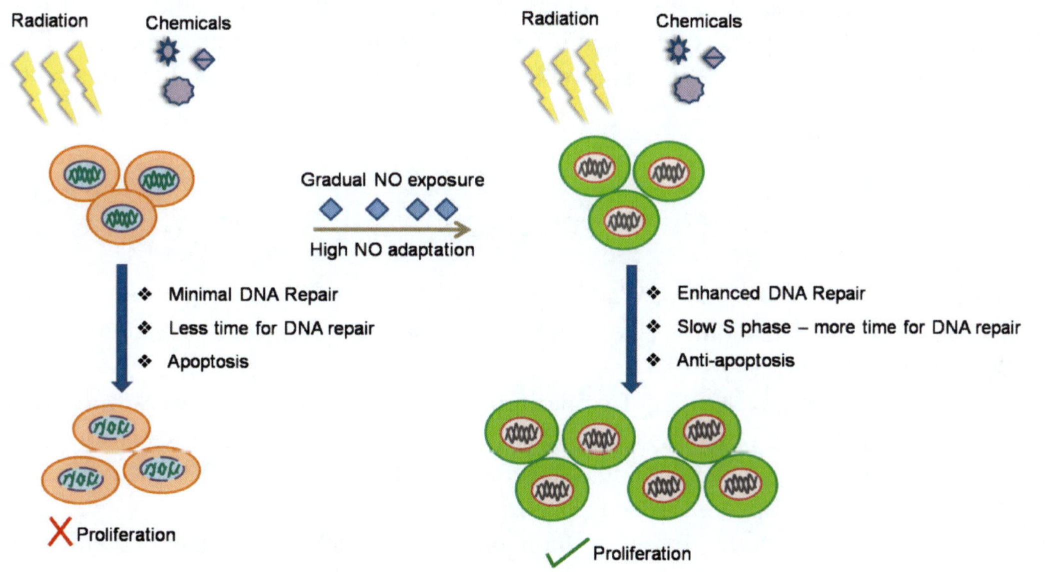

Figure 10. HNO cells show enhanced DNA repair, cell cycle modulation, and anti-apoptotic signaling.

8.7. Future Perspective

Radiotherapy is an important treatment for localized solid tumors. One of the important limitations associated with the dose of radiation to be given is the toxicity to normal surrounding cells. Consequently, it is frequently noted in the clinic that a large number of tumors develop radioresistance and recur after treatment. One of the chemical moieties in tumor microenviroment significantly influencing tumor behavior is the free radical NO. In clinical studies, we and others have reported that patients with high NO expressing tumors show poor prognosis compared to those with low NO expressing tumors. With this background, we developed in vitro model systems to study role of nitric oxide in tumor progression. We adapted 10 human cell lines to high nitric oxide concentrations by gradually increasing NO concentration up to 600μm. These are 5 adenocarcinomas (4 breast cell line, 1 lung cell line) and 5 squamous cell carcinomas (head and neck squamous cell lines). Interestingly these cell lines showed aggressive growth behavior with resistance to radiation and UV exposure. Thus, we carried out detailed biochemical studies to understand the mechanisms underlying this resistance in HNO cells. Our studies show enhanced free radical scavenging enzyme system, GST-pi in HNO cells, induced expression of genes in DNA repair pathway, and deregulated cell cycle with slow progression through S phase. The DDR pathway is a target for intervention to target specific tumor cells depending upon the particular DNA repair pathway being over expressed or to reverse the therapy resistance. Currently several DDR pathway inhibitors are in clinical trials for various cancers. Thus, the molecular studies on DNA repair pathways can be undertaken in patients to identify biomarkers of radiation resistance in tumors and to decipher the response of tumor cells to different drugs. Cell cycle modulation is one of the key events in tumor development. Our studies further highlight the importance of cell cycle perturbations in imparting radiation resistance to HNO tumor cells. Several small molecules against cell cycle regulatory proteins have been tested in the past and are under clinical trials. These molecular studies will enable caregivers to stratify patients with tumors in different stages for better therapy. The mechanisms underlying radiation resistance will also identify patients with higher probability to progress to metastasis.

REFERENCES

[1] Kowalczuk, C.I., et al., (2006), Wavelength dependence of cellular responses in human melanocytes and melanoma cells following exposure to ultraviolet radiation. *Int. J. Radiat. Biol.*, 82(11): p. 781-92.

[2] Runger, T.M., et al., (2012), Comparison of DNA damage responses following equimutagenic doses of UVA and UVB: a less effective cell cycle arrest with UVA may render UVA-induced pyrimidine dimers more mutagenic than UVB-induced ones. *Photochem. Photobiol. Sci.*, 11(1): p. 207-15.

[3] Placzek, M., et al., (2007), Effect of ultraviolet (UV) A, UVB or ionizing radiation on the cell cycle of human melanoma cells. *Br. J. Dermatol.*, 156(5): p. 843-7.

[4] Gentile, M., L. Latonen, and M. Laiho, (2003), Cell cycle arrest and apoptosis provoked by UV radiation-induced DNA damage are transcriptionally highly divergent responses. *Nucleic Acids Res..*, 31(16): p. 4779-90.

[5] Douki, T., (2013), The variety of UV-induced pyrimidine dimeric photoproducts in DNA as shown by chromatographic quantification methods. *Photochem. Photobiol. Sci.*, 12(8): p. 1286-302.

[6] Faraji, S. and A. Dreuw, (2013), Physicochemical Mechanism of Light-Driven DNA Repair by (6-4) Photolyases. *Annu. Rev. Phys. Chem.*, [Epub ahead of print]

[7] Rastogi, R.P., et al., (2010), Molecular Mechanisms of Ultraviolet Radiation-Induced DNA Damage and Repair. *Journal of Nucleic Acids*, vol. 2010, Article ID 592980, 32 pages.

[8] Ito, S., et al., (2011), Tet proteins can convert 5-methylcytosine to 5-formylcytosine and 5-carboxylcytosine. *Science,* 333(6047): p. 1300-3.

[9] Markovitsi, D., et al., (2013), Interaction of UV radiation with DNA. Photochem. Photobiol. Sci., 12(8): p. 1256-8.

[10] Palomera-Sanchez, Z. and M. Zurita, (2011), Open, repair and close again: chromatin dynamics and the response to UV-induced DNA damage. *DNA Repair* (Amst), 10(2): p. 119-25.

[11] Zhang, X., et al., (1997), Identification of possible reactive oxygen species involved in ultraviolet radiation-induced oxidative DNA damage. *Free Radic. Biol. Med.*, 23(7): p. 980-5.

[12] du Penhoat, M.A., et al., (2010), Double-strand break induction and repair in V79-4 hamster cells: the role of core ionisations, as probed by ultrasoft X-rays. *Int. J. Radiat. Biol.,* 86(3): p. 205-19.

[13] Kozmin, S.G., et al., (2009), The formation of double-strand breaks at multiply damaged sites is driven by the kinetics of excision/incision at base damage in eukaryotic cells. *Nucleic Acids Res.*, 37(6): p. 1767-77.

[14] Lo, H.L., et al., (2005), Differential biologic effects of CPD and 6-4PP UV-induced DNA damage on the induction of apoptosis and cell-cycle arrest. *BMC Cancer*, 5: p. 135.

[15] Murray, J.M., T. Stiff, and P.A. Jeggo, (2012), DNA double-strand break repair within heterochromatic regions. *Biochem. Soc. Trans*, 40(1): p. 173-8.

[16] Thoma, B.S. and K.M. Vasquez, (2003), Critical DNA damage recognition functions of XPC-hHR23B and XPA-RPA in nucleotide excision repair. *Mol. Carcinog*, 38(1): p. 1-13.

[17] Latypov, V.F., et al., (2012), Atl1 regulates choice between global genome and transcription-coupled repair of O(6)-alkylguanines. *Mol. Cell*, 47(1): p. 50-60.

[18] Tapias, A., et al., (2004), Ordered conformational changes in damaged DNA induced by nucleotide excision repair factors. *J. Biol. Chem.*, 279(18): p. 19074-83.

[19] Yokoi, M., et al., (2000), The xeroderma pigmentosum group C protein complex XPC-HR23B plays an important role in the recruitment of transcription factor IIH to damaged DNA. *J. Biol. Chem.*, 275(13): p. 9870-5.

[20] Hanawalt, P.C. and G. Spivak, (2008), Transcription-coupled DNA repair: two decades of progress and surprises. *Nat. Rev. Mol. Cell Biol.*, 9(12): p. 958-70.

[21] Hegde, M.L., T.K. Hazra, and S. Mitra, (2008), Early steps in the DNA base excision/single-strand interruption repair pathway in mammalian cells. *Cell Res.*, 18(1): p. 27-47.

[22] Sukhanova, M.V., et al., (2005), Human base excision repair enzymes apurinic/apyrimidinic endonuclease1 (APE1), DNA polymerase beta and poly(ADP-ribose) polymerase 1: interplay between strand-displacement DNA synthesis and proofreading exonuclease activity. *Nucleic Acids Res.*, 33(4): p. 1222-9.

[23] Eot-Houllier, G., et al., (2007), Interplay between DNA N-glycosylases/AP lyases at multiply damaged sites and biological consequences. *Nucleic Acids Res.*, 35(10): p. 3355-66.

[24] Kim, Y.J. and D.M. Wilson, 3rd, (2012), Overview of base excision repair biochemistry. *Curr. Mol. Pharmacol.*, 5(1): p. 3-13.

[25] Li, X. and W.D. Heyer, (2008), Homologous recombination in DNA repair and DNA damage tolerance. *Cell Res.*, 18(1): p. 99-113.

[26] Cousineau, I., C. Abaji, and A. Belmaaza, (2005), BRCA1 regulates RAD51 function in response to DNA damage and suppresses spontaneous sister chromatid replication slippage: implications for sister chromatid cohesion, genome stability, and carcinogenesis. *Cancer Res.*, 65(24): p. 11384-91.

[27] Long, D.T., et al., (2011), Mechanism of RAD51-dependent DNA interstrand cross-link repair. *Science,* 333(6038): p. 84-7.

[28] Matsuzaki, K., A. Shinohara, and M. Shinohara, (2008), Forkhead-associated domain of yeast Xrs2, a homolog of human Nbs1, promotes nonhomologous end joining through interaction with a ligase IV partner protein, Lif1. *Genetics,*179(1): p. 213-25.

[29] Frankenberg-Schwager, M., et al., (2009), Single-strand annealing, conservative homologous recombination, nonhomologous DNA end joining, and the cell cycle-dependent repair of DNA double-strand breaks induced by sparsely or densely ionizing radiation. *Radiat Res.*, 171(3): p. 265-73.

[30] Sperka, T., J. Wang, and K.L. Rudolph, (2012), DNA damage checkpoints in stem cells, ageing and cancer. *Nat. Rev. Mol. Cell Biol.*, 13(9): p. 579-90.

[31] Friedberg, E.C., (2003), DNA damage and repair. *Nature,* 421(6921): p. 436-40.

[32] Lim, S. and P. Kaldis, (2013), Cdks, cyclins and CKIs: roles beyond cell cycle regulation. *Development,* 140(15): p. 3079-93.

[33] Malumbres, M. and M. Barbacid, (2009), Cell cycle, CDKs and cancer: a changing paradigm. *Nat. Rev. Cancer*, 9(3): p. 153-66.

[34] Santamaria, D. and S. Ortega, (2006), Cyclins and CDKS in development and cancer: lessons from genetically modified mice. *Front Biosci.*, 11: p. 1164-88.

[35] Poehlmann, A. and A. Roessner, (2010), Importance of DNA damage checkpoints in the pathogenesis of human cancers. *Pathol. Res. Pract.,* 206(9): p. 591-601.

[36] Smith, J., et al., (2010), The ATM-Chk2 and ATR-Chk1 pathways in DNA damage signaling and cancer. *Adv. Cancer Res.*, 108: p. 73-112.

[37] Shibata, A., et al., (2010), Role of ATM and the damage response mediator proteins 53BP1 and MDC1 in the maintenance of G(2)/M checkpoint arrest. *Mol. Cell Biol.,* 30(13): p. 3371-83.

[38] Lavecchia, A., C. Di Giovanni, and E. Novellino, (2009), CDC25A and B dual-specificity phosphatase inhibitors: potential agents for cancer therapy. *Curr. Med. Chem.,* 16(15): p. 1831-49.

[39] Rudolph, J., (2007), Inhibiting transient protein-protein interactions: lessons from the Cdc25 protein tyrosine phosphatases. *Nat. Rev. Cancer*, 7(3): p. 202-11.

[40] Canepa, E.T., et al., (2007), INK4 proteins, a family of mammalian CDK inhibitors with novel biological functions. *IUBMB Life*, 59(7): p. 419-26.

[41] Chen, Y., L. Zhang, and K.A. Jones, (2011), SKIP counteracts p53-mediated apoptosis via selective regulation of p21Cip1 mRNA splicing. *Genes Dev.*, 25(7): p. 701-16.

[42] Thakur, V.S., et al., (2010), p53-Dependent p21-mediated growth arrest pre-empts and protects HCT116 cells from PUMA-mediated apoptosis induced by EGCG. *Cancer Lett,* 296(2): p. 225-32.

[43] Burke, A.J., et al., (2013), The yin and yang of nitric oxide in cancer progression. *Carcinogenesis,* 34(3): p. 503-12.

[44] Alderton, W.K., C.E. Cooper, and R.G. Knowles, (2001), Nitric oxide synthases: structure, function and inhibition. *Biochem. J.,* 357(Pt 3): p. 593-615.

[45] Paradise, W.A., et al., (2010), Nitric oxide: perspectives and emerging studies of a well known cytotoxin. *Int. J. Mol. Sci.*, 11(7): p. 2715-45.

[46] Forstermann, U. and W.C. Sessa, (2012), Nitric oxide synthases: regulation and function. *Eur. Heart J.*, 33(7): p. 829-37, 837a-837d.

[47] Thomas, D.D., et al., (2008), The chemical biology of nitric oxide: implications in cellular signaling. *Free Radic. Biol. Med.*, 45(1): p. 18-31.

[48] Bryan, N.S., K. Bian, and F. Murad, (2009), Discovery of the nitric oxide signaling pathway and targets for drug development. *Front Biosci.* (Landmark Ed), 14: p. 1-18.

[49] Kots, A.Y., K. Bian, and F. Murad, (2011), Nitric oxide and cyclic GMP signaling pathway as a focus for drug development. *Curr. Med. Chem.,* 18(22): p. 3299-305.

[50] Martinez-Ruiz, A. and S. Lamas, (2004), S-nitrosylation: a potential new paradigm in signal transduction. *Cardiovasc. Res.*, 62(1): p. 43-52.

[51] Klaunig, J.E., et al., (1998), The role of oxidative stress in chemical carcinogenesis. *Environ Health Perspect*, 106 Suppl 1: p. 289-95.

[52] Toledo, J.C., Jr. and O. Augusto, (2012), Connecting the chemical and biological properties of nitric oxide. *Chem. Res. Toxicol.*, 25(5): p. 975-89.

[53] Hirst, D. and T. Robson, (2010), Nitric oxide in cancer therapeutics: interaction with cytotoxic chemotherapy. *Curr. Pharm. Des.*, 16(4): p. 411-20.

[54] Bentz, B.G., et al., (2000), The yin and yang of nitric oxide: reflections on the physiology and pathophysiology of NO. *Head Neck*, 22(1): p. 71-83.

[55] Yang, G.Y., S. Taboada, and J. Liao, (2009), Induced nitric oxide synthase as a major player in the oncogenic transformation of inflamed tissue. *Methods Mol. Biol.*, 512: p. 119-56.

[56] Broholm, H., et al., (2003), Nitric oxide synthase expression and enzymatic activity in human brain tumors. *Clin. Neuropathol.*, 22(6): p. 273-81.

[57] Nussler, A.K. and T.R. Billiar, (1993), Inflammation, immunoregulation, and inducible nitric oxide synthase. *J. Leukoc. Biol.*, 1993. 54(2): p. 171-8.

[58] Hirst, D.G. and T. Robson, (2010), Nitrosative stress as a mediator of apoptosis: implications for cancer therapy. *Curr. Pharm. Des.*, 16(1): p. 45-55.

[59] Epperlein, M.M., et al., (1996), Nitric oxide in cigarette smoke as a mediator of oxidative damage. *Int. J. Exp. Pathol.*, 77(5): p. 197-200.

[60] Miyazawa, M., et al., (2003), Suppressed apoptosis in the inflamed gastric mucosa of Helicobacter pylori-colonized iNOS-knockout mice. *Free Radic. Biol. Med.*, 34(12): p. 1621-30.

[61] Karadayi, N., et al.,(2013), Inducible nitric oxide synthase expression in gastric adenocarcinoma: impact on lymphangiogenesis and lymphatic metastasis. *Diagn Pathol,* 8: p. 151.

[62] Chen, G.G., et al., (2008), Increased inducible nitric oxide synthase in lung carcinoma of smokers. *Cancer,* 112(2): p. 372-81.

[63] Masri, F., (2010), Role of nitric oxide and its metabolites as potential markers in lung cancer. *Ann. Thorac. Med.*, 5(3): p. 123-7.

[64] Kanwar, J.R., et al., (2009), Recent advances on the roles of NO in cancer and chronic inflammatory disorders. *Curr. Med. Chem.*, 16(19): p. 2373-94.

[65] Fukumura, D., S. Kashiwagi, and R.K. Jain, (2006), The role of nitric oxide in tumour progression. *Nat. Rev. Cancer*, 6(7): p. 521-34.
[66] Aranda, E., et al., (2012), Nitric oxide and cancer: the emerging role of S-nitrosylation. *Curr. Mol. Med.*, 12(1): p. 50-67.
[67] Lala, P.K. and A. Orucevic, (1998), Role of nitric oxide in tumor progression: lessons from experimental tumors. *Cancer Metastasis Rev.*, 17(1): p. 91-106.
[68] Nanni, S., et al., (2009), Endothelial NOS, estrogen receptor beta, and HIFs cooperate in the activation of a prognostic transcriptional pattern in aggressive human prostate cancer. *J. Clin. Invest.*, 119(5): p. 1093-108.
[69] Clemons, N.J., et al., (2010), Nitric oxide-mediated invasion in Barrett's high-grade dysplasia and adenocarcinoma. *Carcinogenesis*, 31(9): p. 1669-75.
[70] Oktem, G., et al., (2006), Chemotherapy influences inducible nitric oxide synthase (iNOS) and endothelial nitric oxide synthase (eNOS) activity on 3D breast cancer cell line. *Oncol. Res.*, 16(4): p. 195-203.
[71] Ying, L. and L.J. Hofseth, (2007), An emerging role for endothelial nitric oxide synthase in chronic inflammation and cancer. *Cancer Res.*, 67(4): p. 1407-10.
[72] Anttila, M.A., et al., (2007), Prognostic significance of iNOS in epithelial ovarian cancer. *Gynecol. Oncol.*, 105(1): p. 97-103.
[73] Raspollini, M.R., et al., (2004), Expression of inducible nitric oxide synthase and cyclooxygenase-2 in ovarian cancer: correlation with clinical outcome. *Gynecol. Oncol.*, 92(3): p. 806-12.
[74] Seril, D.N., J. Liao, and G.Y. Yang, (2007), Colorectal carcinoma development in inducible nitric oxide synthase-deficient mice with dextran sulfate sodium-induced ulcerative colitis. *Mol. Carcinog*, 46(5): p. 341-53.
[75] Watanabe, K., et al., (2000), COX-2 and iNOS, good targets for chemoprevention of colon cancer. *Biofactors,* 12(1-4): p. 129-33.
[76] Bentz, B.G., et al., (1990), Nitric oxide synthase type 3 is increased in squamous hyperplasia, dysplasia, and squamous cell carcinoma of the head and neck. *Ann. Otol. Rhinol. Laryngol.,* 108(8): p. 781-7.
[77] Bentz, B.G., G.K. Haines, 3rd, and J.A. Radosevich, (2000), Increased protein nitrosylation in head and neck squamous cell carcinogenesis. *Head Neck*, 22(1): p. 64-70.
[78] Bentz, B.G., et al., (2002), Nitric oxide and apoptosis during human head and neck squamous cell carcinoma development. *Am. J. Otolaryngol.*, 23(1): p. 4-11.
[79] Bentz, B.G., et al., (1998), Endothelial constitutive nitric oxide synthase (ecNOS) localization in normal and neoplastic salivary tissue. *Head Neck*, 20(4): p. 304-9.
[80] Chandra, R., et al., (2001), Expression of nitric oxide synthase type 3 in reflux-induced esophageal lesions. *Otolaryngol Head Neck Surg.*, 124(4): p. 442-7.
[81] Bentz, B.G., et al., (1997), Cytoplasmic localization of endothelial constitutive nitric oxide synthase in endometrial carcinomas. *Tumour Biol..*, 18(5): p. 290-300.
[82] Bentz, B.G., et al., (2004), Nitrosative stress induces DNA strand breaks but not caspase mediated apoptosis in a lung cancer cell line. *J. Carcinog.*, 3(1): p. 16.
[83] Bentz, B.G., et al., (2007), The kinetics and redox state of nitric oxide determine the biological consequences in lung adenocarcinoma. *Tumour Biol..*, 28(6): p. 301-11.
[84] Jenkins, D.C., et al., (1995), Roles of nitric oxide in tumor growth. *Proc. Natl. Acad. Sci. USA,* 92(10): p. 4392-6.

[85] Xu, W., L. Liu, and I.G. Charles, (2002), Microencapsulated iNOS-expressing cells cause tumor suppression in mice. *FASEB J.,* 16(2): p. 213-5.

[86] Juang, S.H., et al., (1998), Suppression of tumorigenicity and metastasis of human renal carcinoma cells by infection with retroviral vectors harboring the murine inducible nitric oxide synthase gene. *Hum. Gene Ther*, 9(6): p. 845-54.

[87] Roberts, P.J., et al., (2001), The physiological expression of inducible nitric oxide synthase (iNOS) in the human colon. *J. Clin. Pathol.*, 54(4): p. 293-7.

[88] Ambs, S., et al., (1998), Frequent nitric oxide synthase-2 expression in human colon adenomas: implication for tumor angiogenesis and colon cancer progression. Cancer Res, 58(2): p. 334-41.

[89] Ambs, S., et al., (1998), Up-regulation of inducible nitric oxide synthase expression in cancer-prone p53 knockout mice. *Proc. Natl. Acad. Sci. USA*, 1998. 95(15): p. 8823-8.

[90] Hussain, S.P., et al., (2004), Nitric oxide, a mediator of inflammation, suppresses tumorigenesis. *Cancer Res,* 64(19): p. 6849-53.

[91] Xu, W., et al., (2002), The role of nitric oxide in cancer. Cell Res, 12(5-6): p. 311-20.

[92] Kim, S.J., et al., (2002), ERK-1/2 and p38 kinase oppositely regulate nitric oxide-induced apoptosis of chondrocytes in association with p53, caspase-3, and differentiation status. *J. Biol. Chem.,* 277(2): p. 1332-9.

[93] Brune, B. and J. Zhou, (2003), The role of nitric oxide (NO) in stability regulation of hypoxia inducible factor-1alpha (HIF-1alpha). *Curr. Med. Chem.*, 10(10): p. 845-55.

[94] Sandau, K.B., J. Fandrey, and B. Brune, (2001), Accumulation of HIF-1alpha under the influence of nitric oxide. *Blood,* 97(4): p. 1009-15.

[95] Sidorkina, O., et al., (2003), Inhibition of poly(ADP-RIBOSE) polymerase (PARP) by nitric oxide and reactive nitrogen oxide species. *Free Radic. Biol. Med.*, 35(11): p. 1431-8.

[96] Yu, Z., et al., (2006), Nitric oxide-dependent negative feedback of PARP-1 trans-activation of the inducible nitric-oxide synthase gene. *J. Biol. Chem.*, 281(14): p. 9101-9.

[97] Brune, B., A. von Knethen, and K.B. Sandau, (2001), Transcription factors p53 and HIF-1alpha as targets of nitric oxide. *Cell Signal*, 13(8): p. 525-33.

[98] Ridnour, L.A., et al., (2008), Molecular mechanisms for discrete nitric oxide levels in cancer. *Nitric Oxide*, 19(2): p. 73-6.

[99] Ambs, S. and S.A. Glynn, (2011), Candidate pathways linking inducible nitric oxide synthase to a basal-like transcription pattern and tumor progression in human breast cancer. *Cell Cycle*, 10(4): p. 619-24.

[100] Glynn, S.A., et al., (2010), Increased NOS2 predicts poor survival in estrogen receptor-negative breast cancer patients. *J. Clin. Invest.*, 120(11): p. 3843-54.

[101] Vesper, B.J., et al., (2009), Long-Term Adaptation of Lung Tumor Cell Lines with Increasing Concentrations of Nitric Oxide Donor, *The Open Lung Cancer Journal*, 2: p. 35-44.

[102] Onul, A., et al., (2012), Long-term adaptation of the human lung tumor cell line A549 to increasing concentrations of hydrogen peroxide. *Tumour Biol..*, 33(3): p. 739-48.

[103] Vesper, B.J., et al., (2010), Long-term adaptation of breast tumor cell lines to high concentrations of nitric oxide. *Tumour Biol..*, 31(4): p. 267-75.

[104] Yarmolyuk, Y.R., et al., (2011), Part I. Development of a model system for studying nitric oxide in tumors: high nitric oxide-adapted head and neck squamous cell carcinoma cell lines. *Tumour Biol.*, 32(1): p. 77-85.

[105] Tarjan, G., et al., (2011), Part II. Initial molecular and cellular characterization of high nitric oxide-adapted human tongue squamous cell carcinoma cell lines. *Tumour Biol.*, 32(1): p. 87-98.

[106] Aqil, M., et al., (2013), Part II-mechanism of adaptation: A549 cells adapt to high concentration of nitric oxide through bypass of cell cycle checkpoints. *Tumour Biol.*

[107] De Vitto, H., et al., (2013), Part III. Molecular changes induced by high nitric oxide adaptation in human breast cancer cell line BT-20 (BT-20-HNO): a switch from aerobic to anaerobic metabolism. *Tumour Biol.*, 34(1): p. 403-13.

[108] Hussein, M.R., (2005), Ultraviolet radiation and skin cancer: molecular mechanisms. *J. Cutan Pathol.,* 32(3): p. 191-205.

[109] Hayes, J.D. and D.J. Pulford, (1995), The glutathione S-transferase supergene family: regulation of GST and the contribution of the isoenzymes to cancer chemoprotection and drug resistance. *Crit. Rev. Biochem. Mol. Biol.*, 30(6): p. 445-600.

[110] Nebert, D.W. and V. Vasiliou, (2004), Analysis of the glutathione S-transferase (GST) gene family. *Hum. Genomics*, 1(6): p. 460-4.

[111] Cullen, K.J., et al., (2003), Glutathione S-transferase pi amplification is associated with cisplatin resistance in head and neck squamous cell carcinoma cell lines and primary tumors. *Cancer Res.*, 63(23): p. 8097-102.

[112] Wang, X., et al., (1997), Overexpression and amplification of glutathione S-transferase pi gene in head and neck squamous cell carcinomas. *Clin. Cancer Res.*, 3(1): p. 111-4.

[113] Bentz, B.G., G.K. Haines, 3rd, and J.A. Radosevich, (2000), Glutathione S-transferase pi in squamous cell carcinoma of the head and neck. *Laryngoscope*, 110(10 Pt 1): p. 1642-7.

[114] Brabender, J., et al., (2002), Glutathione S-transferase-pi expression is downregulated in patients with Barrett's esophagus and esophageal adenocarcinoma. *J. Gastrointest Surg*, 6(3): p. 359-67.

[115] Chandra, R.K., et al., (2002), Expression of glutathione s-transferase pi in benign mucosa, Barrett's metaplasia, and adenocarcinoma of the esophagus. *Head Neck*, 24(6): p. 575-81.

[116] Shi, H., et al., (2008), Expression of multidrug-resistance-related proteins P-glycoprotein, glutathione-S-transferases, topoisomerase-II and lung resistance protein in primary gastric cardiac adenocarcinoma. *Cancer Invest*, 26(4): p. 344-51.

[117] Vesper, B.J., et al., (2013), Part I. Molecular and cellular characterization of high nitric oxide-adapted human breast adenocarcinoma cell lines. *Tumour Biol.,* 34(1): p. 203-14.

[118] Chapman, J.R., M.R. Taylor, and S.J. Boulton, (2012), Playing the end game: DNA double-strand break repair pathway choice. *Mol. Cell*, 47(4): p. 497-510.

[119] Shimada, M. and M. Nakanishi, (2013), Response to DNA damage: why do we need to focus on protein phosphatases? *Front Oncol.*, 3: p. 8.

[120] Serrano, M.A., et al., (2012), DNA-PK, ATM and ATR collaboratively regulate p53-RPA interaction to facilitate homologous recombination DNA repair. *Oncogene,* 32:2452-2462

[121] Aqil, M., et al., (2013), Part I-mechanism of adaptation: high nitric oxide adapted A549 cells show enhanced DNA damage response and activation of antiapoptotic pathways. *Tumour Biol.*

[122] Kuroda, S., Y. Urata, and T. Fujiwara, (2012), Ataxia-telangiectasia mutated and the Mre11-Rad50-NBS1 complex: promising targets for radiosensitization. *Acta Med. Okayama*, 66(2): p. 83-92.

[123] Zhang, J. and S.N. Powell, (2005), The role of the BRCA1 tumor suppressor in DNA double-strand break repair. *Mol. Cancer Res.*, 3(10): p. 531-9.

[124] Curtin, N.J., (2012), DNA repair dysregulation from cancer driver to therapeutic target. *Nat. Rev. Cancer*, 12(12): p. 801-17.

[125] Langerak, P. and P. Russell, (2011), Regulatory networks integrating cell cycle control with DNA damage checkpoints and double-strand break repair. *Philos. Trans. R. Soc. Lond. B Biol. Sci.*, 366(1584): p. 3562-71.

[126] Besson, A., S.F. Dowdy, and J.M. Roberts, (2008), CDK inhibitors: cell cycle regulators and beyond. *Dev. Cell*, 14(2): p. 159-69.

[127] Potapova, T.A., et al., (2011), Mitotic progression becomes irreversible in prometaphase and collapses when Wee1 and Cdc25 are inhibited. *Mol. Biol. Cell,* 22(8): p. 1191-206.

[128] Wu, W., et al., (1998), Overexpression of cdc25A and cdc25B is frequent in primary non-small cell lung cancer but is not associated with overexpression of c-myc. *Cancer Res.,* 58(18): p. 4082-5.

[129] Cangi, M.G., et al., (2008), Constitutive overexpression of CDC25A in primary human mammary epithelial cells results in both defective DNA damage response and chromosomal breaks at fragile sites. *Int. J. Cancer*, 123(6): p. 1466-71.

[130] Nishioka, K., et al., (2001), Clinical significance of CDC25A and CDC25B expression in squamous cell carcinomas of the oesophagus. *Br. J. Cancer*, 85(3): p. 412-21.

[131] Wang, Z., et al., (2010), Overexpression of CDC25B, CDC25C and phospho-CDC25C (Ser216) in vulvar squamous cell carcinomas are associated with malignant features and aggressive cancer phenotypes. *BMC Cancer,* 10: p. 233.

[132] Stone, J.G., et al., (2011), The cell cycle regulator phosphorylated retinoblastoma protein is associated with tau pathology in several tauopathies. *J. Neuropathol. Exp. Neurol.*, 70(7): p. 578-87.

[133] Khidr, L. and P.L. Chen, (2006), RB, the conductor that orchestrates life, death and differentiation. *Oncogene*, 25(38): p. 5210-9.

In: UV Radiation: Properties, Effects, and Applications
Editor: James A. Radosevich
ISBN: 978-1-63321-090-5

Chapter 9

ULTRAVIOLET VISION

Kelly D. Garcia, D.V.M., Ph.D., D.A.C.L.A.M.[1,•]
and James E. Artwohl, M.S., D.V.M., D.A.C.L.A.M.[1]
[1]Biologic Resources Laboratory, UIC, University of Illinois at Chicago, Chicago, IL, US

ABSTRACT

The importance of ultra violet (UV) vision was overlooked by humans for years probably because we lack the ability to detect UV light. In fact, the ability of organisms to detect and respond to UV light was considered a fluke or insignificant. However, as investigators have explored the evolution of the visual system, it has become clear that UV vision is widespread taxonomically in vertebrates and invertebrates and that it plays an important role in many visually guided behaviors.

The following chapter delves into some of the salient features of UV vision and provides an overview on current thoughts on how and why UV vision has evolved in complex organisms.

Keywords: Ultraviolet vision, Opsins, Visual pigments, Spectral tuning, evolution

ABBREVIATIONS

G-protein	guanine-nucleotide-binding protein
LWS	longwave sensitive
MWS	middle wave sensitive
SWS2	short wave sensitive
UV	Ultraviolet

• E-mail: kellygar@uic.edu.

9.0. VISION

The first step in the process of vision is the transduction of light by visual pigments into the movement of ions in the central nervous system. All visual pigments have bell-shaped absorbance spectra. However, the width and height of the curve varies with species and are conferred by variations in the visual pigments. For this reason the absorption spectra of visual pigments are fundamental to their function; only photons that are absorbed play any part in the visual process. Adjusting factors to modify the ability of an organism to detect specific spectrums of light and to adjust their behavior accordingly is referred to as spectral tuning [1].

Visual pigments are comprised of an opsin protein and a chromophore. Since most vertebrates possess only one type of chromophore (11-*cis*-retinal), spectral tuning is achieved through variations in the amino acid sequence of opsin. All opsins possess a Lysine residue at site 296 that is covalently linked to the chromophore via a Schiff base. In vertebrates, this Schiff base is thought to be protonated, with the negatively charged glutamic acid residue at site 113 [2].

Absorption of light causes the isomerization of the chromophore from 11-*cis*- to all-*trans*-retinal, which in turn causes major structural changes that include the displacement of the positively charged Schiff base from its interaction to Glutamic acid113 [3, 4]. This structural change ultimately leads to a neural signal. Each opsin molecule has a wave length of light where there is peak absorption (λ_{max}). Opsins that are tuned to respond to light in the violet to UV range require a larger amount of energy to obtain a structural shift [5].

Opsins are members of the superfamily of G-protein-coupled receptors which function through the activation of a guanine-nucleotide-binding protein (G-protein) and an effector enzyme which changes the level of a second messenger in the cell cytoplasm [6]. Cyclic GMP is an intracellular chemical that is a second messenger which keeps Na^+ channels partly open when rods are in darkness.

Rods are thus slightly depolarized. Exposure to light initiates a cascade of events that deactivates cyclic GMP, closing Na^+ channels, which hyperpolarizes the cell and reduces the release of glutamate. The process of changing the energy of light into a neural signal is termed transduction.

Visual pigments are found in the two types of photoreceptor cells: rods and cones. Phylogenetic analysis of opsin sequences show that vertebrates have four classes of visual pigment: a longwave sensitive (LWS) λmax 500-570 nm, middle wave sensitive (MWS) with λ_{max} 480-520nm, short wave sensitive (SWS2) with λ_{max} 415-470 nm (SWS2), and violet-UV sensitive (SWS1) with λ_{max} 355-435 [1, 7]. Rods function in light:dark detection in dim light and are the basis of scotopic system. They usually have a single class of rhodopsin molecules with an A in the blue-green region of the light spectrum. Cones function in color detection at normal light levels during the day and are the basis of photopic vision[1, 7].

9.1. UV LIGHT

UV light ranges from 10 to 400 nm. Shorter wavelengths destroy nucleic acids and cell membranes [8, 9]. Earth's ozone layer prevents the entry of these shorter UV wavelengths which allows terrestrial life to flourish. Many animals filter medium to low UV wavelengths through protective mechanisms. However, the ability to perceive longer UV wavelengths has been clearly established in a wide range of vertebrate taxa [6].

The ability of a species to perceive UV light is correlated with the amount of UV light in the species' environment [1]. Forty percent of higher wavelength UV only penetrates water more than half a meter, so many animals that live in deep water have lost the ability to detect UV light.

Alternatively, UV light comprises a higher percentage of light at dawn, dusk, and in Polar Regions when the sun is on the horizon due to the high degree of atmospheric scatter [10]. The earth's atmosphere bends light of shorter wavelengths to a greater degree. UV light also reflects to a high degree from snow and ice [11]. Thus, animals which live in Polar Regions or are active dawn and dusk have an advantage if capable of seeing UV light.

9.2. Evolution of Sensory Systems

To understand how UV vision has evolved, it is necessary to first understand how selection pressures lead to the evolution of sensory systems in general. The fundamental or first step in the evolution of the sensory system is the evolution of a sensory task. A sensory task is broadly defined as a systematic behavioral or physiologic response that results from information received [12].

The environment contains an overwhelming amount of information. In order to cope effectively, each organism must filter out non-important from essential information.

The term 'matched filters' was first used in 1987 [13]. This term describes sensory and neural properties that allow an organism to match patterns of information with behavioral responses.

The visual system is the system that gives the organism the ability to discriminate among objects and lights based solely on differences in their spectral properties [7]. The visual spectrum of light represents that spectrum of light organisms have evolved to detect and respond to, and is generally considered to be light with wavelengths between 400 and 700 nms [1]. However, the specific spectrum of light and intensity of light that can be detected varies greatly with species.

Many organisms, including all species of invertebrates thus far investigated and multiple species of fish, amphibians, birds, and mammals have evolved the ability to detect light with wavelengths shorter than 400 nm, or UV light. For these species, UV light plays critical roles in mate selection, foraging and communication. Its importance is critical enough to have originated independently at least four times in frogs, birds, primates, and certartiodactyls [14].

9.3. Evolution of UV Vision

The availability of molecular data has been useful to researchers interested in understanding ecology and behavior. Lessons learned by such researchers actively studying the evolution of UV vision exemplifies the beauty of integrative biology since the data allows researchers to tell a comprehensive story [15].

As early as the 1930s, investigators recognized that insect behavior was altered by UV light [16]. Comparative sequence analysis of visual pigments that absorb UV light, or short wavelength-sensitive pigments, suggested that contemporary pigments evolved from a common ancestor [17, 18].

However, the complexity of the evolutionary path elucidated by modern molecular techniques was substantially underestimated. Recently, researchers assessed the statistical probability of various ancestral opsins serving as a precursor to the modern day proteins [14]. They generated a number of hypothesized opsins and tested their physical characteristics. Their modeling predicts that the ability to detect UV light has evolved independently more than ten times in fish, amphibians, birds, and mammals.

9.4. Function of UV Vision

UV vision plays critical roles in mate selection, foraging, and communication in species that see UV light. Extensive phylogenetic surveys showed that UV vision is present in some, but not all, species of fish, amphibians, reptiles, birds, and mammals [15]. Anoline lizards have a dewlap which reflects UV light significantly and is complemented by UV visions [19]. The presence of UV vision in young salmonids is lost after 2 years and is thought to be correlated with change of diet from zooplankton which is found in shallow water to deep water prey where UV light does not penetrate [20, 21]. Arctic reindeer take advantage of UV light's tendency to scatter and it's reflecting properties in Polar Regions where UV is relatively more abundant [22].

Communication occurs both within species and between species. Voles mark their path with urine and feces, which can be detected much more easily under UV light than visible light [23]. UV reflective vole urine caused increased activity of Eurasian kestrels and rough-legged buzzards [24]. In addition, higher vole mortality was reported in urine treated areas [25]. Work in Australia and Scandinavia does not support this common theme amongst avian predators though [26, 27].

9.4.1. Birds

The first behavioral evidence for UV vision in birds came from operant discrimination experiments on hummingbirds and pigeons. Subsequently, it took 20 more years for the first avian UV-sensitive visual pigment to be characterized. Now, it is known that UV vision is widespread in birds and plays a role in mate selection, avoiding predators, and finding food [28].

In 2003, researchers published results from a study in which they examined gene sequence of SWS1 pigments from 46 different species of birds in 14 avian orders.

Based on single nucleotide substitutions, they concluded that UV vision developed independently at least 14 times in birds [29]. Moreover, their work demonstrates that the development of UV vision is so complex making inferences of spectral sensitivities from closely related taxa is inadvisable.

9.4.1.1. The Mechanism of Color Vision in Birds

The retinae of the least derived avian species have a rod which can absorb light maximally (λ_{max}) with a wavelength of 505 nm; cones containing a visual pigment based on the LWS opsin with a λ_{max} of 565 nm; a MWS pigment with a λ_{max} the same as the rod pigment, and a cone with a SWS pigment with a λ_{max} around 445 nm and, and a cone with a UVS/VS pigment with a λ_{max} between 355 and 420 nm [28, 6].

This pattern may be representative of an ancestral avian retina. However, there is considerably interspecific variation. For example, some bird species lack the full complement of avian visual pigments. Even among those birds with a "typical" set of visual pigments, there is interspecific variation in λ_{max}. The variation is most profound in pigments conferring shortwave sensitivity, including that of ultraviolet (UV) wavelengths [29]. Available data suggest that the UVS/VS class of visual pigments fall into two groups: species having λ_{max} values in the UV (e.g. Budgerigar *Melopsittacus undulatus*, λ_{max} 371 nm; European Starling, λ_{max} 362 nm), and other species having UVS/VS pigments with λ_{max} values in the range 400-420 nm (e.g. Humboldt Penguin *Spheniscus humboldti* λ_{max} 403 nm; Mallard Duck *Anas platyrhynchos* λ_{max} 420 nm;) [29].

One thing to bear in mind is that while it is clear that species will vary in their sensitivity to UV wavelengths due to their visual pigments alone, there are other factors that potentially limit sensitivity to the UV, including intra-ocular pigments in the cornea, lens and ocular media, and intracellular filters, in the form of colored oil droplets in the photoreceptors themselves [30].

9.4.1.2. The Function of UV Vision in Birds

Little is known on the function of UV vision in avian species. Plausible possibilities include foraging, signaling (e.g. in mate choice), and orientation.

In bees, UV and polarization vision are linked because the arrangement of UV receptors in the eye allows detection of the pattern of polarized light in the sky [31]. Therefore, investigators have tried to determine if birds also use polarization patterns and if these patterns assist in orientation, and other compass information. In support of this, it has been noted that there are more short-wavelength receptors in the ventral (skyward-looking) surface the pigeon's retina and the cones in this region are relatively dispersed, which suggests panoramic vision rather than fine spatial resolution. But not all birds show this pattern; indeed starlings have more UV cones in the posterior-dorsal (down and forward-looking) region of the retina [30, 32].

The first study clearly demonstrating the use of UV cues in foraging came when Viitala *et al.* demonstrated Common Kestrels *Falco tinnunculus* used UV to detect active vole trails [33]. They suggested that vole urine absorbs strongly in the UV and this provides a strong contrast against the background. Field experiments showed that raptors tended to hunt near artificial vole trails after urine and feces-soaked straw were experimentally added. Lab experiments showed that Kestrels spent more time above arenas with vole urine, but only under illumination that included UV wavelengths and not when illumination was restricted to human-visible light. The implication is that the birds see the vole trails better when UV information was present [33].

While it does not guarantee a role in visual signaling, the fact that a species' plumage reflects UV is intriguing. For example, in Blue Tits there is pronounced sexual dimorphism in only the UV spectrum of plumage. Male crests are brighter than females and there is significant between-sex variation in both spectral shape and the wavelength of peak reflectance. Furthermore, mate choice trials indicated that this region of plumage may play a role in female mate choice since all females tested chose the male with the brightest crest [32].

Ultraviolet light might also provide higher contrast that makes finding food easier. Environmental factors may play a role in the evolution of UV vision. Open spaces, for example, have more UV light than do forests. Ultimately, the color sensitivity may be a result of other changes that affect the amount of ultraviolet light the birds' eyes receive [32].

9.4.2. UV Vision in Mammals

In multi-system organisms, the visual system is a complex organ system basically comprised of two distinct components: the photopic system and the scoptopic system. Photopic arm of the visual system is mediated by cones and functions under well-lighted circumstances, while scoptopic vision is mediated by rods which mediate low-acuity vision [34, 35].

Consistent with the findings in birds, the relative proportion of rods and cones in the mammalian retina varies with species. However, in general, the retina of most mammals is rod-dominated with the retina of diurnal species ranging between 70 and 95% cones and the retina of nocturnal species being comprised of up to 97% by cones [34]. Furthermore, the wavelength at which light can be maximally absorbed varies with species. Old World primates have trichromatic vision which comes from the combination of an autosomal SWS1 gene and duplicates copies of the LWS gene, one of which encodes an MWS pigment with a lambda max around 535 nm. In contrast, the Pinnepedia (seals, sea-lions, and walruses) have lost color vision altogether by the complete loss of a functional SWS1 gene [6]. Mammals are believed to have evolved as nocturnal creatures. Rodents make up the majority of mammals that have UV vision [35]. Diurnal rodents such as squirrels and ground squirrels sense violet light and rats, mice, certain hamsters such as djungarian hamster and South American rodents such as the degu sense UV light [35].

From a molecular perspective, the hypothesized ancestral opsin is more closely related to mouse and rat opsin when compared to other mammals [14]. Hypothesized ancestral opsin is also more closely related to fish, amphibians, and reptiles when compared to other mammals. Maximum absorbance values are changed in various ways by as many as 9 difference amino acid substitutions within the mammalian phylum.

9.4.2.1. Distribution of UV Photosensitive Cells in the Retina

The distribution of UV photoreceptors varies by species. Generally, UV photoreceptors are less than the LWS photoreceptors. Primates possess UV photoreceptors that are more uniformly distributed throughout the retina. In the mouse, the UV photoreception is in the ventral retina [36]. The background spectrum of the sky would illuminate the ventral retina to a greater degree [6].

9.4.2.2. Rodent Vision

Most mammals, and rodents in particular, have a rod dominated retina with only 3% of photoreceptor cells being cones. Diurnal mammals have between 5 to 30 percent cones [36,

37]. Eutherian mammals lost the SWS2 opsin and a rhopsin gene. Most mammals are dichromats LWS and SWS1 opsins. Cetacean such as whales and dolphins and pinnipeds such as seals, sea-lions, and walruses are monochromats and do not see UV light [38].

The mouse cornea and lens allow 50% of 350 nm UV to reach the retina and some 310 nm light will reach the retina. This is thought to be due to the lesser amount of tissue to penetrate [6] and to the lack of ultra-violet filter which is present in the lens of diurnal mammals like the squirrel and primate [39].

Mice and rats have been the most studied species from a behavioral and neurophysiological perspective. Unlike most other mammals the mouse retina contains a greater total amount of UV sensitive opsin. The amount of UV sensitive opsin is about 3 times greater than the medium wavelength opsin [40]. UV sensitive opsin sensitivity in terms ERG amplitude is roughly 7 times greater than the MWS opsin sensitivity. It is thought that the UV signal is muted somehow in the central nervous. Possibly the cones co-expressing both opsins relay to neurons associated with medium opsin sensation. In addition, mice have dichromatic color visions and can differentiate UV light from light in the 480-520 nm range [41, 42], a characteristic shared with rats [43].

Table. Examples of Organisms with and without the Ability to Detect UV Light

With Ability	Without Ability
Butterflies	African Giant Rat
Dragonflies	Walruses
Bees, Wasps	Sea Lions
Beach Isopod	Dolphin
Fruit Fly	Owl Monkey
Mouse	Old Brown Trout
Rat	Macaque
Degu	Humans
Dwarf Hamster	Cow
Salamander	Pig
Chameleon	Octopus
Goldfish	Cuttlefish
Young Brown Trout	Squid
Honey Opossum	Guinea Pig
Canary	Wallaby
Starling	Ferret
Zebra Finch	Hyena
Common Rhea	Dog
Budgerigar	Cat
African Grey Parrot	Polar Bear
Macaw	Chicken
Cockatoo	Duck
Reindeer	Quail
Bats	Ostrich
	Syrian Hamster

Conclusion

The ability of an organism to filter non-essential sensory information from information requiring a behavioral response is essential for successful existence. Humans lack the ability to detect UV light. Thus, it is tempting to assume UV light does not provide useful behavioral cues to other species. However, the number and breadth of species with the ability to detect UV light suggests the ability represents a distinct advantage at least in certain environments (see table). Notably, the ability to detect UV is most common in species who encounter high levels of UV light in their surroundings. For such organisms, not only is the ability to detect UV light common, pressures have supported the independent evolution of the ability dozens of times in multiple species across multiple classes of organism.

References

[1] Bjron L. O. and Ghiradella H. Spectral Tuning in Biology. (2008) In *Photobiology* (ed L. O. Bjron). pp 155-196.

[2] Nathans J. (1990) Determinants of visual pigment absorbance: identification of the retinylidene Schiff's base counterion in bovine rhodopsin. *Biochemistry*, 29: 9746-9752.

[3] Zvyaga T. A., K. Fahmy, T. P. Sakmar. (1994) Characterization of rhodopsin-transducin interaction: a mutant rhodopsin photoproduct with a protonated Schiff base activates transducin. *Biochemistry*, 33: 9753-9761.

[4] Shieh T., M. Han, T. P. Sakmar, S. O. Smith. (1997) The steric trigger in rhodopsin activation. *J. Mol. Biol.*, 269: 373-384.

[5] Nickle B., P. R. Robinson. (2007) The opsins of the vertebrate retina: insights from structural biochemical, and evolutionary studies. *Cell Mol. Life Sci.*, 64: 2917-2932.

[6] Hunt D. M., Wilkie S. E., J. K. Bowmaker, S. Poopalasundaram. (2001) Vision in the UV. *Cell Mol. Life Sci.,* 58: 1583-1598.

[7] Jacobs G. H. (2012) The evolution of vertebrate color vision. *Adv. Exp. Med. Biol.,* 739: 156-172.

[8] Cadet J., Wagner J. R.. (2013) DNA base damage by reactive oxygen species, oxidizing agents, and UV radiation. *Cold Spring Harb. Perspect. Biol.*, Feb. 1:5(2).

[9] Varma S. D., Kovtun S., Hegde K. R. (2011) Role of UV irradiation and oxidative stress in cataract formation-medical prevention by nutritional antioxidants and metabolic agonists. *Eye Contact Lens*, Jul.;37(4):233-45.

[10] Lythgo J. N. (1979) The Ecology of Vision (Clarendon, Oxford).

[11] Weatherhead B., A. Tanskanen, A. Stevermer. (2007) Ozone and UV radiation. In Arctic Climate Impact Assessment Report (ed JE Walch) pp 151-182. Cambridge: Cambridge University Press.

[12] Nilsson D. E. The evolution of eyes and visually guided behavior. (2009) *Phil. Trans. R.Soc. B,* 364, 2833-2847.

[13] Wehner, R. (1987) Matched filter' – neural models of the external world. *J. Comp. Physiol. A.*, 161, 511-531.

[14] Shi Y. and S. Yokoyama. (2003) Molecular analysis of the evolutionary significance of UV vision in vertebrates. *Proc. Natl. Acad. Sci.*, 100: 8308-8313.

[15] Zhang J. (2003) Paleomolecular biology unravels the evolutionary mystery of vertebrate UV vision. *Proc. Natl. Acad. Sci.*, 100: 8045-8047.

[16] Menzel R., Backhaus W. (1991) Colour Vision in Insects. In: Gouras P (ed) Vision and visual dysfunction: the perception of colour. Mac-Millan, London, pp 262–288.

[17] Yokoyama S. and Shi Y. (2000) Genetics and evolution of ultraviolet vision in vertebrates. *FEBS Lett.,* 486:167-172.

[18] Shi Y., Radlwimmer F. B., Yokoyama S. (2001) Molecular genetics and the evolution of ultraviolet vision in vertebrates. *Proc. Natl. Acad. Sci. USA*, 98:11731-11736.

[19] Fleishman J. L., E. R. Loew. (1993) UV vision in lizards. *Nature,* 365:297.

[20] Bowmake J. K., Y. W. Kunz. (1987) UV receptors tetrachromatic color vision and retinal mosaics in the brown trout (Salmo trutta): age-dependent changes. *Vision Res.,* 27: 2101-2108.

[21] Browman H. I., C. W. Hawryshyn. (1994) The developmental trajectory of UV photosensivity in rainbow trout is altered by thyroxin. *Vision Res.*, 34: 1297-1406.

[22] Hogg C., M. Neveu, K. Stokkan, L. Folkow, P. Cottrill, R. Douglas, D. M. Hunt, G. Jeffery. Arctic reindeer extend their visual range into the UV. *J. Exp. Biol.*, 214: 2014-2019.

[23] Viitala J., E. Korpimaki, P. Palokangas, M. Koivula. (1995) *Nature,* 373: 425-427.

[24] Koivula M., J. Viitala. (1998) Rough-egged buzzards use vole scent marks to assess hunting areas. *J. Avian. Biol.*, 30:329-281.

[25] Koivula M., E. Korpimaki. (2001) Do scent marks increase predation risk of microtine rodents? *Oikos,* 95:275-281.

[26] Kellie A., S. J. Dain, P. B. Banks. (2004) UV properties of Australian mammal urine. *J. Comp. Physiol.*, 190: 429-435.

[27] Lind O., M. Mitkus, P. Olsson, A. Kelber. (2013) UV sensitivity and color vision in raptor foraging. *J. Exp. Biol.*, 15: 1819-1826.

[28] Odeen A., Hastad O., Alstrom P. (2011) Evolution of ultraviolet vision in the largest avian radiation – the passerines. *BMC Evolutional Biology*, 11:313-320.

[29] Odeen A., Hastad O. (2003) Complex distribution of avian color vision systems revealed by sequencing the SWS1 opsin from total DNA. *Mol. Biol. Evol.,* 20:855-861.

[30] Cuthill I. C., Partridge J. C., Bennett A. T. D. (1999) UV vision and its function in birds. http://www.int-ornith union.org/files/proceedings/durban/Symposium/S45/ S45.4.htm

[31] Salcedo E., Zheng L., Phistry M., Bagg E. E., Britt S. G. (2003) Molecular basis for ultraviolet vision in invertebrates. *J. Neursci.*, 23(34):10873-10878.

[32] Rajchard J. (2009) Ultraviolet (UV) light perception by birds: a review. *Veterinarnia Medicina,* 54 (8): 351–359.

[33] Viitala J., Korpimaki E., Palokangas P., Koivula M. (1995)Attraction of Kestrels to vole scent marks visible in ultraviolet-light. *Nature,* 373, 425–427.

[34] Peichl L. (2005) Diversity of mammalian photoreceptor properties: adaptations to habitat and lifestyle? *Anat. Rec. a Discov. Mol. Cell Evol. Biol.*, 287: 1001-1012.

[35] Zhang J. (2003) Paleomolecular biology unravels the evolutionary mystery of vertebrate UV vision. *Proc. Natl. Acad. Sci.,* 100: 8045-8047.

[36] Szel A., P Rohlich, R. Caffe, T. V. Veen. (1996) Distribution of cone photoreceptors in the mammalian retina. *Microsc. Res. Tech.*, 35: 445-462.

[37] Gouras P., B. Ekesten. (2004) Why do mice have ultra-violet vision? *Exp. Eye res.*, 79: 887-892.

[38] Henriksson J. T., J. P. Bergmanson, J. E. Walsh. (2010) UV radiation transmittance of the mouse eye and its individual media components. *Exp. Eye Res.,* 90:382-387.

[39] Cooper G. F., J. G. Robson. (1996) The yellow color of the lens of man and other primates. *J. Physiol.*, 203: 411-417.

[40] Jacobs G. H., G. A. Williams. (2007) Contribution of the mouse UV pigment to the ERG and to vision. *Doc. Ophthal.*, 115: 137-144.

[41] Jacobs G. H., G. A. Williams, J. A. Fenwick. (2004) Influence of cone pigment coexpression on spectral sensitivity and color vision in the mouse. *Vis. Res.,* 44: 1615-1622.

[42] Huberman A. D., C. M. Niell. (2011) What can mice tell us about how vision works? *Trends Neurosci.*, 34: 464-473.

[43] Jacobs G. H., J. A. Fenwick, F. A. Williams. (2001) Cone-based vision of rats for UV and visible lights. *J. Exp. Biol.*, 204: 2439-2446.

In: UV Radiation: Properties, Effects, and Applications
Editor: James A. Radosevich
ISBN: 978-1-63321-090-5

Chapter 10

DEGRADATION OF EMERGING AROMATIC MICROPOLLUTANTS BY UV-BASED OXIDATION PROCESSES

M. Sánchez-Polo[1*], J. Rivera-Utrilla[1], R. Ocampo-Perez[1], J. J. López-Peñalver[1], I. Velo-Gala[1] and M. M. Abdel Daiem[1,2]
[1]Departamento de Química Inorgánica, Facultad de Ciencias, Universidad de Granada, Granada, España
[2]Environmental Engineering Departement, Faculty of Engineering, Zagazig Univeristy, Zagazig, Egypt

ABSTRACT

Ultraviolet (UV) radiation is frequently applied to disinfect water intended for human consumption and wastewater. Due to the greater chemical contamination of water, UV radiation is increasingly proposed as a technology to remove organic micropollutants, underlining its high efficacy to eliminate certain pesticides and pharmaceuticals from water. Significant advances have recently been made in our understanding of the photo-chemical processes undergone by organic contaminants and pharmaceuticals in aqueous medium. However, fewer data are available on their photochemical transformation. The objective of this chapter is to summarize the efficacy of ultraviolet (UV) radiation in the direct or indirect photodegradation of some emergent micropollutants. For this purpose, i) a kinetic study was performed, determining the quantum yield of the process; and ii) the influence of the different operational variables was analyzed [initial concentration of pollutant, pH, presence of natural organic matter compounds, radicals promoters addition ($K_2S_2O_8$, H_2O_2, activated carbon, activated carbon/TiO_2) and chemical composition of water], and iii) the time course of total organic carbon (TOC) concentration and toxicity during micropolluntant photodegradation was studied. The very low quantum yields obtained for the compounds studied are responsible for the low efficacy of the quantum process during direct photon absorption in micropollutant phototransformation. The R_{254} values obtained show that the dose habitually used for water dis-infection is not sufficient

* Corresponding author: M. Sánchez-Polo. Departamento de Química Inorgánica, Facultad de Ciencias, Universidad de Granada, Granada, 18071, España. Tel: 34-958242888, Fax: 34-958248526, E-mail: mansanch@ugr.es.

to remove this type of compounds; therefore, higher doses of UV irradiation or longer exposure times are required for their removal.

The concentration of organic micropolluntants has a major effect on their photodegradation rate. The study of the influence of pH on the values of parameters ε (molar absorption coefficient) and k'E (phodegradation rate constant) showed no general trend in the behavior of emergent pollutants as a function of the solution pH. The components of natural organic matter, gallic acid (GAL), tannic acid (TAN), and humic acid (HUM), may act as promoters and/or inhibitors of $OH^{•}$radicals via photoproduct ion of H_2O_2. It is interesting to note that the addition of radical promoters such as H_2O_2 or $K_2S_2O_8$ markedly increased the effectiveness of UV radiation through the generation of $HO^{•}$ or $SO_4^{•-}$ radicals, respectively.

Regardless of the system considered, the results obtained showed that the micro-pollutant degradation rate was higher with lower concentrations. The solution pH had a major effect on micropollutant degradation with the UV/H_2O_2, and UV/$K_2S_2O_8$ systems. The presence of activated carbon during the pharmacecutical photodegradation process markedly increases the removal rate. The results obtained indicate that activated carbon exerts the greatest synergic effect on diatrizoic acid removal by the UV/AC system, with a synergic contribution >53 % at one minute of treatment. Moreover, the presence of activated carbons with a high carboxyl groups content enhances 2,4-D photodegradation by the UV/TiO_2 system.

Carboxyl groups in the graphene planes of the activated carbon participate in the additional generation of $OH^{•}$ radicals by interacting with the electrons produced by the UV/TiO_2 system.

Keywords: UV radiadion, natural organic matter, total organic carbon, micropollutants

ABBREVIATIONS

CTC	Chlorotetracycline
Cyt	Cytarabine
DMZ	Dimetridazole
DTZ	Diatrizoate
GAL	gallic acid
HUM	humic acid
MNZ	Metronidazole
NOM	natural organic matter
OTC	Oxytetracycline
RNZ	Ronidazole
TAN	tannic acid
TC	Tetracycline
TNZ	Tinidazole
TOC	total organic carbon
UV	Ultraviolet

10.0. INTRODUCTION

Ultraviolet (UV) radiation is frequently applied to disinfect drinking water and wastewater. Due to the greater chemical contamination of water, UV radiation is increasingly proposed as a technology to remove organic micropollutants, underlining its high efficacy to eliminate certain pesticides and pharmaceuticals from water.

Significant advances have recently been made in our understanding of the photochemical processes undergone by organic contaminants and pharmaceuticals in aqueous medium [1-3]. However, fewer data are available on their photochemical transformation [1].

The majority of pharmaceuticals are photo-active, i.e., able to absorb light. This is because their structures generally contain aromatic rings, heteroatoms, and other functional groups that make them prone to absorb UV-VIS radiation (direct photolysis) or to react with photosensitizing species capable of inducing pharmaceutical photodegradation in natural water (indirect photolysis).

During direct photolysis, photon absorption gives rise to compounds in excited electronic states that are susceptible to chemical transformation.

However, indirect or sensitized photolysis leads to the transformation of contaminants by energy transference or by chemical reactions with transitory species formed by the presence of light, such as hydroxyl radicals (HO), singlet oxygen ($^1O_2$), and triplet excited states of natural organic matter ($^3NOM^*$) [4-8].

UV-based advanced oxidation processes (AOPs) have proven highly effective to degrade organic compounds in aqueous solution. AOPs generate species with high oxidizing power (e.g., $HO^\bullet$ radicals) that interact with the pollutant and degrade it into byproducts with lower molecular weight, and they can even achieve its complete mineralization. Hydroxyl radicals have been used to eliminate organic micropollutants from aqueous phase in various studies.

Ultraviolet radiation (UV) combined with H_2O_2 is more effective at BPA degradation than with direct photolysis, obtaining a reaction rate constant of hydroxyl radical ($HO^\bullet$) with BPA of $1.02 \pm 0.23\times10^{10}$ $M^{-1}s^{-1}$ [9]. Unlike the hydroxyl radical, the sulfate and carbonate radicals are selective transient species against organic compounds with a high oxidization potential ($E^0 = 2.6$ V and $E^0 = 1.78$ V, respectively).

The most frequent methods to generate sulfate radicals, $SO_4^{\bullet-}$, are by the photochemical, thermal, or chemical decomposition of $S_2O_8^{2-}$ .Various studies have found the $UV/K_2S_2O_8$ system to be more effective than the UV/H_2O_2 system to treat organic compounds in aqueous solution, offering a higher percentage degradation within a shorter treatment time and reducing the medium toxicity to a greater extent [10, 11]. For its part, the carbonate radical is one of the most extensively studied inorganic radicals, due to its ubiquitous nature in the environment and relatively long lifetime. However, there has been little research on the use of this radical species to degrade organic compounds.

There is currently little knowledge of: i) the mechanism involved in aromatic contaminant photooxidation, ii) the influence of the different operational variables or, iii) the role in this process of the chemical matrix of water, especially dissolved organic matter.

Hence, the objective of the present study was to summarize the efficacy of UV radiation in the direct photooxidation of organic micropollutants (cytarabine, tetracycline, oxytetracycline, chlorotetraciclyne, diatrizoic acid, 2,4-Dichlorophenol metronidazol, dimetridazol, tinidazol, ronidazol). For this purpose: i) a kinetic study was conducted to determine the quantum yield of the process and, ii) the influence of the different operational variables (initial concentration of antibiotic, pH, presence of NOM components and chemical composition of water) was analyzed and, iii) the time course of total organic carbon (TOC) concentration and toxicity during nitroimidazole photodegradation was studied.

Moreover, the effectiveness of $HO^{\bullet}$, $SO_4^{\bullet-}$ radicals in the photodegradation of organic micropollutant in aqueous solution has been investigated by studying the influence of some operational parameters, i.e., initial concentration of micropollutant, initial concentration of H_2O_2 or $K_2S_2O_8$, solution pH, and chemical composition of the water.

A further aim was to determine the variation in the total organic carbon and in the toxicity of emergent micropollutant photodegradation byproducts during these treatments.

Finally, the effectiveness of the systems of UV/activated carbon and UV/TiO_2/activated carbon to remove organic micropollutants has been analyzed. Some of these results have already been published [10-21].

10.1. Results and Discussion

10.1.1. Direct Photodegradation of Organic Micropollutants

Figure 1 depicts, as an example, the kinetics of direct nitroimidazole photodegradation with low-pressure Hg lamp (254 nm). At 3 h of treatment, the concentration of all nitroimidazoles was reduced by 70-85%. A pseudo-first order kinetic model was applied to the above results to determine the nitroimidazole photodegradation rate constant.

Equation (1) [22] was then used to calculate the quantum yield (Φ) for each nitroimidazole. Table 1 shows the results obtained.

$$\Phi_\lambda = \frac{k_\lambda}{2.303 \cdot E_\lambda \cdot \varepsilon_\lambda} \tag{1}$$

In this equation, k_λ is the photodegradation rate constant (s^{-1}); E_λ is the rate of energy emitted, corresponding to the photon flow emitted by the lamp (Einstein$\cdot s^{-1} \cdot m^{-2}$); ε_λ is the molar absorption coefficient at the wavelength in question ($m^2 \cdot mol^{-1}$); and Φ_λ is the quantum yield (mol$\cdot$Einstein^{-1}).

The rate of energy irradiated by the lamp was determined by actimometry, using a solution of 5 μM atrazine as actinometer [5] and obtaining an energy of $1.027 \cdot 10^{-4}$ Einstein$\cdot s^{-1} \cdot m^{-2}$ for the lamp used. For this purpose, the quantum yield of atrazine was considered to be 0.046 mol$\cdot$Einstein^{-1} and the molar absorption coefficient to be 386 $m^2 \cdot mol^{-1}$ at 254 nm wavelength [23].

Table 1 shows the values of the study parameters, showing a very low quantum yield for all the compounds studied, responsible for a low efficacy of the quantum process in nitroimidazole phototransformation. This low efficacy confirms the need for long irradiation times to achieve their complete removal. The values obtained are similar to those reported by other authors for pharmaceutical compounds [9, 18, 24-26].

The value of $(34.7\pm1.8)\cdot10^{-4}$ mol·Einstein^{-1} obtained for MNZ is close to the value of 0.0033 mol·Einstein^{-1} previously reported [27].

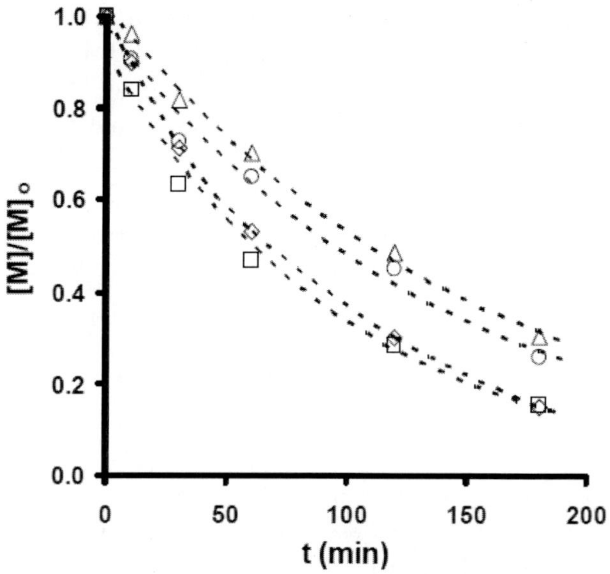

Figure 1. Removal kinetics of the four nitroimidazoles by means of UV photodegradation. [nitroimidazole] o= 200 μM, pH= 5-7, T= 298 K; (◇) MNZ; (□) DMZ; (△) TNZ; (○) RNZ.

Table 1. Parameters obtained from direct irradiation at 254 nm of the pharmaceutical compounds studied

Compound	ε ($m^2\cdot mol^{-1}$)	k x 10^4 (s^{-1})	Φ x 10^4 (mol·$Eins^{-1}$)	k'_E ($m^2\cdot Eins^{-1}$)	R_{254} (%)
Metronidazole (MNZ)	209	1.72 ± 0.09	34.70 ± 1.80	1.68 ± 0.09	0.14 ± 0.01
Dimetridazole (DMZ)	224	1.670 ± 0.20	31.50 ± 3.90	1.63 ± 0.20	0.14 ± 0.02
Tinidazole (TNZ)	233	1.09 ± 0.09	19.60 ± 1.70	1.06 ± 0.09	0.09 ± 0.01
Ronidazole (RNZ)	226	1.18 ± 0.18	22.10 ± 3.40	1.15 ± 0.18	0.10 ± 0.01
Tetracycline (TC)	1203	18.34 ± 2.12	9.20 ± 1.00	15.67± 0.18	0.24 ± 0.01
Chlorotetracycline (CTC)	1216	40.45 ± 3.78	14.30 ± 1.00	38.98 ± 0.18	0.56 ± 0.02
Oxytetracycline (OTC)	1744	36.32 ± 1.34	18.20 ± 0.60	32.76 ± 0.18	0.49 ± 0.01
Cytarabine (Cyt)	968	1.60±0.09	0.06± 0.06	1.4 ± 0.18	0.00 ± 0.01

For comparative purposes, it is essential to consider the apparent photodegradation rate constant normalized by the energy of the lamp, k'_E ($m^2\cdot$Einstein^{-1}), using Equation (2). This constant is independent of the fluctuations in energy irradiated by the lamp and permits direct comparisons among phototransformation rate constants obtained with different photoreactors [24].

$$k'_E = \frac{k_\lambda}{E_\lambda} \tag{2}$$

In this equation, k_λ (s^{-1}) is the photodegradation rate constant of contaminants used in Equation 1, and E_λ ($Einstein \cdot s^{-1} m^{-2}$) is the radiant energy emitted by the lamp at a wavelength of 254 nm, calculated by actinometry [5]. The k'_E values obtained are shown in Table 1. Table 1 also shows the percentage micropollutants removal for an irradiation dose of 400 $J \cdot m^{-2}$ (R_{254}). This parameter determines the applicability of UV radiation in contaminant photo-degradation under the real conditions of a water treatment plant. An irradiation dose of 400 $J \cdot m^{-2}$ was selected as reference because it is the minimum value recommended by different European bodies for water disinfection [28-30]. With the present lamp, the time required to reach an irradiation dose of 400 $J \cdot m^{-2}$ is 8.3 s, and the radiant energy is equivalent to $8.49 \cdot 10^{-4}$ $Einstein \cdot m^{-2}$ for a wavelength of 254 nm. Hence, the percentage of nitroimidazole removed with this irradiation dose, R_{254}, can be calculated by means of Equation (3).

$$R_{254} = 1 - e^{-(k'_E \times 8.49 \cdot 10^{-4})} \tag{3}$$

In this equation, k'_E is the apparent photodegradation rate constant normalized by the energy of the lamp (Equation (2)). As shown in Table 1, a very low percentage of compound removal was achieved at the minimum irradiation dose (400 $J \cdot m^{-2}$). Hence, the usual dose for water disinfection is not adequate to remove this type of pharmaceutical, which requires higher doses of UV irradiation or longer exposure times to be eliminated by direct photolysis.

Two key parameters of the efficacy of any treatment system are: i) the concentration of total organic carbon and, ii) the toxicity of the oxidation subproducts generated. Figure 2 shows, as an example, the time course of TOC and toxicity values during the photodegradation of the four nitroimidazoles under study. According to the results in Figure 2, although the nitroimidazole concentration considerably decreases with radiation time, oxidation subproducts do not transform into CO_2 to the desired extent.

Consequently, they generate fractions of smaller molecular weight than the original nitroimidazole and maintain the TOC concentration constant throughout the treatment time. Moreover, as shown in Figure 2, the mixture of subproducts generated during nitroimidazole photodegradation sometimes shows higher toxicity than the original product. Hence, nitroimidazole photodegradation may give rise to compounds that are pharmacologically active and have higher toxicity than the original compound. We highlight the results obtained for TNZ, which show a substantial increase in toxicity at 3 h of treatment. Similar results have been obtained for the rest of the compounds studied [2, 10, 11, 13-15, 17-20].

10.2. Influence of the Different Operational Variables on Nitroimidazole Photodegradation

This section analyzes the effects on compounds photodegradation performance of the operational parameters studied: micropollutant concentration, solution pH and presence of natural organic matter.

10.2.1. Influence of Nitroimidazole Concentration

Figure 3 depicts, as an example, the quantum yield of nitroimidazoles as a function of their concentration, showing that the initial concentration affects the quantum yield and therefore the photodegradation rate.

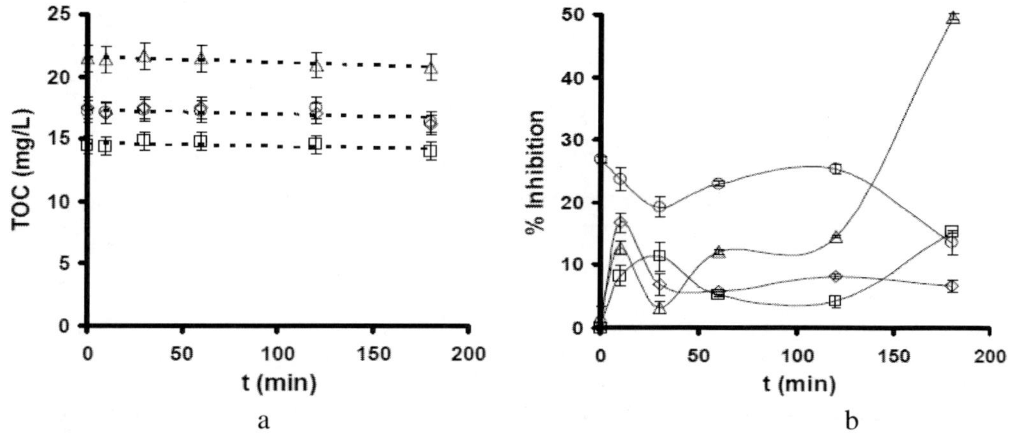

Figure 2. Time course of the concentration of total organic carbon (a) and toxicity (b) during UV irradiation of the four nitroimidazoles in ultrapure water. [nitroimidazole]$_o$= 200 μM, pH= 5-7, T= 298 K; (◇) MNZ; (□) DMZ; (△) TNZ; (○) RNZ.

The decrease in photodegradation rate in nitroimidazoles with the increase of their concentration is related to the energy absorbed by each nitroimidazole molecule.

Hence, given that the radiation energy deposited in the medium per unit volume is constant, nitroimidazole molecules can accept more radiant energy at lower concentrations, explaining the behavior observed. Similar results have been observed for the rest of the compounds studied [10-15, 18-20].

10.2.2. Influence of Solution pH

The influence of the solution pH in nitroimidazole photodegradation was analyzed in experiments with pH values ranging from 2 to 9. Based on the pK_a of each nitroimidazole and the distribution of species that each presents, the four nitroimidazoles are in their cationic and/or neutral form in the pH range studied. Figure 4 shows the variation in global molar absorption coefficient (ε) and global photodegradation rate constant (k'_E) as a function of the solution pH for each nitroimidazole. Interestingly, the changes observed are determined by the pK_a of the antibiotic. According to previous Equation, an increase in the molar absorption coefficient (ε) would produce an increase in the k'_E value, but the results show no general tendency that describes the influence of pH on the ε and k'_E values of these nitroimidazoles.

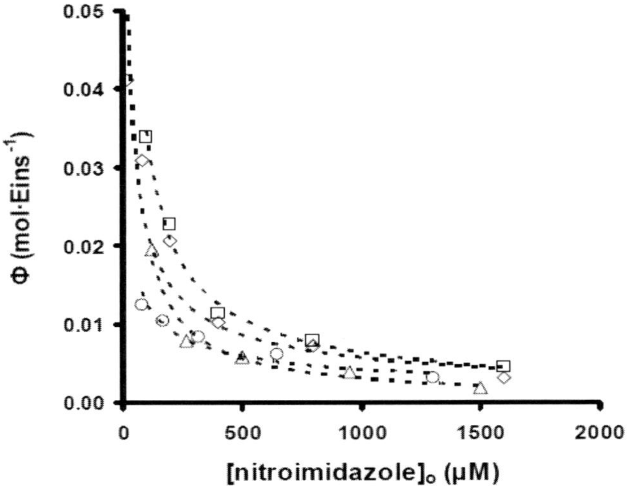

Figure 3. Quantum yield of the photodegradation process as a function of initial nitroimidazole concentration. pH= 5-7, T= 298 K; (◇) MNZ; (□) DMZ; (△) TNZ; (○) RNZ.

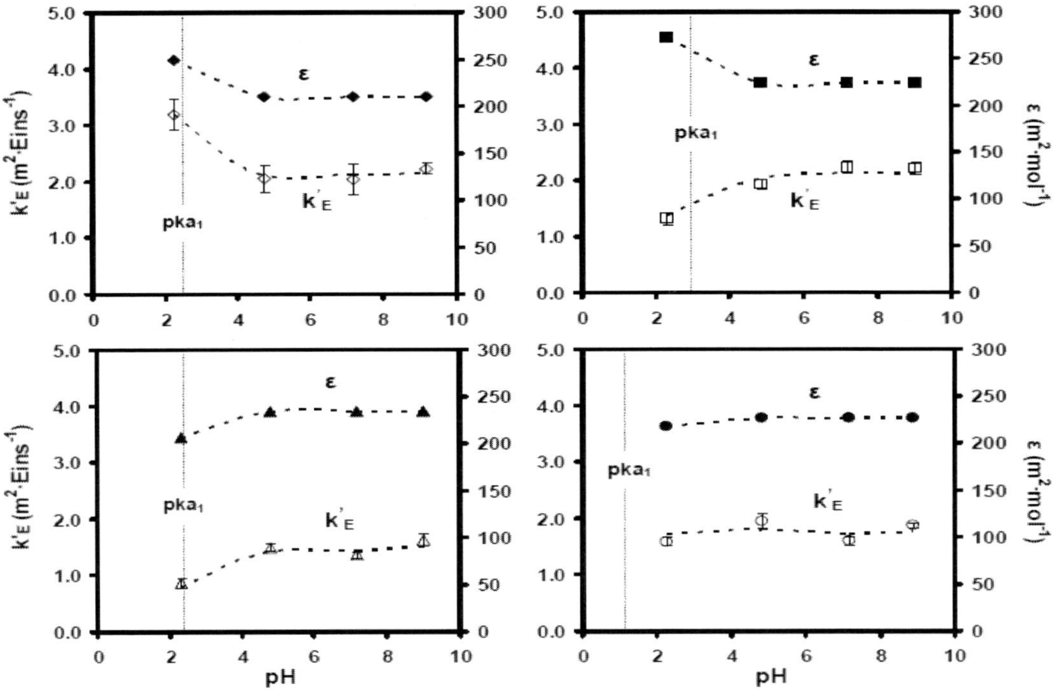

Figure 4. Molar absorption coefficient (ε) and apparent photodegradation constant normalized by the energy emitted by the lamp (k'_E) as a function of solution pH for the four nitroimidazoles studied. T= 298 K; (◇) MNZ; (□) DMZ; (△) TNZ; (○) RNZ.

In fact, the variation in trends among these nitroimidazoles differentiates four distinct behaviors:

a MNZ: increase in ε and k'_E at pH< 4
b DMZ: increase in ε and decrease in k'_E at pH< 5
c TNZ: decrease in ε and k'_E at pH< 5
d RNZ: ε and k'_E remain virtually unchanged in the studied pH range.

The variations in these parameters as a function of the medium pH are similar to those observed for other pharmaceutical compounds [24].

This global behavior was studied in greater depth by quantitative analysis of the photodegradation of each nitroimidazole species in the medium as a function of the pH, using Equation 4, which is obtained by combining Equations (1) and (3).

$$k'_{Ej} = 2.303 \cdot \varepsilon_j \cdot \Phi_j \tag{4}$$

In this equation, k'_{Ej} is the apparent rate constant for each nitroimidazole species "j" present in the solution. The contribution of each of each species towards total k'_E can be represented by Equation (6):

$$k'_E = \sum_j \alpha_j \cdot k'_{Ej} \tag{5}$$

In this equation, α_j represents the molar fraction of each of the species ($\Sigma_j \alpha_j = 1$). Once the species distribution and pK_a of each nitroimidazole is known, the rate constant and molar absorption coefficient can be calculated by means of linear regression:

$$k'_E = (1-\alpha_2) \cdot k'_{E1} + \alpha_2 \cdot k'_{E2} = k'_{E1} + (k'_{E2} - k'_{E1})\alpha_2 \tag{6}$$

$$\varepsilon = (1-\alpha_2) \cdot \varepsilon_1 + \alpha_2 \cdot \varepsilon_2 = \varepsilon_1 + (\varepsilon_2 - \varepsilon_1)\alpha_2 \tag{7}$$

After calculating k'_{Ej} and ε_j for the ionic (j=1) and neutral (j=2) species of each nitroimidazole, the quantum yield of the corresponding species can be obtained by means of Equation (8).

$$\Phi_j = \frac{k'_{Ej}}{2.303 \cdot \varepsilon_j} \tag{8}$$

Table 2 shows the results obtained by applying these equations. Using these parameters, the values of k'_E and Φ can be determined for any pH value in the studied range. As shown in Table 2, the rate constant (k'_E) and quantum yields (Φ) values obtained for the species of each nitroimidazole vary by only one order of magnitude, with minimum values for the ionic form of TNZ and maximum values for the ionic form of MNZ.

Table 2. Kinetic parameters of direct phototransformation of ionic (j = 1) and neutral (j = 2) species of nitroimidazoles

Nitro-imidazole	pk_{a_1}	ε_1 ($m^2 \cdot mol^{-1}$)	k'_{E_1} ($m^2 \cdot Eins^{-1}$)	$\Phi_1 \cdot 10^3$ ($mol \cdot Eins^{-1}$)	ε_2 ($m^2 \cdot mol^{-1}$)	k'_{E_2} ($m^2 \cdot Eins^{-1}$)	$\Phi_2 \cdot 10^3$ ($mol \cdot Eins^{-1}$)
MNZ	2.58	271.18	3.78	6.05	209.72	2.10	4.35
DMZ	2.81	288.12	1.33	2.01	224.45	1.73	3.35
TNZ	2.30	178.88	0.28	0.69	233.91	1.42	2.64
RNZ	1.32	171.31	0.31	0.79	226.13	1.80	3.46

10.2.3. Influence of the Presence of Gallic, Tannic or Humic Acid

The influence of natural organic matter (NOM) components during nitroimidazole photodegradation was analyzed with experiments of MNZ photodegradation in the presence of three components of NOM (gallic acid (GAL), tannic acid (TAN) and humic acid (HUM)). The concentration of NOM in waters ranges from approximately 0.1 mg to >100 mg of total organic carbon (TOC) per liter, depending on their origin [31-35]. Knowledge of the amount of light absorbed by the different components of NOM is necessary to study the influence of humic matter during UV irradiation of MNZ and its effects on the removal rate. According to the transmittance data obtained, the amount of UV light absorbed by NOM can be known, hence reducing the number of photons that reach the MNZ molecules [24].

The little transmittance shown by gallic (GAL), tanic (TAN) and humic (HUM) acid at high concentrations will have a very negative effect on the direct photodegradation of MNZ. The results obtained show that the presence of GAL has a markedly different effect on MNZ degradation in comparison to the presence of TAN or HUM. MNZ removal is favored only at a GAL/MNZ concentration ratio < 1.4. Moreover it was observed that low concentrations of GAL in the medium favor MNZ photodegradation, despite the lesser transmittance of GAL versus TAN or HUM. '

Therefore, GAL may act as a promoter of OH· radicals, which oxidize MNZ molecules. In contrast, the behavior in the presence of TAN and HUM suggests a pre-dominant $OH^{\bullet}$radical inhibition effect, due to their complex structure and the high reactivity of NOM against $OH^{\bullet}$ radicals ($k_{OH^{\bullet}} = 10^8\ M^{-1}s^{-1}$). More information about the role of natural organic matter on pharmaceutical compounds degradation can be found elsewhere [15].

10.3. MICROPOLLUTANTS DEGRADATION BY UV-BASED ADVANCED OXIDATION PROCESSES

10.3.1. UV/H_2O_2 Process

The effectiveness of the UV/H_2O_2 system was assessed studying the influence of the different operational variables (micropollutant concentration, pH, H_2O_2 concentration, and chemical composition of water) on micropollutant degradation kinetics. The reaction rate constant of each compound with HO radicals was also determined.

The results obtained are presented in Table 3. It can be observed that the values obtained are very similar regardless of the chemical compound considered.

Table 3. Reaction rate constants of some pharmaceutical compounds and HO·radicals

Compound	k x 10^{10} (M^{-1} s^{-1})
Metronidazole (MNZ)	4.40
Dimetridazole (DMZ)	5.60
Tinidazole (TNZ)	4.50
Ronidazole (RNZ)	13.90
Tetracycline (TC)	0.63
Chlorotetracycline (CTC)	0.56
Oxytetracycline (OTC)	0.38
Cytarabine (Cyt)	3.50
Diatrizoate (DTZ)	1.20

Table 4. Experimental conditions and apparent reaction rate constants for cytarabine degradation by UV/H_2O_2 in Milli-Q water at T = 25°C

[Cyt] (mg L^{-1})	[H_2O_2] (μM)	pH	k_{Ap}x10^3 (min^{-1})	% degradation of Cyt at 120 min.
10	0	7	0.9	10
10	100	7	4.2	40
10	200	7	8.1	62
10	300	7	11.0	73
10	400	7	18.6	90
10	500	7	19.3	92
10	1000	7	33.9	98
20	200	7	4.0	38
5	200	7	14.1	80
3	200	7	25.8	95
1	200	7	45.3	99
10	400	2	5.5	48
10	400	4	18.3	89
10	400	6	18.6	92
10	400	8	21.0	91

Table 4 shows, as an example, the experimental conditions for each study variable in the case of cytarabine micropollutant. The degradation kinetics of cytarabine was interpreted by a pseudo first order kinetic model, represented by the following equation:

$$\frac{d[Cyt/Cyt_0]}{dt} = -k_{Ap}[Cyt/Cyt_0] \tag{9}$$

Table 4 also depicts the values of apparent reaction rate constants obtained by fitting the Equation (9) for the experimental data.

The results show that the addition of increasing amounts of H_2O_2 (Table 4) increased the cytarabine degradation. Thus, the addition of 300 μM of H_2O_2 increased the apparent reaction rate constant k_{Ap} around 12-fold from 0.9×10^{-3} to 11×10^{-3} min^{-1}. This is due to the generation of $HO^\bullet$ radicals, which attack cytarabine by degrading it into byproducts of smaller molecular weight. The cytarabine removal rate also increased at higher H_2O_2 concentrations. This improves the effectiveness of the system, since UV capture by H_2O_2 is greater at higher H_2O_2 concentration, favoring the generation of $HO^\bullet$ radicals. However, results in Table 1 show that the addition of very high concentrations of H_2O_2 does not produce an increase in cytarabine degradation rate of the same magnitude. This is because $HO^\bullet$ radicals have low selectivity and, at high H_2O_2 concentrations, initiate secondary reactions in which the $HO^\bullet$ radical inhibits or, recombines in accordance with Equations (10) to (13).

$$HO^\bullet + H_2O_2 \rightarrow H_2O + HO_2^\bullet \tag{10}$$

$$HO_2^\bullet + H_2O_2 \rightarrow HO^\bullet + H_2O + O_2 \tag{11}$$

$$2HO_2^\bullet \rightarrow H_2O_2 + O_2 \tag{12}$$

$$HO_2^\bullet + HO^\bullet \rightarrow H_2O + O_2 \tag{13}$$

The effect of the initial cytarabine concentration was studied by performing experiments at different initial cytarabine concentrations and at the same initial H_2O_2 concentration. Table 4 depicts the results, which show that degradation kinetics accelerated with the decrease in cytarabine concentration due to an increase in the $[HO^\bullet]/[Cyt]_0$ ratio. Table 4 also depicts the effect of medium pH on cytarabine degradation rate, showing a considerably reduced cytarabine degradation at pH = 2, with only 48% removal after 2 h of irradiation, compared to values > 89% at pH values of 4, 6, 7, and 8.

This can be attributed to the inhibiting character of chloride ions present in the system due to acidification of the medium. In accordance with Equation (14), these ions can react with $HO^\bullet$ radicals and reduce the concentration available for cytarabine degradation.

$$Cl^- + HO^\bullet \rightarrow HClO^{-\bullet} \quad k = 4.3\times10^9\ L\ mol^{-1}\ s^{-1} \tag{14}$$

The radical formed in Equation (14) presents lower reactivity than the $HO^\bullet$ radical, implying that it does not participate in cytarabine degradation, explaining the decrease in degradation rate. Moreover, the $HO^\bullet$ radicals can recombine at acidic pH and form H_2O_2, as indicated in Equations (12) and (13), reducing their effectiveness in cytarabine oxidation. Furthermore, most the oxidation reactions of organic compounds with $HO^\bullet$ in aqueous phase are disadvantaged at acidic pH due at the generation of H^+.

In order to analyze the applicability of UV/H_2O_2 system for cytarabine degradation, experiments were conducted with ultrapure water, groundwater, and wastewater.

The properties and characteristics of these types of water and the apparent reaction rate constant for the removal of cytarabine and $HO^{\bullet}$ inhibition rates are listed in Table 5.

It is showed that the apparent reaction rate constants (cytarabine removal) were 50% lower for groundwater and wastewater than for ultrapure water. In order to interpret the results in Table 5, we determined the inhibition rates of $HO^{\bullet}$ radicals by the species present in the medium, using Equation (15).

$$r_{HO^{\bullet}} = k_{H^+}[H^+] + k_{TOC}[TOC] + k_{HCO_3^-}[HCO_3^-] \qquad (15)$$

where $r_{HO^{\bullet}}$ is the inhibition rate of $HO^{\bullet}$ radicals in s^{-1}. $k_{H^+} = 7\times10^9\ M^{-1}\ s^{-1}$, $k_{TOC} = 2\times10^8\ M_c^{-1}\ s^{-1}$ $k_{HCO_3^-} = 8.5\times10^6\ M^{-1}\ s^{-1}$ [32-34]. $[H^+]$, [TOC], and $[HCO_3^-]$ are initial concentrations of each species present in the water. M_c is the molarity of natural organic matter, based on the moles of carbon, assuming 12 g mol^{-1}.

Results in Table 5 show that: i) the wastewater had the greatest radical inhibiting capacity, reducing the concentration of $HO^{\bullet}$ radicals available to react with cytarabine and, ii) the wastewater had the lowest light transmittance, absorbing the UV radiation and considerably reducing the number of photons reaching the H_2O_2.

These findings confirm that the organic matter in wastewater acts as a filter of UV light, reducing the effectiveness of the treatment to remove cytarabine from the medium.

The reduced k_{Ap} value of groundwater in comparison to ultrapure water was mainly due to the greater $HO^{\bullet}$ radical inhibition capacity of the species that it contains.

The effectiveness of a system to treat water polluted with organic compounds is dependent on: i) the transformation of dissolved organic carbon into CO_2 (pollutant mineralization) and, ii) the low or nil toxicity of degradation byproducts. Figure 5 shows the time course of TOC as a function of treatment time for the three types of water studied. Regardless of the chemical composition of the water, the TOC concentration in the system reduced with longer treatment time, confirming that the UV/H_2O_2 system has sufficient oxidizing power to mineralize part of the dissolved organic matter. In absolute terms, the greatest reduction in TOC concentration was observed in the wastewater, indicating that a 55% of the initial organic matter was transformed into CO_2 after 120 min of treatment.

Figure 6 shows the inhibition percentage of *Vibrio Fischeri* bacteria as a function of UV/ H_2O_2 treatment time for the three water types. Regardless of the chemical composition of the water, the percentage bacteria inhibition was greater after treatment, evidencing the formation of degradation byproducts more toxic than cytarabine.

Table 5. Chemical characteristics of the three types of water used, apparent reaction rate constants for the removal of cytarabine, and $HO^{\bullet}$ inhibition rates

Water type	pH	TOC ($mg\ L^{-1}$)	$[HCO_3^-]$ ($meq\ L^{-1}$)	$T^{a)}$ %	$k_{Ap}\times10^3$ (min^{-1})	$r_{HO^{\bullet}}$ (s^{-1})
Ultrapure	7.0	0.0	0.0	100	18.6	7.00×10^2
Groundwater	8.1	0.0	8.9	99	9.6	1.30×10^3
Wastewater	7.8	12.2	10.5	64	8.6	2.05×10^5

[a]Transmittance at 271 nm.

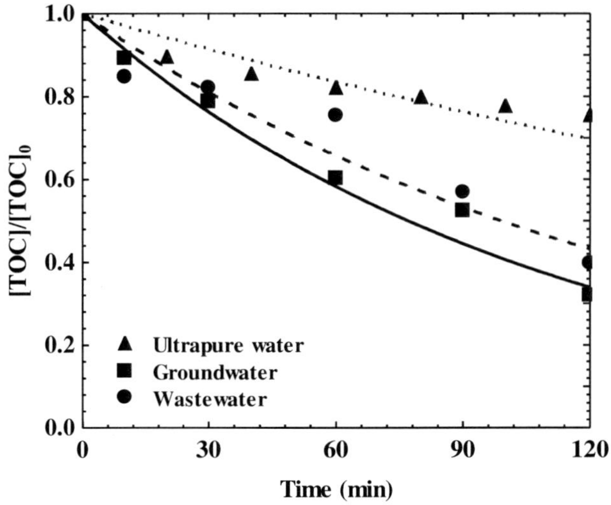

Figure 5. Effect of water type on the time course of TOC during treatment with UV/H_2O_2. $[Cytarabine]_0 = 10\ mg\ L^{-1}$, $[H_2O_2]_0 = 400\ \mu M$.

10.3.2. $UV/K_2S_2O_8$ Process

The reaction rate constants of each compound with sulfate radicals have been determined previously to understand better the efficiency of $UV/K_2S_2O_8$ process.

The results are presented in Table 6. It can be observed that the values obtained are very similar regardless of the chemical compound considered.

More information related to the use of $UV/K_2S_2O_8$ process for the degradation of pharmaceutical compounds can be found elsewhere [10, 11, 14, 19, 20].

In this manuscript the data obtained for cytarabine will be presented as an example of emerging pollutant degradation by $UV/K_2S_2O_8$ process.

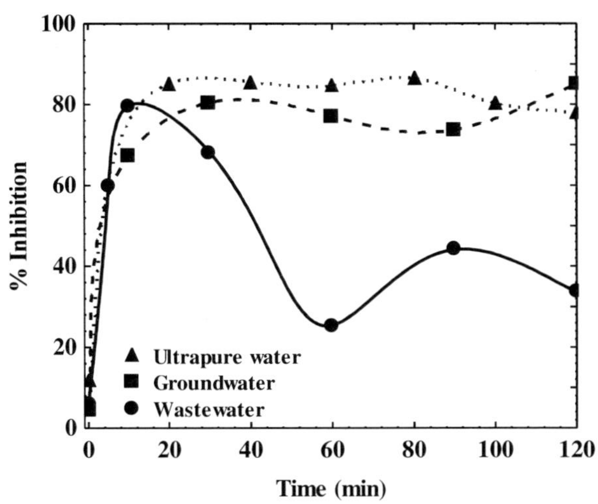

Figure 6. Time course of system toxicity during UV/H_2O_2 treatment of the different water types. $[Cytarabine]_0 = 10\ mg\ L^{-1}$, $[H_2O_2]_0 = 400\ \mu M$.

Table 6. Reaction rate constants of some pharmaceutical compounds and $SO_4^{2-}\cdot$ radicals

Compound	k x 10^9 (M^{-1} s^{-1})
Metronidazole (MNZ)	3.45
Dimetridazole (DMZ)	5.56
Tinidazole (TNZ)	6.43
Ronidazole (RNZ)	6.89
Tetracycline (TC)	3.67
Chlorotetracycline (CTC)	2.89
Oxytetracycline (OTC)	4,56
Cytarabine (Cyt)	1.61
Diatrizoate (DTZ)	0.65

10.3.2.1. Cytarabine Degradation by the UV/K2S2O8 System: Influence of Operational Variables

The effect of the initial $K_2S_2O_8$ concentration was assessed by performing experiments at an initial cytarabine concentration of 10 mg L^{-1} (40 μM) and $K_2S_2O_8$ concentrations of 100, 200, 300 400, 500, and 1000 μM (Table 7). The results obtained show that the addition of a small amount of $K_2S_2O_8$ produced a major increase in the degradation rate, obtaining 95% cytarabine degradation after 2 h of treatment with 100 μM of K_2S_2O. It can also be observed that, under the same conditions, the percentage cytarabine degradation was 2.5-fold higher with this system than with UV/H_2O_2. Table 7 also lists the values of the apparent reaction rate constants obtained. Under the above experimental conditions, these rate constants were on average nine-fold higher with this system than with UV/H_2O_2 although the reaction rate constant of cytarabine with the sulfate radical was slightly lower than the reaction rate constant with the hydroxyl radical. These results confirm the higher concentration of radicals generated for this system due to the formation of both radicals $SO_4^{\bullet-}$ and $HO^{\bullet}$.

The results presented in Table 7 show that the excessive addition of $K_2S_2O_8$ does not substantially increase the cytarabine degradation rate, attributable to the excessive generation of $SO_4^{\bullet-}$ radicals at high concentrations of $K_2S_2O_8$, which can recombine according to Equations (16, 17). Furthermore, as commented above, $HO^{\bullet}$ radicals generated in this system are also inhibited at high concentrations of $HO^{\bullet}$.

$$SO_4^{\bullet-} + SO_4^{\bullet-} \rightarrow S_2O_8^{-2} \; k = 4\times10^8 \; M^{-1} \; s^{-1} \quad (16)$$

$$SO_4^{\bullet-} + S_2O_8^{-2} \rightarrow SO_4^{-2} + S_2O_8^{\bullet-} \; k = 6.1\times10^9 \; M^{-1} \; s^{-1} \quad (17)$$

The effect of the initial cytarabine concentration was studied by performing experiments at different initial cytarabine concentrations (1, 2, 5, 10, and 20 mg L^{-1}) with the same initial $K_2S_2O_8$ concentration (200 μM).

The results obtained show that the degradation kinetics became much faster with a decrease in the cytarabine concentration, as in the UV/H_2O_2 system. However, at both low and high cytarabine concentrations, the $UV/K_2S_2O_8$ system was much more effective than the UV/H_2O_2 system, as demonstrated by the k_{Ap} values.

Table 7. Experimental conditions and apparent reaction rate constants for cytarabine degradation by UV/$K_2S_2O_8$ at T = 25 °C

[Cyt] (mg L^{-1})	[$K_2S_2O_8$] (μM)	Water type	pH	$k_{Ap}\times10^3$ (min^{-1})
10	0	Milli-Q	7.0	0.9
10	100	Milli-Q	7.0	26.0
10	200	Milli-Q	7.0	63.8
10	300	Milli-Q	7.0	166.5
10	400	Milli-Q	7.0	152.6
10	500	Milli-Q	7.0	164.0
10	1000	Milli-Q	7.0	304.0
20	200	Milli-Q	7.0	33.5
5	200	Milli-Q	7.0	157.0
2	200	Milli-Q	7.0	193.0
1	200	Milli-Q	7.0	366.0
10	200	Milli-Q	2.0	10.7
10	200	Milli-Q	4.0	56.3
10	200	Milli-Q	6.0	47.3
10	200	Milli-Q	8.0	40.2
10	200	Groundwater	8.1	22.9
10	200	Wastewater	7.8	17.1

The effect of the solution pH on cytarabine degradation by the UV/$K_2S_2O_8$ system was studied by performing experiments at pH = 2, 4, 6, 7, and 8 at the same initial cytarabine concentration and same $K_2S_2O_8$ concentration. Table 7 depict the results obtained showing that the cytarabine degradation was strongly affected by the pH. Thus, the lowest percentage cytarabine degradation was obtained, as with the UV/H_2O_2 system, at pH = 2, with only 50% removal after 1 h of irradiation. The drastic decrease in reaction rate at pH = 2 was attributable to several facts: i) reactions, directly related to the attack of cytarabine with radicals, are disadvantaged as a result of proton formation; ii) formation of additional acidic inorganic products, which are favorable at lower pH, such as H_2SO_5, H_2SO_4, $HS_2O_8^-$, HSO_4, which have a lower oxidation potential than $HO^•$ and $SO_4^{•-}$ radicals [24] and, iii) inhibition of $HO^•$ which are favored at acidic pH due the greater stability of H_2O_2.

The applicability of the UV/$K_2S_2O_8$ system for cytarabine degradation was analyzed by conducting experiments with groundwater, wastewater and ultrapure water (as with the UV/H_2O_2 system). The highest percentage degradation of cytarabine was obtained with ultra-pure water (98% after 1 h), followed by groundwater (75%) and wastewater (65%) (results not shown).

Table 7 also lists the apparent reaction rate constants, showing that as in the UV/H_2O_2 system the lowest degradation rate was obtained for wastewater and as commented above, this may be due to the screening effect produced by dissolved organic matter (absorbing part of the UV radiation) and to the radical scavenging capacity of some species contained in wastewater.

Figure 7 depicts TOC values as a function of treatment time, showing a reduction in TOC concentration for all three types of water. As in the UV/H_2O_2 system, the mineralization of organic compounds was highest in wastewater, followed by groundwater and ultrapure water.

These results indicate that the oxidizing power of UV/$K_2S_2O_8$ and UV/H_2O_2 systems is adequate to mineralize the dissolved organic matter.

Figure 8 depicts the percentage inhibition of *Vibrio Fischeri* bacteria during UV/$K_2S_2O_8$ treatment of the different water types, showing that the percentage inhibition remained virtually constant with longer treatment time, regardless of the type of water studied. These results indicate that the degradation byproducts generated are not excessively toxic. We highlight the much lower percentage inhibition of bacteria in this system than in the UV/H_2O_2 system, confirming its potential applicability in the removal of organic pollutants from water.

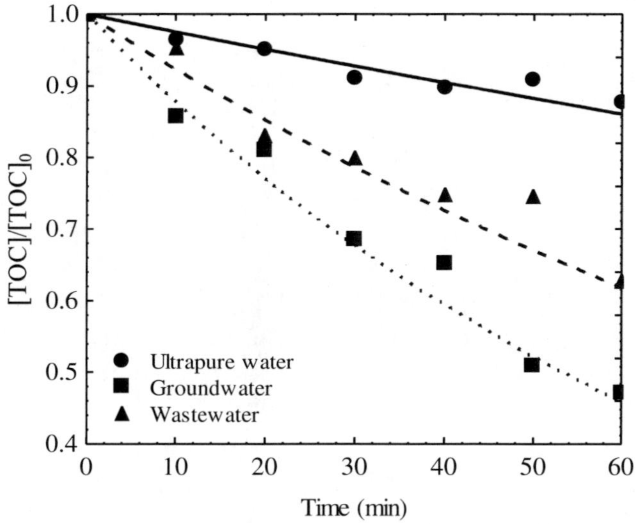

Figure 7. Time course of TOC with UV/$K_2S_2O_8$ treatment of the different types of water. $[\text{Cytarabine}]_0 = 10$ mg L^{-1}, $[K_2S_2O_8]_0 = 200$ μM.

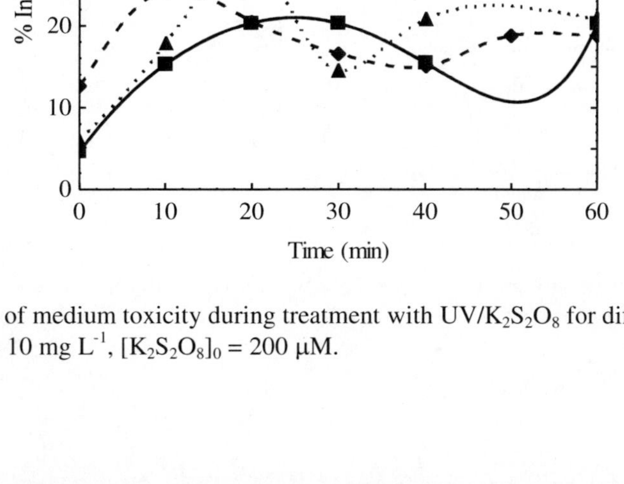

Figure 8. Time course of medium toxicity during treatment with UV/$K_2S_2O_8$ for different types of water. $[\text{Cytarabine}]_0 = 10$ mg L^{-1}, $[K_2S_2O_8]_0 = 200$ μM.

10.3.3. UV/TiO₂/Activated Carbon System

Figure 9 depicts the variation in percentage 2,4-D degradation as a function of treatment time with the UV/TiO_2 system in the presence of carbons S, M, and W, while Figure 9 depicts this variation in the presence of carbons $W_{O3\text{-}30}$ and $W_{O3\text{-}120}$.

Figure 9 shows that the highest percentage 2,4-D removal after 60 min of treatment was obtained with carbon S (80%), followed by carbons M (72%) and W (59%).

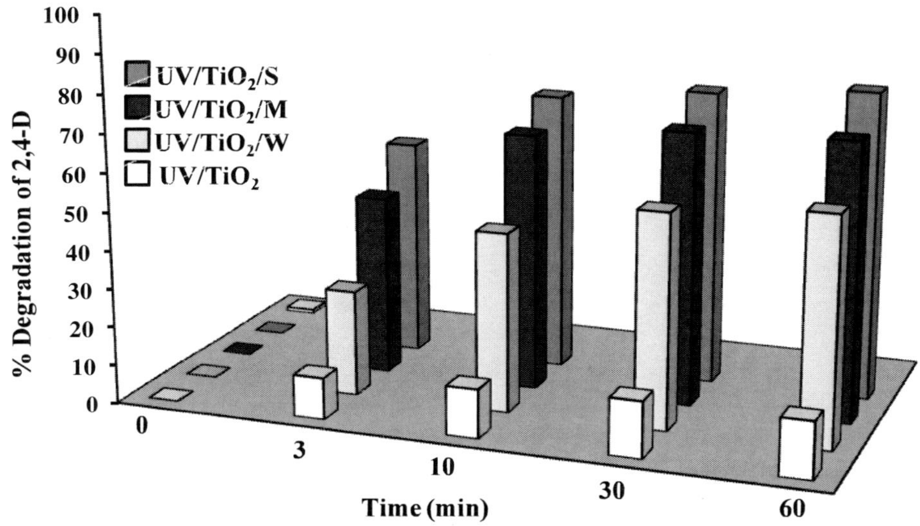

a

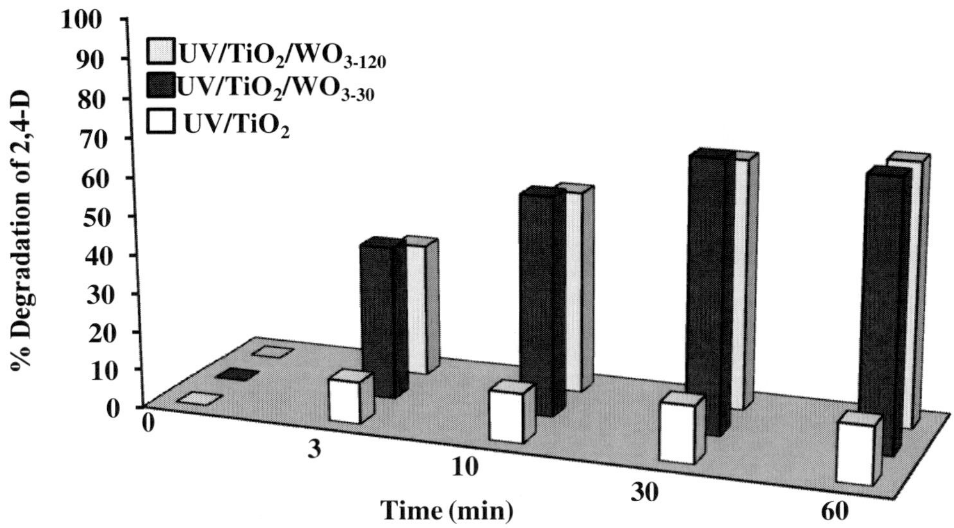

b

Figure 9. 2,4-D photodegradation by means of the UV/TiO_2 system in the presence of different activated carbons. T = 25° C, pH = 7, [2,4-D]= 50 mg/L, mass of TiO_2 and activated carbon = 5 mg, and V = 30 mL.

Comparing with the results in Figure 2b, the removal of 2,4-D was 1.1, 1.38, and 2.57-fold higher, respectively, with the combined UV/TiO_2/activated carbon process than with the adsorption process. The role of activated carbon in the UV/TiO_2/activated carbon system was analyzed by determining the contributions of adsorption and photocatalysis (UV/TiO_2) to the removal of 2,4-D at 60 min of the combined treatment.

The results in Table 3 show that: i) the greatest adsorptive contribution to the global removal process is made with carbons S and M, ii) with these two carbons (S and M), the contribution of the photocatalytic process (UV/TiO_2) is very low in comparison to the contribution of the adsorptive process and, iii) with carbon W, the total percentage removal obtained by the UV/TiO_2/activated carbon system is markedly higher than the result of summing the adsorptive and catalytic contribu-tions, indicating that its presence has a synergic effect on the 2,4-D removal process.

The results in Table 8 were compared with the chemical and textural properties of the activated carbons. No clear relationship was observed between the surface area and the synergic contribution to the global removal process, given that the three activated carbons have a similar surface area ($\cong$1200 m^2/g).

However, the content of carboxyl and phenol groups is higher in carbon W than in carbons S or M. Hence, the synergic activity of carbon W in the removal of 2,4-D by the UV/ TiO_2/activated carbon system may be related to the elevated carboxyl group content on its surface (0.32 meq/g vs. 0.00 meq/g on carbon S and 0.04 meq/g on carbon M).

The role of carboxyl groups in the photocatalytic process was examined by increasing the carboxyl group content of carbon W through oxidation with ozone for 30 min ($W_{O3\text{-}30}$) and 120 min ($W_{O3\text{-}120}$) and then texturally and chemically characterizing these samples. The ozone treatment reduced the surface area value by 50%, increased the content of carboxyl and lactone groups, and reduced the content of phenol groups.

Figure 9 depicts the photocatalytic degradation of 2,4-D in the presence of these ozonated samples ($W_{O3\text{-}30}$ and $W_{O3\text{-}120}$), demonstrating that the degradation was markedly enhanced by their presence, achieving 70% removal of 2,4-D after 60 min in both cases. This is much higher than the percentage removal obtained with the untreated carbon (W). As shown in Table 3: i) these ozonated carbon samples ($W_{O3\text{-}30}$ and $W_{O3\text{-}120}$) make a very low adsorptive contribution (<4%) to the global 2,4-D removal process and, ii) they make the highest synergic contribution to this global process (>50 %).

In order to identify the mechanism by which ozonated carbon samples enhance 2,4-D photocatalysis, we chemically characterized sample $W_{O3\text{-}30}$ by exposing it to UV and UV/ TiO_2 radiation in the absence of 2,4-D. The results in Table 8 show: i) similar values of surface acid groups between UV-treated and untreated samples and, ii) a lower concentration of carboxyl groups and markedly higher content of phenol and lactone groups in the UV/ TiO_2-treated *versus* untreated samples.

These results verify the proposed mechanism (reactions [1-3]), indicating that the main pathway by which activated carbons enhance 2,4-D removal by the UV/TiO_2/carbon system is the reduction of superficial carboxyl groups to aldehyde groups and finally to alcohol groups. Alcohol groups formed on the carbon surface may in turn react with adjacent carboxyl groups in the graphene planes of the activated carbon, generating lactone groups. This would explain the increase in lactone groups in sample $W_{O3\text{-}30}$ with the UV/TiO_2/$W_{O3\text{-}30}$

system. The increase in phenol groups in sample $W_{O3\text{-}30}$ would result from the decarboxylation of carboxyl groups by $OH^{\bullet}$ radicals, transforming them into phenol groups.

Table 8. Adsorptive and photocatalytic contributions to the removal of 2,4-D with the UV/TiO₂/activated carbon system

Activated Carbon	Total (%)	Adsorption (%)	UV/TiO_2 (%)	Synergic contribution (%)
S	80	72	14	0
M	72	56	14	2
W	59	23	14	22
$W_{O3\text{-}30}$	70	3	14	53
$W_{O3\text{-}120}$	70	4	14	52

More information can be found elsewhere [16, 17].

$$\text{-COOH} + H_2O \xrightarrow{e^-_{aq}} \text{-COH} + H_2O_2 \tag{18}$$

$$H_2O_2 + h\nu(UV) \rightarrow 2OH^{\bullet} \tag{19}$$

$$\text{-COH} + 2H_2O \xrightarrow{2e^-_{aq}} \text{-CH}_2\text{OH} + 2OH^{\bullet} \tag{20}$$

10.3.4. UV/Activated Carbon System (UV/AC)

After studying DTZ direct photolysis by UV radiation, we investigated the degradation of this compound when activated carbon was present during the photolytic process (UV/AC).

Figure 10 depicts DTZ removal by UV radiation and by the simultaneous use of UV radiation and activated carbons C, S, M, or W. It is observed that the removal rate is markedly increased by the presence of activated carbon.

Based on the degradation kinetics obtained, we calculated the values of the reaction rate constants in Table 3. The values of these constants confirm the increase in reaction rate observed in the degradation kinetics. This may be attributable to the contribution to the overall removal process of DTZ adsorption on the activated carbons.

For this reason, we obtained the DTZ adsorption kinetics on all of the activated carbon samples. Figure 11 shows, as an example, the results obtained for carbon W. We can observe that there is virtually no DTZ adsorption on W, although this is the activated carbon that produced the greatest increase in DTZ removal by the UV/AC system (Figure 11).

These results indicate that DTZ adsorption on activated carbon is not the sole cause for the increase in DTZ removal in the UV/AC system and that the activated carbon must make other types of contribution to the overall removal process.

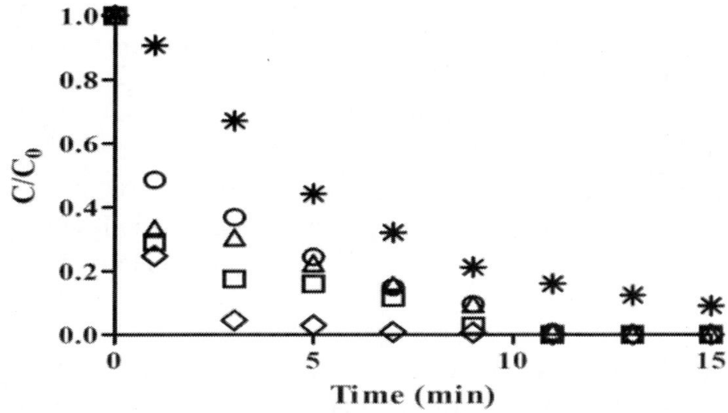

Figure 10. DTZ degradation by direct photolysis (✳) and by the UV/AC system with activated carbons C (○), M (△), S (□), or W (◇). $[DTZ]_0 = 25$ mg L^{-1}; pH = 6.5; T = 298 K.

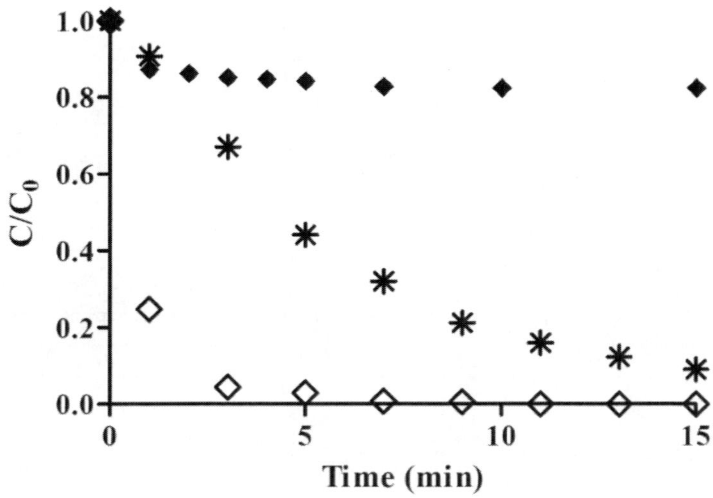

Figure 11. DTZ degradation by the UV/AC system. Direct photolysis (✳), adsorption on activated carbon W (◆), UV/W (◇). $[DTZ]_0 = 25$ mg L^{-1}; pH = 6.5; T = 298 K.

The role of activated carbon in DTZ removal process by UV/AC was further explored by determining the percentage removal (designated "synergic effect") of the carbons, which was calculated by subtracting the adsorptive and photolytic contributions from the global removal percentage in the UV/AC system (Equation (21)). Results are listed in Table 9.

$$\% S_{UV/CA} = \% D_{UV/CA} - \% D_{UV} - \% A_{CA} \tag{21}$$

In this equation, $\% S_{UV/CA}$ is the percentage DTZ removal due to the synergic effect produced by the presence of activated carbon during exposure to UV radiation, $\% D_{UV/CA}$ is the total percentage DTZ removal in the photocatalytic process UV/AC, $\% D_{UV}$ is the percentage DTZ degradation by direct photolysis, and $\% A_{CA}$ is the percentage DTZ removal by adsorption on the activated carbon.

Among the different commercial activated carbons used, the results in Table 3 show that carbon S made the highest adsorptive contribution to the global removal process and carbon W the lowest. However, we highlight that carbon W exerts the greatest synergic effect on DTZ removal by the UV/AC system, with a synergic contribution of >53% after the first minute of treatment.

Figure 12 depicts the time course of the synergic contribution to the overall DTZ removal by the UV/AC system for the four original activated carbons. The results obtained indicate that, regardless of the activated carbon used, the synergic contribution decreases with longer treatment time.

Table 9. Removal fractions and reaction rate constants for DTZ removal by UV/AC. $[DTZ]_0 = 25$ mg L^{-1}; pH = 6.5; T = 298 K

Exp. Num.	Activated carbon	k_{OB} (15 min) (min^{-1})	% UV degradation (1 min)	% Removal by adsorption (1 min)	% Removal by UV/AC (1 min)	% Synergic removal (1 min)
4	C	0.47 ± 0.01	9.29	14.20	51.50	28.01
5	C-H•	0.75 ± 0.04	9.29	11.70	49.79	28.80
6	C- e_{aq}^-	0.69 ± 0.03	9.29	11.60	44.61	23.72
7	C-HO•	1.05 ± 0.05	9.29	17.00	53.99	27.70
8	C-0	2.05 ± 0.00	9.29	12.90	87.09	64.90
9	M	0.53 ± 0.04	9.29	18.70	66.60	38.61
10	M-H•	1.06 ± 0.02	9.29	30.40	65.38	25.69
11	M- e_{aq}^-	1.05 ± 0.04	9.29	0.86	65.04	54.89
12	M-HO•	0.93 ± 0.05	9.29	17.00	62.36	36.07
13	M-0	0.69 ± 0.04	9.29	12.40	53.05	31.36
14	S	0.59 ± 0.03	9.29	28.80	71.00	32.91
15	S-H•	0.42 ± 0.05	9.29	8.49	45.00	27.22
16	S- e_{aq}^-	0.28 ± 0.02	9.29	9.07	32.49	14.13
17	S-HO•	1.03 ± 0.04	9.29	14.80	64.10	39.91
18	S-0	1.07 ± 0.02	9.29	7.67	47.00	30.04
19	W	1.42 ± 0.04	9.29	12.80	75.20	53.11
20	W-H•	1.02 ± 0.04	9.29	1.04	69.47	59.14
21	W- e_{aq}^-	5.06 ± 0.06	9.29	1.02	100.00	89.69
22	W-HO•	5.26 ± 0.06	9.29	1.67	92.81	81.85
23	W-0	0.99 ± 0.01	9.29	8.23	62.87	45.35

This may be attributable to: i) adsorption of DTZ and degradation compounds on the activated carbon surface that hinders the access of UV radiation to its active sites and/or, ii) modifications by UV radiation of the chemical functionalities on the surface of the activated carbon, reducing its synergic activity.

In order to explain the results in Figure 11, we conducted DTZ degradation experiments with the UV/AC system using DTZ-saturated activated carbons. Figure 13 shows, as an example, the results obtained for carbon S (carbon with higher adsorptive contribution to the global removal process). There is a decrease in the reaction rate at the beginning of the UV/saturated AC process due to the absence of DTZ removal by adsorption; however, the percentage degradation values continue to be higher than those obtained by direct photolysis, confirming that DTZ adsorption is not the most important effect in the reduction of the synergic activity of the activated carbon.

We studied DTZ photodegradation in the presence of carbons that had undergone different gamma irradiation treatments. The treatments modified the surface chemistry of the carbons. Thus, we obtained carbons treated with reducing agents (solvated electrons, e-aq, and hydrogen atoms, H$^{\bullet}$) and carbons treated with oxidizing agents (HO$^{\bullet}$ radicals). Their chemical and textural characteristics are given elsewhere [21].

Table 9 exhibits the kinetic constants obtained and the synergic contribution to the global DTZ removal process.

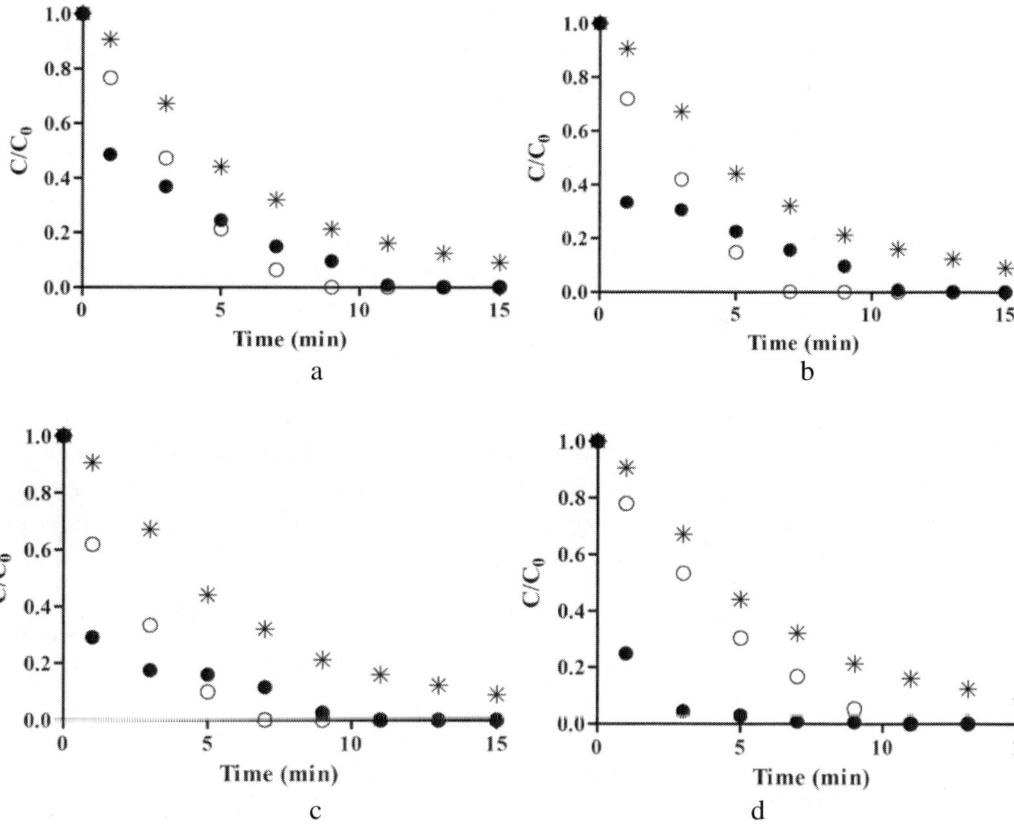

Figure12. DTZ removal by the UV/AC system. (✳), Direct photolysis; (●), UV/AC; (○), UV+ adsorption. A) Carbon C; B) Carbon M; C) Carbon S; D) Carbon W. $[DTZ]_0 = 25$ mg L^{-1}; pH = 6.5; T = 298 K.

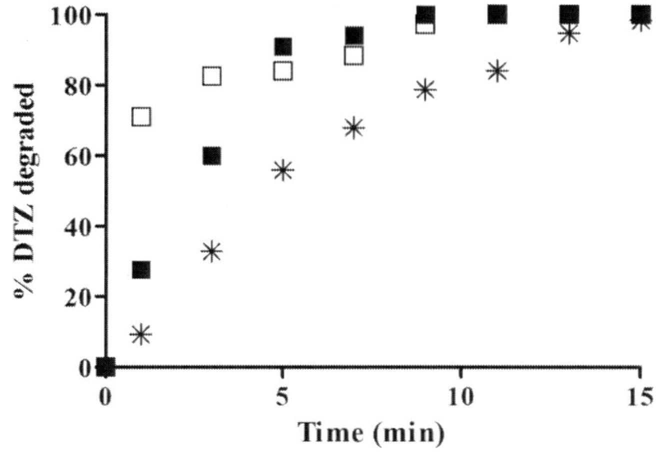

Figure 13. DTZ degradation by the UV/AC system using saturated activated carbon. (✳), Direct photolysis; (□), UV/Carbon S; (■), UV/Saturated carbon S. $[DTZ]_0$ = 25 mg L^{-1}; pH = 6.5; T= 298K.

We shall first discuss the results of carbon W, the activated carbon showing the greatest synergic activity. According to the results, the carbons producing the greatest increase in DTZ removal are W-e_{aq}^- and W-$HO^\bullet$, whereas no substantial increase in DTZ removal is observed using the sample treated with hydrogen atoms (W-$H^\bullet$). In addition, Table 9 shows that the synergic contribution of samples W-e_{aq}^- and W-$HO^\bullet$ to the overall removal process for one minute of treatment is markedly higher than the contribution of the original carbon (W).

These results appear to indicate that, regardless of the treatment applied (oxidizing or reducing agents), irradiation of the activated carbon increases its synergic activity. The chemical properties of the treated carbons (Table 4S, in supporting information) show that both treatments increase the percentage of -Ph-OH, C-O (ester/anhydride) groups in comparison to the original carbon.

In order to confirm that these groups participate in the synergic activity of the activated carbon, the remaining activated carbons studied were treated under the same experimental conditions.

The results show that, regardless of the activated carbon sample considered, its synergic activity is enhanced by the gamma radiation treatment, although this behavior depends on the baseline material. Thus, the reaction rate varies in the order W-$HO^\bullet$>W-e^-_{aq}>W>W-$H^\bullet$ >W-0 for carbon W but in the order M-$H^\bullet$>M-e^-_{aq}>M-$HO^\bullet$>M-0> M for carbon M.

When the experimental data were related to the textural and chemical properties of the activated carbons, no general trend was found among the 20 samples studied. However, in general, the samples that most enhance the synergic activity of the activated carbon are those with a higher percentage of surface oxygen and, among these, those with higher percentages of ester/anhydride groups and of carbon atoms with sp^2 hybridization [21].

In order to dig deep into the activated carbon characteristics affecting its photocatalytic activity, we have determined the band gap (E_g) of the activated carbons (Table 10).

According to the results in Table 10, the activated carbons behave as semiconductor materials and therefore as photoactive materials in the presence of UV radiation, because all E_g values are < 4 eV. It can also be observed that the gamma radiation treatment reduces the band gap energy values of the materials and that, in the same series of activated carbons, lower E_g values correspond to higher k_{OB} values (Table 9), especially in series with greater band gap reductions.

Table 10. Band gap values (E_g) of the activated carbons, calculated according to the Kubelka-Munk method

Carbon	E_g (eV)	Carbon	E_g (eV)
C	3.65 ± 0.03	S	3.58 ± 0.02
C-H$^•$	3.36 ± 0.02	S-H$^•$	3.63 ± 0.02
C- e_{aq}^-	3.14 ± 0.02	S- e_{aq}^-	3.16 ± 0.02
C-HO$^•$	3.00 ± 0.02	S-HO$^•$	2.92 ± 0.02
C-0	3.04 ± 0.02	S-0	2.98 ± 0.02
M	3.50 ± 0.02	W	3.68 ± 0.02
M-H$^•$	3.13 ± 0.02	W-H$^•$	3.35 ± 0.02
M- e_{aq}^-	3.20 ± 0.02	W- e_{aq}^-	3.15 ± 0.02
M-HO$^•$	3.23 ± 0.02	W-HO$^•$	3.10 ± 0.02
M-0	3.33 ± 0.02	W-0	3.23 ± 0.02

The W series show the highest performance in synergic removal (Table 9), although their E_g values are not the lowest among the activated carbons. This may be due to the presence of sulfur heteroatoms in the composition of carbon W, which has higher sulfur concentrations than the other carbons [21]. The presence of sulfur produces an increase in carbon atoms with sp^2 hybridization of the carbons, which facilitates the transit of electrons, the main triggers of the photocatalytic process that degrades the compound. We found that sp^2 hybridization is increased in the materials modified with gamma radiation, explaining their usually superior behavior in the process under study. Hence, the photocatalytic activity of the carbon is favored not only by the decrease in E_g values but also by the amount of carbon atoms with sp^2 hybridization of the material. Thus, although, in general, the W series has not the lowest E_g values, it obtains the best performance because of its higher fraction of sp^2 hybridization.

These findings indicate the possible mechanism underlying the photoactive activity of the activated carbons. We propose that the photons from UV light would fall on the activated carbons and generate electron-hole pairs through their irradiation with a sufficient amount of energy to promote electrons from the valence band to the conduction band. The photogenerated electrons would spread throughout the graphene layers and reach molecules of the absorbed DTZ and oxygen molecules.

The electrons reduce the adsorbed O_2 to form superoxide radicals ($O_2^{\bullet -}$), which can react with the water molecule and trigger the formation of oxidizing radical species that will interact with the compound, contributing to its degradation (Reactions [22-24]). Additionally, the presence of adsorbed oxygen avoids recombination of the electron with the positive hole (Reaction [23]), allowing interaction between the water molecule and the free hole and increasing the effectiveness of the photocatalytic process.

$$e_{CB}^- + O_2 \rightarrow O_2^{\bullet -} \quad (22)$$

$$O_2^{\bullet -} + H_2 \rightarrow HO_2^{\bullet} + HO^- \quad (23)$$

$$2HO_2^{\bullet} \rightarrow O_2 + H_2O_2 \rightarrow O_2 + 2HO^{\bullet} \quad (24)$$

The positive holes are directly responsible for the generation of hydroxyl radicals by interaction with OH^- groups of the carbon surface and by capture of water molecules (Reaction [26]).

$$h^+_{BV} + H_2O \rightarrow HO^\bullet + H^+ \tag{25}$$

CONCLUSION

Quantum yields obtained for the pharmaceutical compounds studied are very low, which is responsible for the low efficacy of the process during direct absorption of photons in nitroimidazole phototransformation.

R_{254} values obtained show that the dose habitually used for the disinfection of waters is not adequate to remove this type of pharmaceutical, demonstrating the need for higher doses of UV irradiation or long exposure times for their removal.

The time course of TOC and of toxicity during direct photodegradation of pharmaceutical compounds, in both ultrapure and real waters, shows that oxidation subproducts do not transform into CO_2 to the desired extent.

Therefore, they generate fractions of lower molecular weight than the original molecule, maintaining a constant TOC concentration throughout the treatment time and possibly giving rise to pharmacologically active compounds with higher toxicity than the original nitroimidazole.

The concentration of pharmaceutical compounds has a major effect on its photodegradation rate. The study of the influence of pH on the values of parameters ε and k'_E shows no general tendency for the behavior of pharmaceutical compounds as a function of the pH.

The NOM components GAL, TAN, and HUM may act as promoters and/or inhibitors of $OH^\bullet$ radicals generated by H_2O_2 photoproduction. Results show that the presence of GAL has a markedly different effect on the MNZ degradation rate from that of TAN or HUM, with an increase in this rate at low GAL concentrations.

These results appear to show that, under these conditions, GAL mainly acts as a promoter of $OH^\bullet$radicals, which oxidize MNZ molecules. In contrast, the presence of TAN or HUM decreases MNZ degradation rate, suggests a predominant effect of their $OH^\bullet$radical inhibiting capacity, due to their complex structure and high reactivity against $OH^\bullet$ radicals.

Differences in pharmaceutical compounds degradation rate among the studied waters, which have different chemical compositions, are not very marked, although there is a slight decrease in wastewaters, mainly because of the UV radiation filter effect of this type of water. The MNZ photodegradation rate constant is slightly higher in groundwater, suggesting a small contribution of indirect photodegradation due to the type of NOM present in this water.

The UV/H_2O_2 system was adequate to degrade pharmaceutical compounds, yielding a reaction rate constant value with $HO^\bullet$ radicals of $k_{HO^\bullet} = 10^{10}$ M^{-1} s^{-1}. The percentage pharmaceutical compound degradation depended on the solution pH. The applicability of the UV/ H_2O_2 system was demonstrated by the cytarabine degradation and TOC removal obtained with its use in groundwater and wastewater. However, the toxicity results indicated that the degradation byproducts of the organic matter in the medium had a higher toxicity in comparison to the original matter; therefore this treatment system should be used with caution for cytarabine degradation.

The $UV/S_2O_8^{-2}$ system was more effective than UV/H_2O_2 to degrade pharmaceutical compounds, achieving higher removal percentages in shorter times due to the generation of $SO_4^{\bullet -}$ and $HO^{\bullet}$ radicals. The solution pH considerably affected the micropollutant degradation, and the lowest percentage degradation was at pH = 2, due to consumption of $SO_4^{\bullet -}$ in the formation of anions with low oxidizing power. The maximum degradation percentage was at neutral pH. The applicability of the $UV/S_2O_8^{-2}$ system was demonstrated by the cytarabine degradation and TOC removal obtained with its use in groundwater, and waste water. TOC results showed that $UV/S_2O_8^{-2}$ was slightly more effective than UV/H_2O_2 to mineralize the organic matter present in the different types of water.

Furthermore, toxicity studies of this system found no formation of degradation byproducts with higher toxicity than cytarabine and/or the organic compounds in natural water. The $UV/S_2O_8^{-2}$ system is therefore recommended for the treatment of polluted water.

The presence of activated carbons with a high carboxyl groups content enhances 2,4-D photodegradation by the UV/TiO_2 system.

Carboxyl groups in the graphene planes of the activated carbon participate in the additional generation of $OH^{\bullet}$ radicals by interacting with the electrons produced by the UV/ TiO_2 system. Consequently, the contribution of $OH^{\bullet}$ radicals to the global 2,4-D degradation process is greater at the beginning of the photocatalytic treatment. The $UV/TiO_2/W_{O3\text{-}30}$ system mineralized 40 % of the organic matter present in the medium, and the toxicity of the degradation byproducts was considerably lower than that of 2,4-D.

The presence of activated carbon during the DTZ photodegradation process markedly increases the removal rate, regardless of the activated carbon used. The results obtained indicate that activated carbon W exerts the greatest synergic effect on DTZ removal by the UV/AC system, with a synergic contribution >53 % at one minute of treatment. Regardless of the activated carbon sample considered, its synergic activity is, in general, enhanced by the gamma radiation treatment. The textural and chemical properties of the activated carbons used show no clear relationship with their synergic contribution.

However, the synergic activity of the activated carbon is more greatly enhanced by the samples with higher percentages of surface oxygen and, among these, the samples with higher percentages of carbon atoms with sp^2 hybridization.

ACKNOWLEDGMENTS

The authors are grateful for the financial support provided by MEC-DGI, FEDER (Project: CTQ2011-29035-C02-02) and Junta de Andalucía (Project: RNM3823; RNM7522).

REFERENCES

[1] Boreen, A. L., Arnold, W. A., McNeill, K. (2004) Photochemical fate of sulfa drugs in then aquatic environment: Sulfa drugs containing five-membered heterocyclic groups. *Environmental Science and Technology* 38(14), 3933-3940.

[2] Latch, D. E., Stender, B. L., Packer, J. L., Arnold, W. A., McNeill, K. (2003) Photochemical fate of pharmaceuticals in the environment: Cimetidine and ranitidine. *Environmental Science and Technology* 37(15), 3342-3350.

[3] Packer, J. L., Werner, J. J., Latch, D. E., McNeill, K., Arnold, W. A. (2003) Photochemical fate of pharmaceuticals in the environment: Naproxen, diclofenac, clofibric acid, and ibuprofen. *Aquatic Sciences* 65(4), 342-351.

[4] Schwarzenbach, R. P., Gschwend, P. M., Imboden, D. M. (2003) *Environmental Organic Chemistry*. John Wiley and Sons: New York, 1311.

[5] Canonica, S., Jans, U., Stemmler, K., Hoigné, J. (1995) Transformation kinetics of phenols in water: Photosensitization by dissolved natural organic material and aromatic ketones. *Environmental Science and Technology* 29(7), 1822-1831.

[6] Canonica, S. and Tratnyek, P. G. (2003) Quantitative structure-activity relationships for oxidation reactions of organic chemicals in water. *Environmental Toxicology and Chemistry* 22(8), 1743-1754.

[7] Gerecke, A. C., Canonica, S., Muller, S. R., Scharer, M., Schwarzenbach, R. P. (2001) Quantification of dissolved natural organic matter (DOM) mediated phototransformation of phenylurea herbicides in lakes. *Environmental Science and Technology* 35(19), 3915-3923.

[8] Zepp, R. G., Baughman, G. L., Schlotzhauer, P. F. (1981b) Comparison of photochemical behavior of various humic substances in water: II. Photosensitized oxygenations. *Chemosphere* 10(1), 119-126.

[9] Rosenfeldt, E. J. and Linden, K. G. (2004) Degradation of endocrine disrupting chemicals bisphenol A, ethinyl estradiol, and estradiol during UV photolysis and advanced oxidation processes. *Environmental Science and Technology* 38(20), 5476-5483.

[10] Méndez-Díaz, J. D., Sánchez-Polo, M., Rivera-Utrilla, Canónica S., von Gunten, U. (2010) Advanced oxidation of the surfactant SDBS by means of hydroxyl and sulphate radical, *Chemical Engineering Journal*, 163, 300-306.

[11] Ocampo–Perez, R., Sánchez-Polo, M., Rivera-Utrilla, J., Leyva-Ramos, R. Degradation of antineoplastic cytarabine in aqueous phase by advanced oxidation processes bases on ultraviolet radiation, *Chemical Engineering Journal* 165, 581-588, 2010.

[12] Gómez-Pacheco, C. V., Sánchez-Polo, M., Rivera-Utrilla, J., López-Peñalver, J. J. (2012) Tetracycline degradation in aqueous phase by ultraviolet radiation, *Chemical Engineering Journal*, 187, 89-95.

[13] Lopez-Peñalver, J., Sánchez-Polo, M., Gómez-Pacheco, C., Rivera-Utrilla, J. (2010) "Photodegradation of tetracyclines in aqueous solution by using UV and UV/H2O2 oxidation processes", *J. of Chemical Technoloy and Biotechnology*, 85, 1325-1333.

[14] Medellin-Castillo, N. A., Ocampo-Pérez, R., Leyva-Ramos, R., Sanchez-Polo, M., Rivera-Utrilla, J., Méndez-Díaz, J. D. (2013) Removal of diethyl phthalate from water solution by adsorption, photo-oxidation, ozonation and advanced oxidation process (UV/H_2O_2, O_3/H_2O_2 and O_3/activated carbon), *Science of the Total Environment*, 442, pp. 26-35.

[15] Prados-Joya, G., Sánchez-Polo, M., Rivera-Utrilla, Ferro-García M. (2011), Photodegradation of the antibiotics nitroimidazoles in aqueous solution by ultraviolet radiation", *Water Research* 45, 393-403.

[16] Rivera-Utrilla, J., Sánchez-Polo, M. Ocampo-Perez, R. (2011) Enhancement of the catalytic activity*Applied Catalysis B: Environmental* 104, 177-184.
[17] Rivera-Utrilla, J., Sánchez-Polo, M., Abdel Daiem, M. M., Ocampo-Pérez, R., (2012) Role of activated carbon in the photocatalytic degradation of 2,4-dichlorophenoxyacetic acid by the UV/TiO_2/activated carbon system, *Applied Catalysis B: Environmental*, 126, 100-107.
[18] Sánchez-Polo, M., Rivera-Utrilla, J., Méndez-Díaz, J. D., Canonica, S., von Gunten, U. (2007) Photooxidation of naphthalenesulfonic acids: Comparison between processes based on O_3, O_3/activated carbon and UV/H_2O_2. *Chemosphere* 68(10), 1814-1820.
[19] Sánchez-Polo, M., Abdel Daiem, M. M., Rivera-Utrilla, J., Ocampo-Pérez, R., (2013) Comparative study of the photodegradation of bisphenol A by $HO^{\bullet}$, $SO_4^{\bullet-}$ and $CO_3^{\bullet-}$ /$HCO_3^{\bullet}$ radicals in aqueous phase, *Science of the Total Environment*, 463, 423-431.
[20] Torres-Luna, J. R., Ocampo-Pérez, R., Sánchez-Polo, M., Rivera Utrilla, J., Velo-Gala, I., Bernal-Jacome, L. A. (2013) Role of $HO^{\bullet}$ and $SO_4^{\bullet-}$radicals on the photodegradation of remazol red in aqueous solution, *Chemical Engineering Journal*, pp. 155-163, 223.
[21] Velo-Gala, I., Lopez-Peñalver, J., Sánchez-Polo, M., Rivera-Utrilla, J., (2013) Activated carbon as photocatalyst of reactions in aqueous phase, *Appliced Catalysis B: Environ.* 142 694-704.
[22] Sharpless, C. M. and Linden, K. G. (2003) Experimental and model comparisons of low - and medium-pressure Hg lamps for the direct and H_2O_2 assisted UV photodegradation of N-nitrosodimethylamine in simulated drinking water. *Environmental Science and Technology* 37(9), 1933-1940.
[23] Hessler, D. P., Gorenflo, V., Frimmel, F. H. (1993) Degradation of aqueous atrazine and metazachlor solutions by UV and UV/H_2O_2 - Influence of pH and herbicide concentration. *Acta Hydrochimica Hydrobiology* 21(4), 209-214.
[24] Canonica, S., Meunier, L., von Gunten, U. (2008) Phototransformation of selected pharmaceuticals during UV treatment of drinking water. *Water Research* 42(1-2), 121-128.
[25] López, A., Bozzi, A., Mascolo, G., Ciannarella, R., Passino, R. (2002) UV and H_2O_2/ UV degradation of a pharmaceutical intermediate in aqueous solution. *Annali di Chimica* 92(1-2), 41-51.
[26] Andreozzi, R., Raffaele, M., Nicklas, P. (2003) Pharmaceuticals in STP effluents and their solar photodegradation in aquatic environment. *Chemosphere* 50(10), 1319-1330.
[27] Shemer, H., Kunukcu, Y. K., Linden, K. G. (2006) Degradation of the pharmaceutical Metronidazole via UV, Fenton and photo-Fenton processes. *Chemosphere* 63(2), 269-276.
[28] DVGW (1997) W 294. UV-Desinfektionsanlagen für die Trinkwasserversorgung - Anforderungen und Prüfung.
[29] ÖNorm (2001) M 5873-1. Anlagen zur Desinfektion von Wasser mittels Ultraviolett-Strahlen.
[30] Cabaj, A., Sommer, R., Schoenen, D. (1996) Biodosimetry: Model calculations for u.v. water disinfection devices with regard to dose distributions. *Water Research* 30(4), 1003-1009.
[31] Artinger, R., Buckau, G., Geyer, S., Fritz, P., Wolf, M., Kim, J. I. (2000) Characterization of groundwater humic substances: Influence of sedimentary organic carbon. *Applied Geochemistry* 15(1), 97-116.

[32] Steelink, C. (2002) Investigating humic acids in soils. *Anal. Chem.* 74(11), 327A-333A.

[33] Choudhry, G. G. (1984) Humic substances. Structural aspects, and photophysical, photochemical and free radical characteristics. *The handbook of environmental chemistry.* Vol. 1C: the natural environment and the biogeochemical cycles, 1-24.

[34] Cooper, W. J., Song, W., Gonsior, M., Kalnina, D., Peake, B. M., Mezyk, S. P. (2008) Recent advances in structure and reactivity of dissolved organic matter in natural waters. *Water Science and Technology: Water Supply* 8, 615-623.

[35] Xu, G., Li, F., Wang, Q. (2007) Effect of humic acids on photodegradation of chloroacetanilide herbicides under UV irradiation. *Journal of Environmental Science and Health.* Part B, Pesticides, food contaminants, and agricultural wastes 42(2), 165-171.

In: UV Radiation: Properties, Effects, and Applications
Editor: James A. Radosevich
ISBN: 978-1-63321-090-5

Chapter 11

The Universe in Ultraviolet Light

Ana I. Gómez de Castro*
Universidad Complutense de Madrid, Spain

Abstract

Ultraviolet (UV) light pervades the Universe. Many astronomical sources emit electromagnetic radiation in the wavelength range from 20 to 320 nanometers among them, the brightest objects in the Universe. UV radiation is efficiently scattered by dust grains and interacts strongly with gas in space due to the atomic structure of matter. For this reason, the UV range is ideally suited to detect subtle astronomical structures and thin gas clouds.

From the ground, the UV night sky is dark. The Earth atmosphere blocks the penetration of UV photons from space. Solar radiation with wavelengths below 280 nanometers does not reach the surface of the Earth.

UV space missions and astronomical satellites show an amazing view of the Universe. This chapter is an introduction to the processes and sources of UV radiation in space. The characteristics of the UV radiation from the Sun, planets, comets in the Solar System, from stars and interstellar clouds in the Galaxy and from the Universe, at large, are outlined.

Keywords: Ultraviolet astronomy, The Sun, The Solar System, Planets atmospheres and Magnetospheres, Stars, White Dwarfs, Star Formation, Starbursts, Interstellar Medium, Intergalactic medium, Quasars, Jets, Black holes, Chemistry

Abbreviations

EVE	Extreme Ultraviolet Explorer
GALEX	Galactic Evolution Explorer
HST	Hubble Space Telescope

* Corresponding author: Ana I. Gómez de Castro. Universidad Complutense de Madrid, Facultad de Ciencias Matemáticas, Plaza de Ciencias 3, 28040 Madrid, Spain. E-mail: anai_gomez@ucm.es.

IGM	Intergalactic Medium
ISM	Interstellar Medium
LMC	Large Magellanic Cloud
nm	Nanometer
PAH	Polycyclic Aromatic Hydrocarbon
SDO	Solar Dynamics Observatory
UV	UltraViolet

11.0. Introduction

Our eyes are sensors adapted to the strongest radiation field on the surface of the Earth, the light from a rather cool star, the Sun, around which our planet orbits. Astronomers denote this radiation field as the visible range of the electromagnetic spectrum. Visible stands for wavelengths between some 350 and 800 nanometers (nm). Electromagnetic radiation with wavelengths below 350 nm and above 20 nm is defined as ultraviolet (UV) radiation [1].

UV photons are absorbed by molecules in the Earth's atmosphere and, very especially, by Ozone. Below 280 nm everything is dark on the Earth; no stars are observed in the night sky, neither the Moon nor even the Sun. For comparison, the UV bands commonly used to describe the effects of UV irradiation in human skin, UVA, UVB and UVC bands, correspond to wavelengths of 365 nm, 313 nm and 254 nm, respectively. Only UVA rays can penetrate into the subcutaneous tissue; UVC rays are fully absorbed (>> 99%) within 100 microns of the epidermis surface, destroying the tissue.

The Ozone blockage allows the large organic molecules, and the human beings, to survive on the Earth surface. As a result, there is a rich wealth of information about the Universe or the very same Earth atmosphere that can only be obtained from space.

The relevance of UV radiation is associated with the basic constitution of matter. The electronic structure of the most abundant atoms and molecules displays its strongest footprints at UV wavelengths. Atoms are resonators to electromagnetic fields. Depending on the distribution of electronic levels, the electromagnetic radiation produced by atoms and molecules is more or less energetic and can be observed anywhere from high energies (X-ray or UV) to low energies (infrared or radiofrequencies). The strongest resonances are those exciting the atoms in their lowest energy states, also known as fundamental states. The same occurs with ions and with the electronic transitions of molecules. The number of electrons at each energy level depends on the temperature. Up to temperatures of about 500,000 K, the most abundant species in the Universe either radiate or absorb radiation in the UV range. Very common molecules such as molecular Hydrogen, Carbon Monoxide, Water or Sulfur hydride interact strongly with UV radiation.

Also dust is very sensitive to UV photons. Space dust is mainly built of carbonate or silicate cocoons covered with a mantle of ice and organic compounds. Large organic molecules, such as the Poly-cyclic Aromatic Hydrocarbons (PAHs), are widespread in the Universe absorbing strongly the UV radiation at 217.5 nm.

Though some PAHs are toxic for human beings – the toxicity of PAHs is structure-dependent – they are considered as candidate molecules for the earliest forms of life to be based on.

For all these reasons, astronomers have sought actively access to the UV range since the beginning of Space exploration. In 1972, the Apollo XVI mission carried to the Moon the small UV telescope in Figure 1, to observe bright stars. As today, there is only one space telescope to observe the Universe in the UV range, the Hubble Space Telescope (HST). Hubble is in a low Earth orbit, at about 560 km above the Earth surface and completes its orbit in 96-97 minutes [2]. This low orbit has allowed for in-orbit maintenance and repair work, as well as for new instruments be set on-board in several occasions. The Space Shuttle programme formally ended the 31st of August of 2011. Hence, Hubble will not be serviced any longer.

The UV Universe shines powerfully above the protecting cover of the Earth's Ozone layer. The Earth glow, the Sun, Jupiter and the Jovian planets, the comets, all radiate in the UV range. However, they have not the same appearance as in the visible range. Strongly variable nests of bright loops growing and erupting on the surface of the Sun are seen in UV light. UV radiation from planets is mainly produced in the upper atmospheric layers, as well as in their extended magnetic envelopes. Beyond the Solar System, the UV radiation from the very luminous hot stars is scattered by the dust grains pervading the Galaxy, producing a diffuse ultraviolet background (see Figure 2). Thin laces of plasma are observed among the stars, produced by the left overs of supernovae explosions and the action of the galactic magnetic field. The subtle connection between galaxies and the intergalactic medium is clearly apparent in the UV.

Credit: NASA.

Figure 1. The UV telescope set in the Moon during the Apollo 16 mission.

Courtesy: GALEX team.

Figure 2. The night sky seen by the GALactic Evolution eXplorer mission. Every circle has a radius of about 0.6 deg. Notice the networks of filaments radiating in the UV.

This chapter is a brief primer to the UV night sky and to the phenomena in the Universe that are best studied by means of UV radiation.

11.1 Generation of UV Radiation from Astronomical Sources

There are two main processes to produce UV radiation from astronomical sources: black body emission and radiation from diffuse (thin) gas in space.

Radiation is produced by all kind of species (atoms, ions, molecules) whenever an electron moves to a lower energy level releasing a photon. If this photon is reprocessed many times (absorbed and re-emitted) within a single body, for instance a star, the output radiation provides direct information on the temperature of the body.

In thermal equilibrium the number of particles at any given velocity (or with a given kinetic energy) is prescribed by a simple function, the Boltzmann distribution that is parametrized in terms of the average kinetic energy of the particles, in the body/system. The absolute or Kelvin scale of temperatures directly measures this average kinetic energy; for instance, at 0 K (0 Kelvin degrees) all particles are at rest.

In a dense body, the many photons produced by the multiple interactions between the particles and the photons generate a well-defined spectrum named as black body spectrum; spectrum is the technical term to define the histogram of the number of photons radiated per wavelength unit.

Let us denote as λ_{max}, the wavelength at which most of the photons are emitted, e.g. the peak of the energy distribution. A simple law was found back in the XIXth century relating the wavelength of the maximum and the temperature of the body, T, in Kelvin degrees. This law is known as Wien's law and states that,

$$\lambda_{max}\,(\mathrm{nm})\ T\,(\mathrm{K}) = 2.9\ 10^{6}$$

Thus a solar type star, with a surface temperature of 5700 K, has its maximum at 509 nm, i.e., in the visible range. However, a 10,000 K star, like Deneb, the brightest star in the Cygnus constellation, has its maximum at 290 nm, i.e., in the UV range. Stellar surface temperatures may reach 50,000 K thus, there are plenty of sources in the night sky that are brighter in the UV than in the visible range.

In the Universe, there are objects much hotter than the stars as, for instance, the white dwarfs. The term "white dwarf" is used to name the naked stellar cores containing the exhausts of the nuclear fuel left over after the life of a Solar-like star ends; their surface temperatures reach as high as few hundredth thousands Kelvin.

But UV radiation is not only produced in dense bodies. UV photons can also be produced in diffuse environments and nebulae. In these objects, radiation escapes directly to space without being scattered or absorbed, as a result, it carries detailed information of the media originating it. Basic parameters such as temperature but also density, velocity, abundance of the various possible species can be measured with a clever selection of the spectral lines, also known as spectroscopic tracers.

In some sources, this information is susceptible to be used to build 3D maps of the radiating nebula; the UV spectrum is the fingerprint of the ionized gas to be analyzed in depth with the full remote sensing artillery.

11.2. UV Radiation in the Solar System

The dominant source of UV radiation in the Solar System is the Sun. The nuclear fusion reactor that generates the solar energy occupies roughly the central quarter of the Sun. An envelope of ionized gas that transports to the surface the heat generated in the reactor cools this core. In the upper layers of the Sun, the transport is made through convective currents that give its granular aspect to the solar surface.

The Sun does not rotate like a rigid body; the rotation rate is faster at the equator than at high latitudes. The combination of differential rotation and convection works like a dynamo that feeds the solar magnetic field. The excess magnetic energy is released in the solar surface through phenomena like coronal loops, eruptions, flares and Solar storms but also in a quieter manner, through waves that heat the Solar atmosphere to reach about 1 million Kelvin in the corona (see Figure 3). Moreover, the Sun produces a continuous flow of particles, the Solar wind, which propagates through the interplanetary medium reaching the planetary atmospheres and beyond.

Credit: K Schrijver and A. Title.

Figure 3. On August 1, 2010, there was a big eruption in the Sun. This is a false-colour image obtained with the Solar Dynamics Observatory (SDO) in the extreme UV during the eruption. The areas where the key events took place are labeled. SDO permitted to trace the topology (white lines) of the Solar magnetic field. Compare them, with the field lines of a simple magnet.

As a result, the transition region cannot be described as a well-defined layer, at a given height above the surface of the Sun (see Figure 4). The location of the transition region itself is set by the full ionization of Helium [3]. Above it, Helium is fully ionized, and cannot help to the thermal control of the atmosphere. The temperature jumps from about 35,000 K to nearly one million Kelvin in about 100 km, which is tiny compared with the solar radius: 696,000 km.

In the UVA, UVB and UVC bands the Sun appearance is similar to the disk seen in visible light from the Earth's surface; these bands capture the UV tail of the blackbody spectrum that dominates the visible radiation from the Sun. In the far UV range (90-200 nm), the solar appearance depends on the wavelength. For instance, at 103 nm, the five times ionized oxygen provides a view to the Sun rather similar to the loops and laces observed in the extreme UV (see Figure 4). However at 130 nm, cool oxygen in the lowest atmospheric layers is detected.

The Sun's magnetic activity affects the Earth's biosphere through a variety of phenomena that include the solar wind or the showers of high velocity particles reaching the Earth after eruptions, flares and coronal mass ejections.

A single ultraviolet spectrum provides simultaneous information on the thermal properties of a wide range of atmospheric components in the high atmosphere of the Sun.

The Solar magnetic activity cycle lasts for 11 years during which the Sun evolves from magnetically quiet to magnetically active. The maximum degree of activity may change from one cycle to the next.

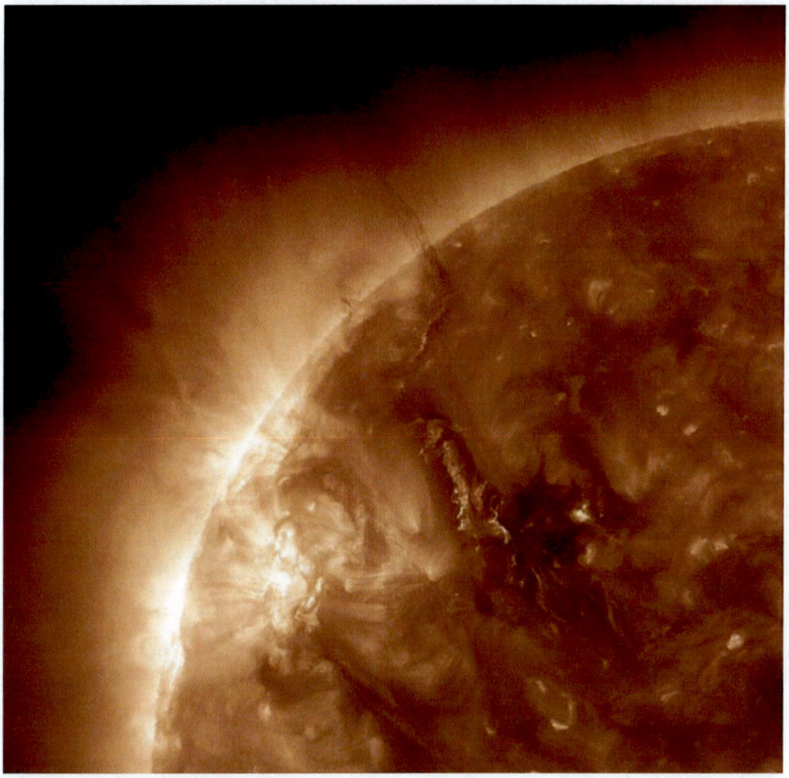

Credit: Solar Dynamic Observatory/NASA.

Figure 4. Extreme UV image of the Sun obtained with the Extreme Ultraviolet Variability Explorer (EVE) on board the NASA mission SDO. The transition region is visible as a low, bright fog over the surface of the Sun. The large, bright structures are magnetic loops in the solar corona. The dark areas are prominences; cool gas lifted up from lower atmospheric levels by the magnetic field.

Furthermore, the Sun's activity has varied during the Solar System life, affecting the evolution of the primitive planetary atmospheres and the environment in which life began on Earth [4]. At present, measurable effects are, for instance, the decrease of the temperature and density of the Earth's thermosphere as the Solar magnetic activity decreases.

The thermosphere is the Earth's layer where the Sun's radiation makes first contact with the atmosphere; it expands and shrinks with the level of extreme UV radiation and hence with the Solar magnetic activity [5].

As relevant as the energy input from solar UV radiation, it is the input in terms of particles and magnetic energy from the Sun. The Solar wind is basically made of two components: fast and slow.

The fast (800 km/s) wind leaves the Sun through coronal holes that are funnel-like regions of open field lines in the Sun's magnetic field (see for instance the open lines in the upper right corner of figure 3). For the ionized gas in the solar surface, the magnetic field lines act as a sort of rigid bars that channel the flow being pushed outwards by the centrifugal force. In a sense, the Sun works like a gigantic sprinkler system filling the interplanetary space with matter.

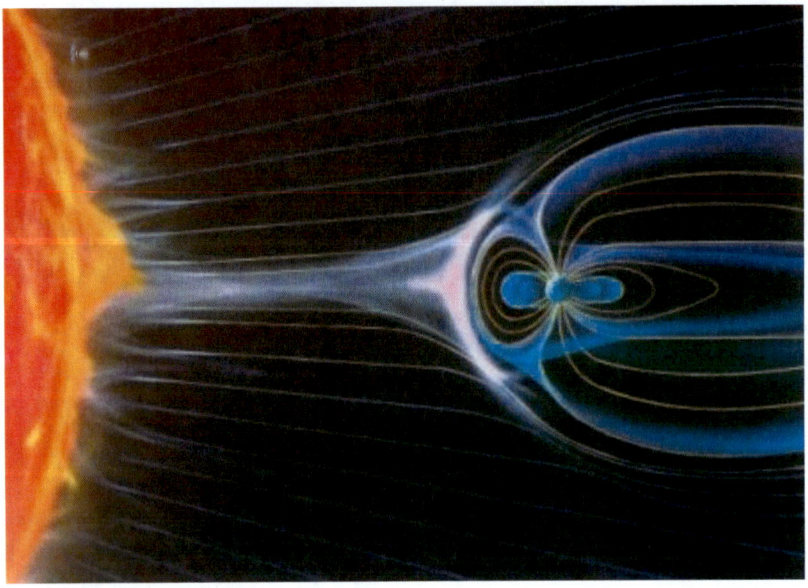

Credit: NASA.

Figure 5. Artist concept of the dynamic conditions in space and the Earth's magnetic field shielding action.

The slow component (400 km/s) to the wind originates from regions above and below the solar equator, at latitudes of 30°-35° during the solar minimum and expands to the pole during the solar magnetic activity cycle; at maximum, the Sun's poles are ejecting the slow wind. The slow wind has turbulent regions and large-scale structures [6].

Both fast and slow winds pass by the Earth sequentially and can produce storms in the Earth's magnetosphere.

The magnetic field of the Earth acts like a shield to the supersonically flowing solar wind. A standing bow shock is formed in the solar wind ahead of the Earth's magnetosphere [7]. The bow-shock is similar to a paraboloid of revolution with its nose being located at about 14 Earth radii from the center of the Earth, Sunwards. The particles in the Solar wind brake to subsonic speeds, so that Solar wind can flow into the magnetosphere. Particles are channeled and accelerated by the Earth magnetic to the magnetic poles producing the aurorae.

Aurorae take different morphologies: homogenous arcs, rays corona, drapery… The radiation from the aurorae traces the location where atmospheric constituents are excited by collision with the precipitating particles from the interplanetary space. These emissions are primarily due to a two steps process. Firstly, the precipitating energetic auroral particles (electrons and ions) collide with the atoms and molecules in the Earth's upper atmosphere exciting these species; their kinetic energy is partially stored in the chemically excited states of atmospheric species. At the end, radiation is produced when these species relax to less energetic configurations. Thus, the auroral emissions can be considered as fingerprints of the atmospheric constituents.

The first UV view of the Earth was obtained with the small lunar telescope in Figure 1.

As shown in Figure 6a, the Sun illuminated hemisphere shines in the UV. In addition, some bands are observed in the night hemisphere caused by the solar wind particles trapped by the Earth's magnetic field.

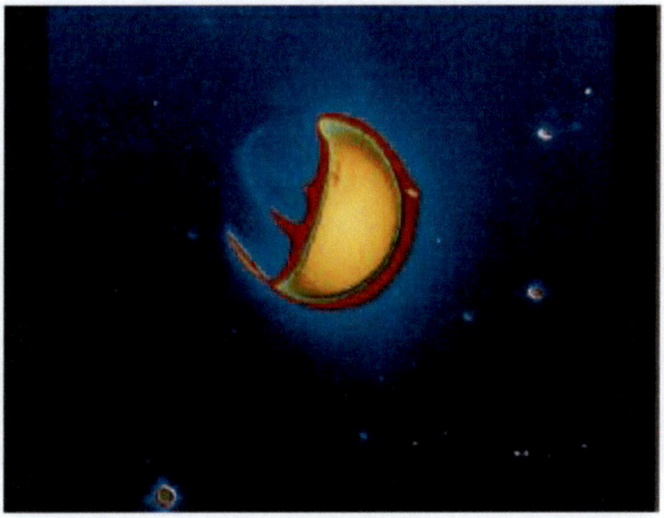

a

Credit: G. Carruthers (NRL) et al., Far UV Camera, Apollo 16, NASA.

Figure 6a. UV image of the Earth obtained from the Moon by the Apollo 16 mission.

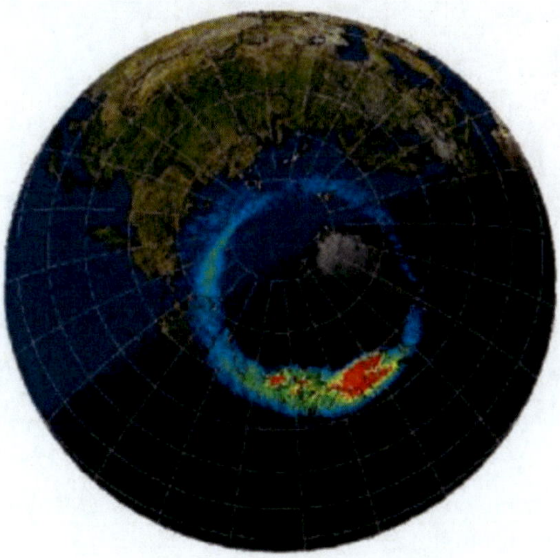

b

Credit: THEMIS mission, NASA.

Figure 6b. Ultraviolet image of the Earth Aurora seen from space by the Cluster mission. The satellite image is overlaid on top of a visible image of the Earth.

The UV radiation from all solar system bodies follows a similar behavior. Planets with tiny or absent magnetic fields reflect the UV radiation from the Sun. Also, the high atmospheric layers emit radiation after collisions with Solar wind particles. In planets hosting magnetic fields, aurorae are observed. For the astronomers, aurorae are an invaluable tool to study the planetary magnetic fields, and thus the internal structure of planets and the process that leads to planet formation [8].

In some objects, like Jupiter, other phenomena may come in. Jupiter's far UV auroral emission results from the currents driven between Jupiter's magnetosphere and Io's ionosphere [9]; Io is a volcanically active satellite.

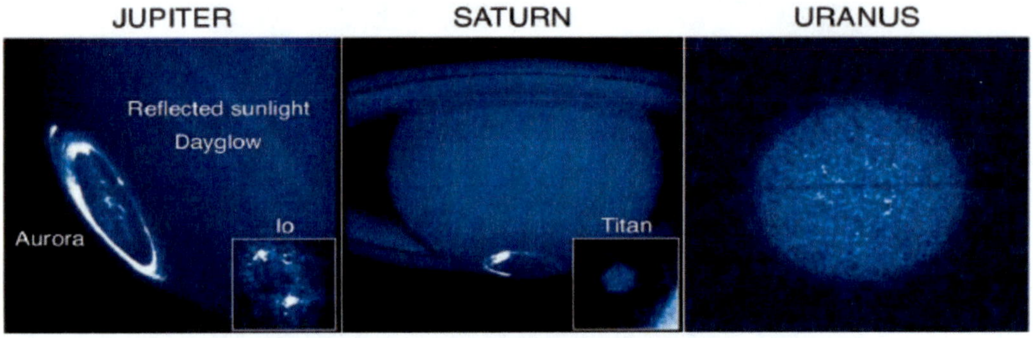

Courtesy: Jonathan Nichols. Based on NASA Hubble Space Telescope images.

Figure 7. A collage of some solar system UV targets. From left to right, these include the atmospheres, auroras and airglow of Jupiter, Io, Saturn,Titan, and Uranus.

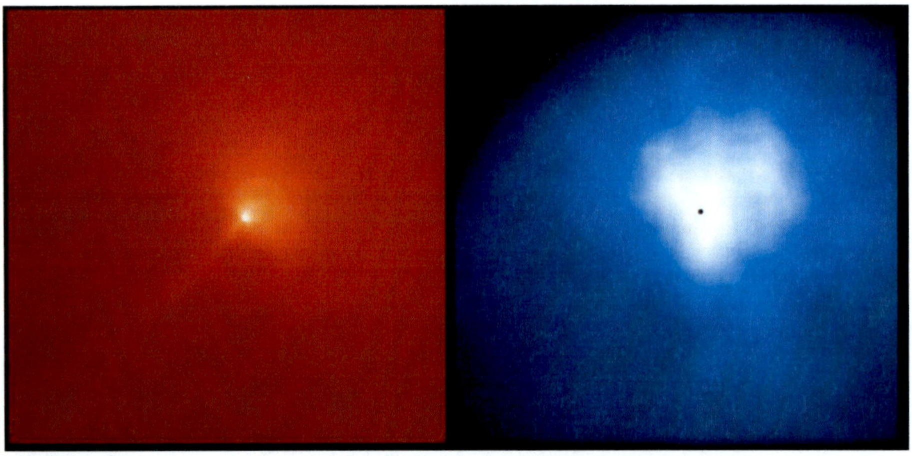

Credit: M.R. Combi.

Figure 8. Comet Hyakutake observed by the Hubble. Left, visible image tracing the dust location, right UV image tracing the Hydrogen location.

Most of the minor bodies in the Solar System have an appearance in the UV similar to that observed in the visible since they are observed through the reflection of the Sun's radiation.

Comets however, are a special case. They are built from rock, dust and ice. As comets approach the Sun, the evaporation of the water ice in the surface creates a tail of debris, known as coma, which reflects the Sun's light [10].

The comas contain water products such as Hydrogen, Oxigen and OH that radiate strongly in the UV producing spectacular images. This effect is nicely illustrated for comet Hyakutake in Figure 8.

11.3. UV Radiation from Stars and Galaxies

The Milky Way also named as the Galaxy, is made of stars, gas, and dust. The main source of UV radiation is hot stars with surface temperatures above 10,000.

Stars produce black body radiation. An important characteristic of black body radiation is that the higher the temperature, the higher the energy radiated per unit surface and time. This relation goes as a power of four, i.e., a star with a surface temperature of 30,000 K radiates 81 times more energy per unit surface than a 10,000 K star. As a result, the most luminous sources in the Galaxy are the hottest stars and they radiate mainly in the UV between 5.8 nm and 290 nm. These luminous stars are readily recognized in any galaxy, as shown in Figure 9 for the 30 Doradus Nebula in the Large Magellanic cloud (LMC). The LMC is a satellite galaxy to the Milky Way located 170,000 light years away from the Sun. At the heart of the nebula, there are more than 200 stars with masses from 10 to 50 times the Sun's mass.

Stars are gaseous bodies where the gravitational contraction force is balanced by the outwards thermal pressure. During the stellar phase, gas heating is produced by the energy released in the nuclear reactions taking place in the stellar core. Therefore, the balance is conditioned by the total mass of the star that, in turn, regulates the rate of nuclear energy production, as well as the efficiency of its transport through all the stellar layers. For the same set of nuclear reactions, stellar surface temperature and radius increase as the stellar mass does it. So does the stellar luminosity.

Credit: NASA, ESA, and F. Paresce, R. O'Connell, and the Wide Field Camera 3 Science Oversight Committee.

Figure 9. Composite image of 30 Doradus obtained by combining several HST images in various bands. The blue color is associated with the near UV band, the green with the visible band and the red with the red and near infrared bands [11].

The efficiency of the nuclear reactions sets the stars lifetime. Nuclear reactions are more efficient in massive stars than in solar type stars (the CNO cycle in massive stars versus the proto-proton chain reaction in solar type stars). As a result, the most massive stars live some hundred thousand years. The least massive stars have lifetimes comparable to the age of the Universe. Thus the massive, UV stars in Figure 9 are young, recently formed stars, in a starburst.

Stars form from giant clouds of gas and dust known as molecular clouds. Any given cloud forms many more low mass stars than high mass stars; the distribution of initial masses goes roughly like $M^{-2.35}$ [12], i.e., for any 10,000 Suns being formed in a massive cloud, only one 50 solar masses star is produced.

In summary, massive stars are scarce, since only very few of them are produced, and their life is short. They are very luminous though, and can be easily recognized even in distant galaxies.

The short life of massive (UV) stars makes of them ideal tracers to locate where star formation occurs in galaxies and which processes trigger it.

In Figure 10, there is a set of images of galaxies obtained with NASA's GALEX mission that has changed our understanding of star formation triggering in the Universe [13].

Galaxies come in few basic morphologies associated with their color, gas content and ability to form stars. Star formation is scarce in elliptical galaxies, as a result, old red giant stars dominate the stellar population. Hence, elliptical galaxies are bright in the visible range (where the radiation from the red stars peaks) and significantly weaker in the UV. Spiral and irregular galaxies, like the Milky Way or the Magellanic Clouds, are live entities where star formation proceeds at a good pace and gas is abundant.

The study of the Galaxy during the last century drove to a coherent scenario where star formation occurs in the dense disk of gas and dust, triggered by the passage of spiral density waves; gas in the disk is compressed by the waves. The subsequent gravitational collapse of the dense molecular clouds drives to the onset of star formation. The galactic magnetic field assists the concentration of matter in isolated pockets, large complexes of molecular gas containing about 100,000 times the mass of the Sun. These complexes are located along the spiral arms at relative distances of about 2,500 light years [14]; from outside, from the intergalactic space, these giant star forming regions look like UV beads on a necklace (see Figure 10).

The GALEX mission has shown that star formation may be triggered in apparently dead elliptical galaxies. Also it has shown that the spiral arms are able to trigger star formation in the extended galactic planes of the spiral galaxies, where the population of visible red giants, is tiny or undetectable. Thus, the paradigm has changed: galaxies are now understood as bodies surrounded by invisible pools of gas (the intergalactic medium) that submitted to the proper triggering can give rise to new generations of stars [15]. Galaxies are not perceived any longer as lost entities, "island universes", in the intergalactic space but as the brightest parts, the tip of the iceberg, of a much larger and subtle network of matter-energy, the cosmic web.

GALEX was also able to map the evolution of the star formation history for about 80% of the life of the Universe. Notice that since the cosmological expansion introduces a redshift (by the Doppler effect) in the wavelength of the radiation measured in the Earth, young, UV bright, stars need to be observed in the optical or the infrared range for redshifts, z, above 2;

z=2 roughly corresponds to looking backwards about 10 billion years in the life of the Universe.

These measurements have pointed out that overall, the star formation rate is apparently decreasing since z=1 to present times [16]. The Universe does not seem to be rejuvenating any longer, as a whole.

Credit: NASA/ESA/JPL-Caltech/STScI/UCLA.

Figure 10 a. Rings and arcs of star formation regions seen in ultraviolet light around a selection of galaxies. The blue color represents the GALEX UV image, superimposed on a visible image obtained with the Hubble.

Credit: NASA/JPL-Caltech/MPIA.

Figure 10b. GALEX image of the M83. Young stars can be seen way beyond the main spiral disk up to 140,000 light-years from its center. The image is one of the longest-exposure, or deepest, images ever taken of a nearby galaxy in ultraviolet light. Near-ultraviolet light (or longer-wavelength ultraviolet light) is colored yellow, and far-ultraviolet light is blue.

11.3.1. The Cosmic Web. Mixing and Chemical Evolution

The cosmic web is a large scale mesh created by the primordial density fluctuations in the Universe; it is a network of filaments that leave large voids of empty space with characteristic diameters of about 0.3 billions of light years.

A 26.8% of the content of the cosmic web is dark matter. Baryonic matter, like the gas, stars or dusty clouds, only represents approximately a 4.3% of the total matter-energy density of the Universe [17].

Only about a fourth of the baryonic mass is detectable in terms of stars or galaxies and this is concentrated in the filaments of the web, where the gravitational potential is strong and galaxies do form.

About 50% of the baryons are predicted to be in low density, highly ionized gas in the intergalactic space heated during the process of galaxy formation and structure generation; the gravitational collapse of the intergalactic medium (IGM) leads to shocks that heats it to temperatures from 100,000 to 1,000,000 Kelvin [18]. This gas radiates primarily in the far and extreme UV and in soft X-rays but unfortunately, it is so tenuous that its UV emission cannot be detected with the current instruments.

However it can be detected and analyzed by absorption spectroscopy since it produces a detectable signature in the spectra of any strong UV source located behind it, typically a QSO. QSOs are supermassive black holes that get their extraordinary luminosity by converting into heating the gravitational energy lost when a flow of matter of about a solar mass per year crosses the events horizon (see Sect. 12.5). From these studies, it is inferred that still about 30% of the baryonic mass in the Universe remains undetected [19].

There is a feedback loop between the IGM and the galaxies. Galaxies are formed from the IGM and massive stars in galaxies drive very powerful winds that blow matter into the interstellar and intergalactic space; these winds may reach velocities of 2000-3000 km/s. Of special relevance are the super winds produced by the starbursts: regions with a very high, abnormally high, star formation rate [20]. Moreover, some massive stars become supernovae that are especially efficient in releasing gas into the IGM.

In this feedback loop, the gas content of the universe is enriched through successive stellar generations. Only very simple atoms were formed in the Big Bang: Hydrogen, Helium and some small amount of Lithium and Berillium. The rapid expansion of the Universe drove it to be cooled to below 1 billion Kelvin, in about three minutes after the Big Bang. At these "low" temperatures, electrons, neutrons and protons recombined to form atoms. Roughly 25% of the mass of the Universe was Helium and 75%, Hydrogen.

Elements of biological interest such as Oxygen, Nitrogen or Carbon, were produced massively much later on, by nucleosynthesis in the stars. Every stellar generation has enriched the Universe with these and many more elements as to cover the whole periodic table. The metallicity of the gas (the logarithm of the abundance of Iron with respect to Hydrogen compared to the solar value) measures this enrichment and it has been found to increase steadily since z=5, on the average [21].

However, the UV observations of the IGM absorption have also shown that the metal enrichment of the Universe is neither uniform nor homogeneous; for instance, chemically processed material has been found in the voids of the Cosmic Web, where the gas density is very low and the star formation rate is expected to be small. Important clues on the metal enrichment spreading in the Universe depend on inter-galactic transport processes. When the time comes that high sensitivity UV imaging of the galaxies and the IGM becomes feasible, the warm/hot plasma emission from galactic halos will be detected, as well as the currents driving matter from/to the galaxies to/from their haloes and the IGM, in general.

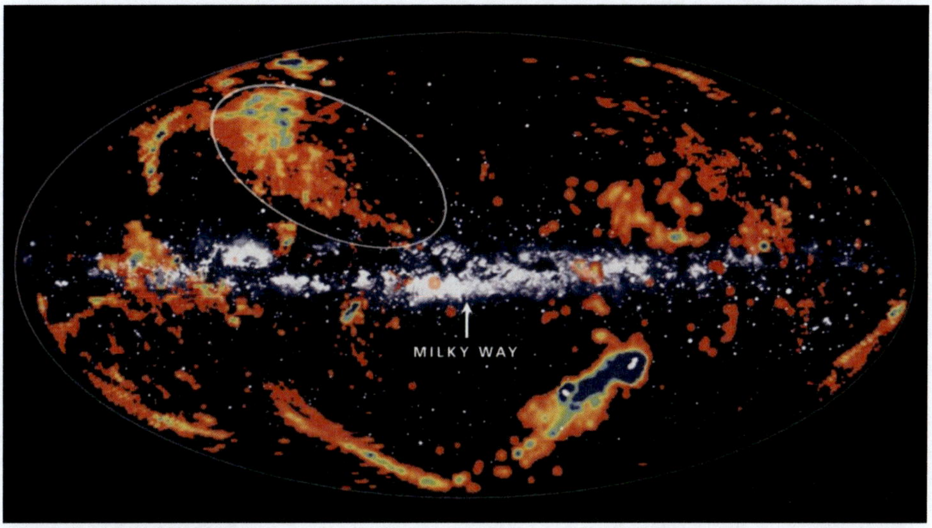

Credit: NASA and B. Wakker.

Figure 11. This image shows the projection of the sky in a plane. The central white band marks the location of the disk of the Milky Way, the "equator" of the galactic coordinates system. The colored structures are high velocity clouds falling onto the galactic disk from the Halo. The map is made in the 21 cm line of atomic hydrogen at radio wavelengths. In the UV, these clouds would be shadowing the UV background [22, 23].

In the Milky Way, this interaction can already be observed in the so-called galactic fountain. The galactic disk is surrounded by a large UV halo of hot gas that must be replenished as the gas cools [24]. In fact, most of the UV emission comes from the flows generated by supernovae and massive star forming regions that open from the stellar disk into the halo. The fountain is a dynamic process that involves a flow time scale that can be much shorter than any of the microphysical time scales due to ionization and recombination. Thus, the gas escaping into the halo has an initial temperature well in excess the million Kelvin.

As the hot plasma expands away from the disk, it cools adiabatically thereby reducing its temperature and density (to be about 10 particles per cubic decimeter) and rains back to the galactic disk, as shown in Figure 11. While matter in the disk is concentrated in dense shells and filaments, the halo acts as a pressure release valve for the hot gas in the disk, thereby controlling the fraction of the galactic disk volume, it occupies.

11.4. UV RADIATION FROM ASTRONOMICAL ENGINES

An engine is a device that transforms any kind of energy into mechanical energy. In astronomy, the widespread gravitational energy is converted into mechanical energy by different processes. The simplest examples are the stellar winds, either those driven by radiation pressure or those solar-like described in Section 12.3. These winds are driven by the excess energy (thermal, radiative…) of a star. There is however, another family of interesting outflows that are produced from sources surrounded by accretion disks. Accretion disks are formed when matter is attracted by a compact source of gravitational field be it a star, a white dwarf, a neutron star or a black hole. Infalling matter has an intrinsic angular momentum, if excessive, matter keeps orbiting around the source until the excess is transferred away. If the flow of infalling matter is high, an accretion disk is formed.

For this to occur, either the source of gravity is a young object surrounded by gas, as it is the case of the young stars or the supermassive black holes, or matter is being transferred within a binary stellar system from a giant star into the companion compact object.

Young stars and their accretion disks form simultaneously, giving rise to the formation of planetary systems [25]. Matter in the disk orbits around the parent star following Johannes Kepler's 3rd law: the orbital period is proportional to the orbital radius to the (3/2) power. For matter to fall onto the star, it needs to brake from this natural Keplerian velocity and hence, transport angular momentum to other scales.

In the micro-scale, the angular momentum is transported into turbulence and finally, into heating. In the macro-scale, rotating bipolar outflows are generated that carry away part of the excess angular momentum with them. These flows are some of the most spectacular astronomical objects (see Figure 12). In some systems, the connection between the accretion disk and the star is made through a magnetic interface, the stellar magnetosphere, and large structures radiating at UV wavelengths are generated; they may reach Mercury's orbit in solar-type stars. About 85% of the energy released by this magnetic interface is radiated in the Lyman-α line at 121.6 nm. The inner area of the disk heats by absorbing this UV radiation. As a result, the gas thermal velocity is high enough to escape the gravitational field. A photoevaporative flow is produced that evacuates the gas content from the inner region of the disk. This inner hole spreads with time. The disk lifetime is some few million years, after planetary systems life begins.

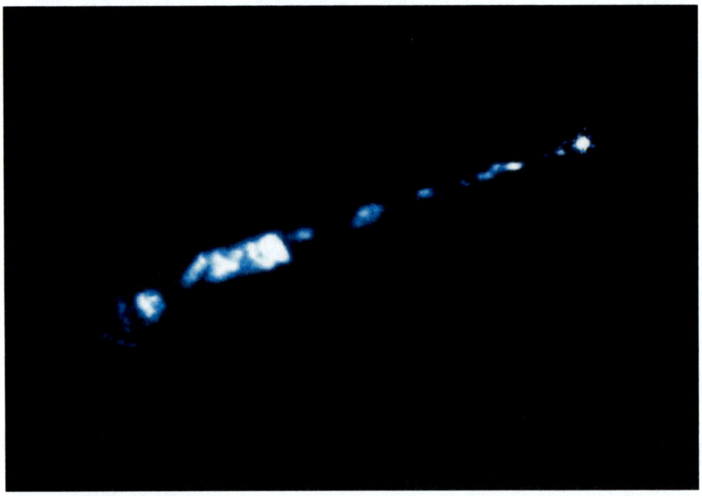

Credit: F. Duccio Macchetto/NASA/ESA

Figure 12. Ultraviolet light image of a 4,000 light-year long jet of plasma emanating from the bright nucleus of the giant elliptical galaxy M87 obtained with the HST and the instrument FOC.

At early times in the history of the Universe, a similar process took place but at a much larger scale; matter concentrated in the center of galaxies reached the critical mass and formed black holes that kept growing by accreting more gas and stars until they reached a mass of millions or even, hundreds of millions times the mass of the Sun, a QSO. Though the formulation of the physical problem is much harder in the black hole context, the general pattern repeats: the excess angular momentum of the matter falling into the black hole is transported into turbulence and heating, as well as into generating large scale bipolar outflows and jets. The inner region of the accretion disk heats to hundreds thousands of Kelvin and radiates in the UV [26]. They are the most luminous sources in the Universe; a QSO radiates about 10^{12} times the luminosity of the Sun, roughly a hundred times the radiation produced by the Milky Way, as a whole. The first QSO was discovered back in 1963, when Schmidt published the distance to the radiosource 3C 273 to be 2.44 billions of light years [27]. As today, about 200,000 quasars are known covering a very broad range of distances (from 600 million to 29 billion light years), though quasars were much more common in the early universe. Quasars radiate in all ranges of the electromagnetic spectrum but they are the brightest in the UV, in the Lyman-α line at 121.6 nm. Quasars contribute to the photoionization of the intergalactic medium and are the background sources against which the haloes of the galaxies and the IGM distribution are studied.

The general physics controlling accretion and outflow is non-stationary and highly non-linear since magnetic fields and relativity are involved. This implies a significant mathematical complicacy that only can be addressed in two manners: either by working with simplified models or by designing good numerical experiments (which, in turn, requires the simplified models to be properly understood).

This theoretical complexity, together with the limited spectral and time resolution in many previous studies, has defeated some of the bravest efforts to tackle the problem of accretion physics.

11.5. The Role of UV Radiation in Astrochemistry

The interstellar medium and the circumstellar environment are major chemical laboratories though the kinetics and energy exchange processes differ significantly from the Earth standards. Astrochemistry is vacuum chemistry; the compounds are either in gas phase in the interstellar space or in solid phase in the icy mantle of the interstellar dust grains.

In the *gas phase*, Coulomb fields drive the reactions since the ISM density is very low. In this environment, the UV radiation field becomes a major chemical agent leading to photodissotiation and photoionization.

In the *solid phase*, on the surface of the interstellar dust grains, UV radiation releases bonds of the chemical compounds accelerating the chemical processes. Moreover, the photoelectric effect produces hot photoelectrons from the dust that inject energy into the gas phase.

Strong UV radiation fields are able to photodissociate and destroy all the molecules and suppress chemical processes.

Accretion disks around young stars are the most interesting sites to study astrochemical processes. Young planetary systems grant the opportunity to assess the astrochemical conditions in environments where live can emerge.

At early phases, the composition of accretion disks is dominated by molecular hydrogen. Dust however, controls the thermal properties of disks. The gas to dust mass ratio is about a hundred.

The disk is surrounded by a thin atmosphere of atomic gas produced by the photodissotiation of molecules induced by the UV radiation from the star. However, molecular hydrogen self-shields efficiently. UV photons are trapped by the H_2 in the disk atmosphere (in the Lyman and Werner bands) protecting the disk interior, where dust grains and planetesimals grow.

Dust in the disk is mainly made of small amorphous carbon grains, large silicate grains (in two basic compounds olivine and orthopyroxene), and silicate grains coated with "dirty ice." "Dirty ice" is mainly constituted of H_2O, CO, and other simple molecules. However, its exact composition is very uncertain and depends on the physical conditions and the local star forming history in the parent molecular cloud. The most abundant ice in protostellar disks is H_2O ice.

Understanding disk chemistry requires considering both gas phase and dust grain-surface processes. Dust grains are key ingredients in the disk chemistry since they control the degree of ionization in the most opaque areas of the disk; dust is an efficient electron donor for recombining ions and a sink for neutrals. At low temperatures ($T \simeq 10$ K) large molecules are stationary on the grain surface and grow by reaction with mobile species like hydrogen. However, as the temperature rises, diffusion and thermal evaporation produce desorption of the light species such as hydrogen, thus slowing down the hydrogenation-driven chemistry on the surface. Later, heavier compounds move from the grain surface to the gas phase.

Desorption processes depend exponentially on the grain temperature, gas phase abundances and chemistry; thus chemical modeling is very sensitive to the processing of the grain mantles.

In general, chemical models of disk structure and evolution are developed with reduced sets of equations that are defined selecting only those species necessary to compute the abundances of the compounds under study.

Unfortunately, there is not yet a fully consistent disk model that solves for the chemical and dynamical structures simultaneously. As the chemistry of the disk evolves on timescales comparable to the dynamical timescale, this is a relevant source of uncertainty [28].

Conclusion

This chapter outlines the richness and relevance of the UV range for astronomical studies. As such, I expect that it is wide enough to provide a proper perspective and deep enough to give some clues about the extraordinary scientific rewards that UV astronomical instrumentation provides.

REFERENCES

[1] Gómez de Castro, A. I., (2012) *"Ultraviolet astronomy",* McGraw-Hill encyclopedia of Science and Technology.

[2] *"Hubble Space Telescope: HST overview"* retrieved from www.stsci.edu/hst.

[3] Schrijver, C. J., Title, A. M., Berger, T. E., et al. (1999). "A new view of the solar atmosphere by the Transition Region and Coronal explorer", *Solar Physics*, 187, 261.

[4] Lamer, H., Bredehoft, J. H., Coustenis, A., et al. (2009). "What makes a planet habitable?", *Astronomy and Astrophysics Review*, 17, 181.

[5] Knipp, D. J., Tobiska, W. K., Emery, B. A., (2004). "Direct and indirect thermospheric heating source from solar cycles", *Solar Physics*, 224, 2506.

[6] Gosling, J. T., (2007). "The Solar Wind". *Encyclopedia of the Solar System*, Second Edition, edited by L. A. McFadden, P. R. Weissman, T. V. Johnson, Academic Press, Chapter 5, 99.

[7] Wolf, R. A. (1997). *"Magnetospheric configuration", Introduction to Space Plasmas,* Eds. M. Kivelson and C.T. Russell, Cambridge University Press, p. 288.

[8] Carlson Jr., H. C., Egeland, A. (1997)). *"The aurora and the auroral ionosphere", Introduction to Space Plasmas*, Eds. M. Kivelson and C. T. Russell, Cambridge University Press, p. 459.

[9] Bonfond, B., Grodent, D., Gerard, J. C., et al. (2012). "Auroral evidence of Io's control over the magnetosphere of Jupiter", *Geophysical Research Letters*, 39, 1105.

[10] Brosch, N., Davies, J., Festou, M., et al. (2006). "A view of the future: ultraviolet studies of the Solar System", *Astrophysics and Space Science*, 303, 103.

[11] Sabbi, E., Anderson, J., Lennon, D. J., et al. (2013). "Hubble Tarantula Treasury Project: Unraveling Tarantula's Web. I. Observational overview and first results", *Astronomical Journal*, 146, 53.

[12] Salpeter, Edwin (1955). "The luminosity function and stellar evolution". *Astrophysical Journal,* 121, 161.

[13] Martin, D. C., Fanson, J., Schiminovich, D., et al. (2005). "The Galaxy Evolution Explorer: a space ultraviolet survey mission", *Astrophysical Journal*, 619, L1.

[14] Elmegreen, B. C., (1982). "The formation of giant molecular cloud complexes by the Parker-Jeans instability", *Astrophysical Journal*, 253, 655.

[15] Thilker, D. A., Bianchi, L., Meurer, G., et al., (2007). "A search for extended ultraviolet disks (XUV-disks) galaxies in the local Universe", *Astrophysical Journal* Suppl. Series, 173, 538.

[16] Martin D. C., Small, T., Schiminovich, D., et al. (2007). "The Star Formation and Extinction Coevolution of UV-Selected Galaxies over 0.05<z<1.2", *Astrophysical Journal Suppl. Series*, 173, 415.

[17] Wamsteker, W., Prochaska, J. X., Bianchi, L., et al. (2006). "The need for ultraviolet to understand the chemical evolution of the Universe, and Cosmology". *Astrophysics and Space Science*, 303, 69.

[18] Werk, J. K., Prochaska, J. X., Thom, C., et al. (2013). "The COS-Halos Survey: An Empirical Description of Metal-line Absorption in the Low-redshift Circumgalactic Medium", *Astrophysical Journal Suppl. Series*, 204, 17.

[19] Stocke, J. T., Keeney, B. A., Danforth, C. W., et al. (2013) "Characterizing the circumgalactic médium of nearby galaxies with HST/COS and HST/STIS absorption-line spectroscopy", *Astrophysical Journal*, 763, 148.

[20] Heckman, T. M., Roberts, C., Leitherer, C. (1998) "The ultraviolet spectroscopic properties of Local Starburst: Implications at high redshift", *Astrophysical Journal*, 503, 646.

[21] Prochaska, J. X., Wolfe, A. M., (2002). "The UCSD HIRES/Keck I damaped Lyman alpha abundance data base. II. The implications.", *Astrophysical Journal*, 566, 68.

[22] Giovanelli, R. (1981) "Very high velocity H-I clouds – an intergalactic population", *Astronomical Journal*, 86, 1468.

[23] Lehner, N., Howk, J. C., Thom, C., et al. (2012). "High-velocity clouds as streams of ionized and neutral gas in the halo of the Milky Way", *Monthly Notices of the Royal Astronomical Society,* 424, 2896.

[24] Shapiro, P. R., Field, G. B. (1976). "The consequences of a new hot component of the interstellar médium", *Astrophysical Journal*, 205, 762.

[25] Gómez de Castro, A. I. (2013). *"Young stellar objects and protostellar disks", Planets, Stars and Stellar Systems* Vol. 4, by Oswalt, Terry D.; Barstow, Martin A., ISBN 978-94-007-5614-4. Springer Science+Business Media Dordrecht, 2013, p. 279.

[26] Peterson, B. M., (2010). "Active galactic nuclei in the ultraviolet", *Astrophysics and Space Science*, 320, 69.

[27] Schmidt, M., (1973). "3C 273: a star-like object with large redshift"; *Nature*, 197, 1040.

[28] Bergin, E. A., Aikawa, Y., Blake, G. A., et al., (2007). *"The chemical evolution of protoplanetary disks", Protostars and Planets V.*, B. Reipurth, D. Jewitt, K. Keil (eds.), University of Arizona Press, Tucson, 751.

In: UV Radiation: Properties, Effects, and Applications
Editor: James A. Radosevich
ISBN: 978-1-63321-090-5

Chapter 12

THE USE OF UV LASERS IN FLUORESCENT ACTIVATED CELL SORTER ANALYSIS

Berna Demircan[1], Zane Deliu[3], Ozan A. Alkoc[2] and James A. Radosevich[3,*]

[1]Istanbul Medeniyet University, Medical School, Department of Medical Biology, Istabul, Turkey
[2]Instanbul Medeniyet University, Medical School, Department of Anatomy, Istanbul, Turkey
[3]University of Illinois, College of Dentistry, Oral Medicine and Diagnostic Sciences, Chicago, Illinois, US

ABSTRACT

Flow cytometry and cell sorting are well-established technologies to identify, analyze and/or sort different populations of cells in basic biological research and clinical diagnosis in pathology. This technology makes use of the principles of light scattering by particles crossing a beam of light, and excitation and fluorescence emission of fluorochromes attached to specific molecules or expressed by cells. All flow cytometers are able to interrogate cells and generate data, but only cell sorters can sort cells in a second stage where electrostatic cell sorting takes place.

Flow cytometers and cell sorters use combined systems of fluidics, optics, electronics, and computers. The fluidics system in a typical bench top flow cytometer is used to introduce and focus the particles of interest into the light source (laser) for interrogation. The excitation optics consists of a laser and lenses to shape and focus the beam. The numbers of lasers available in flow cytometry vary. Argon gas is commonly used as an excitation laser. The energized light produced by these lasers emit at 488nm. Many different fluorochromes absorb at 488nm and emit light of longer wavelength in the green (535 nm), orange (585 nm) and red spectrum (600 nm). However, ultraviolet (UV) sensitive or far red fluorochromes require excitation lasers (UV or red diode) that emit in the UV (300-400 nm) or far red (630 nm) range. Typical bench top systems have

* Corresponding Author: James A. Radosevich, Ph.D., Oral Medicine and Diagnostic Sciences, College of Dentistry, 801 S. Paulina Street, Chicago, IL 60612-7213. Email: jrados@uic.edu.

two lasers (argon and red diode). Addition of a Krypton laser provides additional color capabilities in the violet range. All of these lasers will be broadly addressed in this chapter.

This chapter will begin with a discussion of basic principles of flow cytometry and cell sorting, including a technical description of factors, intensively focused on UV lasers, which contribute to the performance of these instruments. The remaining sections will focus on clinical- and research-based applications of flow cytometry and cell sorting.

Keywords: Flow cytometer, Cell sorting, UV lasers

Abbreviations

Cdk2	cyclin-dependent kinase 2
DPSS	diode-pumped-solid state lasers
FACS	fluorescence activated cell sorter
PCR	Polymerase Chain Reaction
PE	phycoerythrin
UV	Ultraviolet radiation
YAG	yttrium aluminum garnet
YVO4	yttrium vanadate (YVO4)

12.0. Definitions and History

The word *flow* indicates some sort of plumbing or fluidics. "Cytometer" is derived from two Greek words, *cyto-*, which means cell, and *–metry*, which refers to counting. Combining these three words gives the term flow cytometry, a technology that measures multiple parameters of cells or particles as they pass through a light source in a fluid stream [1, 2]. Flow cytometry techniques have also been utilized to analyze individual particles that are not cells at all (e.g., viruses [3]. nuclei [4], chromosomes [5], DNA fragments, and latex beads [6, 7].

The technology was invented in the late 1960s by Bonner, Sweet, Hulett, Herzenberg, and others to do flow cytometry and cell sorting of viable cells. Becton Dickinson Immunocytometry Systems introduced the commercial machines in the early 1970s, using the Stanford patent and expertise supplied by the Herzenberg Laboratory and a Becton Dickinson engineering group under Bernie Shoor [8]. The Herzenberg group later coined the term fluorescence activated cell sorter (FACS) in their publications that emerged in 1972. In these publications, FACS was used to sort lymphocytes that had been labeled with a specific antibody [9, 10].

Rapid developments in highly specific monoclonal antibodies combined with huge growth in the availability of many fluorescent molecules to conjugate to these antibodies resulted in massive growth in the field of flow cytometry.

12.1. Principles of Flow Cytometry

A flow cytometer is essentially a fluorescence microscope that gathers light signals emitted from particles flowing across an objective lens. It integrates fluidics, optics, electronics, computer, software, and laser technologies in a single platform [11].

The fluidics system in a typical flow cytometer is used to introduce and focus the particles of interest into the light source for examination. A reservoir of liquid, called sheath fluid is pressurized with room air and shifted toward the illumination point [2]. The cells or particles to be analyzed require being placed in suspension and pressurized to be directed to a fluid stream in all cytometers. The sample is injected as a core stream into a larger, surrounding column of flowing sheath fluid to guarantee that particles appear in single file for analysis. The sample stream becomes compressed to roughly one cell in diameter; this is also known as hydrodynamic focusing [11]. The operator controls the amount of pressure applied to the sample stream, and this in turn affects the uniformity of the interrogation point, where the laser and the sample intersect. The increase in the sample pressure is directly proportional to the increase in the flow rate. In many applications, high pressure or high flow rates are used, however for DNA measurements, the size of the core stream is very important and low flow rates narrow the stream of examination for more precise analysis [2].

It is important that the particles pass through the light source (laser) one at a time. As the particles pass through the laser, they will emit light in form of scattered light or fluorescence. The detectors receive the scattered light and convert the signals into digital data. Some particles emit endogenous fluorescence while others do not. In the latter case, the particles can be stained with fluorescent molecules. The fluorescent molecules absorb light at specific wavelength and emit light at specific wavelength (color) depending on the energy level to which they have been excited. The fluorescent molecules are very useful in flow cytometry because they can be used to label antibodies that are specific to a protein/antigen. When the fluorescent-labeled antibody binds to the protein of interest, the fluorescence signal is a read-out of the amount of protein/antigen to which the antibody has bound. Sometimes, the fluorescent molecules emit light when they are bound to a cellular component [7].

Lasers and lenses to shape and focus the beam form the excitation optics. Lasers in today's cytometers are either gas lasers (e.g., argon ion, krypton ion, or helium–neon lasers) or solid state lasers (e.g., blue solid state laser, red or green diode lasers, violet or near-UV diode lasers, or the relatively new orange fiber laser) [1, 2, 7].

In this chapter, we will review the history of lasers and their use widely in clinical- and research-based applications of flow cytometry and cell sorting.

12.2. Gas Lasers in Flow Cytometry

The earliest flow cytometers used mercury arc lamps and took advantage of the ultraviolet, blue, green and yellow lines. Lasers were rapidly integrated into flow cytometers and these lasers were almost entirely gas-filled ion lasers, including argon, krypton, and helium-neon [13].

12.2.1. Argon- and Krypton-Ion Lasers

The advantage of using Argon and Krypton-Ion lasers lies in the fact that they have long lifetimes. In addition, they produce high quality beams accompanied by low noise and good power stability. However, these lasers are large and they demand a complex cooling system since they produce abundant amounts of waste heat [13]. Singly ionized argon has several laser transitions at wavelengths ranging from the blue-violet (454 nm) to green (528 nm); the highest gain lines are those at 515 and 488 nm, with the 488-nm line predominating in smaller systems and the 515-nm line in larger ones. Argon-ion lasers also produce strong series of lines in the ultraviolet, ranging from 351 to 364 nm [14]. This makes the use of UV-excited dyes such as DAPI, possible which had first been utilized in mercury arc lamp systems. Argon-ion lasers have formed the excitation core of most commercial flow cytometers in the past 25 years [13].

In comparison to Argon-ion laser, the Krypton-ion lasers are less powerful, however, they emit at much broader range of wavelengths. This range of wavelengths produced include UV at 351 and 356 nm, violet at 407 and 413 nm, blue at 520nm, green at 531 nm, yellow at 568 nm, and red at 647 and 676 nm. These lasers first started being used in flow cytometry for two-color immunofluorescence measurements. Despite the fact that they were less powerful and more expensive than argon-ion lasers, the 568-nm line from a krypton-ion laser allows detecting antibodies labeled with Texas Red as well as a variety of fluorescent probes [13, 14].

Mixed-gas argon-krypton lasers have been incorporated into cell sorters, which can produce a wide range of wavelengths from the UV to red, although they are less common.

12.2.2. Helium-Neon Lasers

The most common He-Ne lasers emit red light at 633 nm; other visible wavelengths at which He-Ne lasers are now available include 543, 594, and 611 nm. He-Ne lasers do not require water cooling and thus easier to operate. These lasers were important since they allowed multilaser flow cytometers to be introduced and, hence additional fluorochromes to be used in cytometers. Investigators could excite three fluorochromes [fluorescin, phycoerythrin (PE), and a PE tandem] and APC with a red He-Ne or diode, enabling four-color flow cytometry.

Red He-Ne lasers are very efficient, inexpensive, least complicated, and are the most commonly used lasers available in flow cytometry, and however, they have been displaced with diode lasers in many cases.

12.2.3. Helium-Cadmium Lasers

He-Cd lasers can emit in the blueviolet at 441 nm and in the UV at 325 nm, and like He-Ne lasers, they do not need water-cooling. However, these lasers are not common on flow cytometers due to noise, tube life, and their limited applications. He-Cd lasers have been utilized for only a few flow cytometric applications such as excitation of the fluorescence of the DNA stain DAPI at 325 nm and for the excitation of the chromomycin family of DNA stains (olivomycin, chromomycin A3, and mithramycin) at 441 nm; but these dyes can also be excited by the 457-nm argon-ion laser line [13, 14].

12.3. Dye Lasers

These lasers are used mainly as excitation sources for antibody labels. The dye which will be used is chosen depending on the wavelengths at which the application requires. The laser fluorescent dye rhodamine 6G can produce strong laser lines in the 570 to 620 nm range for excitation of Texas Red, Cy5 or allophycocyanin.

Dye lasers have recently been substituted with solid-state lasers for flow cytometry due to rapid advances in technology.

12.4. Laser Diodes are Light Emitting Diodes (LEDs) Created in the Form of a Laser Cavity

Small red lasers were the first developed diodes with monochromatic wavelengths in the range of 635-645 nm, and incorporated into commercial cytometers in the early 1990s.

The first, blue and violet LEDs and laser diodes were introduced by Shuji Nakamura in 1997. Violet lasers are also monochromatic but they may also emit in the range from 395 to 415 nm.

They are used in DNA content analysis and in exciting several fluorescent probes. The concurrent analysis of fourteen or more specific fluorochromes can be achieved by combining blue-green 488 nm, red He-Ne and violet laser diodes. The wavelengths produced by blue laser diodes (440-450 nm) excite DNA dyes such as mithramycin and chromomycin; hence they are used in chromosome analysis. UV laser diodes that emit at 370-390nm are used for cell cycle analysis, since both Hoeschst and DAPI dyes (which bind to DNA) are excited at these wavelengths. Lastly, green laser diodes were developing for similar application in flow cytometry

It is probable that diode and solid-state lasers, which will be discussed in the next section, will greatly replace ion, He-Ne, and He-Cd lasers as light sources for flow cytometry in the near future.

12.5. Solid State Lasers

The majority of solid-state lasers used in flow cytometry are diode-pumped-solid state lasers (DPSS). The laser diode is used to pump a crystalline lasing medium, resulting in laser emission. An example of this is the use of an 808 infrared laser that pumps an yttrium aluminum garnet (YAG) or yttrium vanadate (YVO4) lasing medium leading in a 1064 nm laser emission.

Green neodymium Nd:YAG and Nd:YVO_4 lasers are good excitation sources for several dyes.

However, these lasers cannot excite fluorescein, a label that has been incorporated into many probes for analysis of a large number of structural and functional components. Fluorescein is excited at the 515 nm wavelength. Thus, the ytterbium YAG (Yb:YAG) lasers can be used to excite fluorescein since their emission wavelength is 1029 nm which can be doubled to 514.5 nm.

In terms of ease of pumping and stability, YAG lasers have advantages over Nd:YAG, suggesting that green Yb:YAG lasers would have less problems than green Nd: YAG lasers. However, the rise of 488-nm solid-state lasers was likely set back by the development of a green Yb:YAG laser for cytometry, and is likely to set back the development of 473-nm, frequency-summed Nd:YAG lasers, which have been offered in some cytometers. [13, 14].

12.6. Applications of UV Lasers in Flow Cytometry

UV lasers are very useful in flow cytometry. The wavelengths produced by these lasers can excite several fluorochromes associated with cell cycle analysis (DAPI and Hoechst), intracellular calcium measurement (chelating probe indo-1), and chromosome sorting.

Modern solid-state UV sources and CW UV lasers are also available for flow cytometry. Some 355 nm lasers are commercially available and produced for industrial purposes as well as laboratory uses.

12.6.1. Cell Sorting

Cell sorters are flow cytometers that select individual particles of interest and divert them from the fluid stream into a collection vessel. Any flow cytometer can be modified to perform cell sorting. This is accomplished by vibrating the sheath stream causing it to break into drops. The sheath stream flows past two charged plates, and the drops can be charged or not depending on the properties of any contained cell, as detected at the interrogation point. Similar to a non-sorting cytometer, the cells in a sorting cytometer flow through the interrogation point where they are hit by a laser beam, which results in the emission of scattered light or fluorescence. In a sorting cytometer, the cells continue to flow downstream, as the stream breaks up into drops, they become enclosed in individual drops.

Cells sorted by flow cytometry are routinely used for functional assays, for PCR replication of cell-type specific DNA sequences, for artificial insemination by sperm bearing X or Y chromosomes, and for cloning of high-expressing transfected cells. In addition, sorted chromosomes are used for the generation of DNA libraries [2, 7, 11].

12.6.1.1. Research-Based Applications

Flow cytometry is used widely in stem cell research and cell type-specific developmental analysis [15, 16].

Sorting has referred to as reading through a highly complex cell-based or solid support molecular library in order to select genetic clones and/or molecules of interest for further chemical, cellular, or molecular studies. With protein expression library screening, cells provide a physical link between expressed phenotype and genotype [17]. Ingenious fluorescent labeling schemes are used to indicate binding or activity characteristics of a specific outcome both qualitatively and quantitatively [18, 19]. These cells can then be repeatedly sorted and expanded, thereby enriching for desired clones. The cloning of fluorescent reporter genes such as green fluorescent protein, dsRed, and their homologs have revolutionized the field of fluorescence microscopy and flow cytometry through their ability to tag individual cells for the expression of single genes or a combination of genes [20-23].

The expression of mammalian and bacterial proteins can be directly monitored by fluorescent reporters easily detectable with mainstream flow cytometers.

Another use of flow cytometers includes their use to screen for G-protein-coupled receptors [24], to decipher immunologic signaling networks with phosphospecific flow cytometry [25], and other enzymatic activity. High-throughput drug screening, signaling through the use of bar-coding, use of a cell sorter as a DNA sequencing device, and transcriptional profiling are but a few of the new uses for cytometers as a molecular tool [26, 27].

12.6.2. Clinical Diagnostics

Flow cytometry has expanded over the past 30 years from a laboratory technique to a routine tool used by clinical pathologists and immunologists for diagnosis and monitoring of patients with cancer and immune deficiencies as well as transplantation medicine and genetics. The fact that high-speed sorters have increased capabilities and flexibility, has made it possible to measure many parameters in parallel using multiple lasers with high sensitivity and accuracy [28-33]. Rapid advances in monoclonal antibodies that specifically bound to cell surface markers [34] and the discovery of a variety of fluorescent dyes with narrow excitation and emission spectra [8] permitted the application of this technology to basic immunology research, clinical immunology diagnostics, and cell selection for preclinical models and clinical trials of transplanting phenotypically defined cell subsets [35].

Prenatal diagnostics is another clinical area where flow cytometry will keep growing. Current methods are amniocentesis or chorionic villus sampling, procedures which carry certain risks. More recently, there have been reports of using flow cytometry to identify and isolate fetal cells within maternal blood [36, 37]. This provides easier, relatively safer, and inexpensive approach for prenatal diagnostics. Analysis of sperm quality by flow cytometry has also been reported [37].

12.6.2.1. Cancer Diagnosis and Tumor Biology

A number of literature reports support the application of flow cytometry in disease diagnosis, prognosis, and monitoring of patients with hematological malignancies [38, 39], as well as in clinical studies of immune status in patients with cancer. Flow cytometry analysis can provide vital diagnostic and prognostic information and identify with high accuracy the distinctive immunophenotype of malignancies and cell DNA content that are used to select specific therapies. For example, various CD201 B cell malignancies are now commonly treated with anti-CD20 Abs (rituximab), and a subset of CD52-expressing T cell and B cell malignancies may be treated with anti-CD52 (alemtuzumab). Flow cytometry is used to distinguish mantle cell lymphoma, an aggressive disorder that requires more intensive therapy, from other more indolent mature CD201 B cell lymphomas with similar morphology. Because the Ag-expression profiles of these B cell neoplasms can be subtly different, flow cytometry can facilitate diagnosis by identifying reproducible expression patterns of multiple Ag is associated with specific neoplasms and relatively quantify Ag density on malignant cells [40].

The power of the flow cytometer to detect minute quantities of specific cells in heterogeneous cell mixtures is used to identify residual malignant cells after therapy, known as minimal residual disease monitoring, in common malignancies for which disease persistence indicates a worse prognosis [41].

12.6.2.2. Cell Cycle Analysis and Ploidy

The measurement of DNA content (ploidy) of tumor cells after intercalation of fluorescent dyes such as propidium iodide into the DNA is one of the common applications in human oncology. Tumor cells from fixed tissues embedded in paraffin blocks can be analyzed by flow cytometry for DNA content which made a considerable contribution to diagnosis and prognosis of common solid tumors [2].

Two basic approaches exist for cell cycle analysis; one involves the administration of thymidine analogues to cancer patients, and the second makes use of endogenous markers involved in specific phases of cell cycle progression [42].

Thymidine analogues have substitutions at the fifth-position of the pyrimidine ring that is sufficiently like the normal methyl group that the nucleotide is incorporated into DNA during S-phase like thymidine. Halogen substitutions can be detected by flow cytometry using specific antibodies. The dynamic analysis of cell cycle progression by *in vivo* labeling with thymidine analogues is a highly versatile technique [43].

Antibodies are also available in flow cytometry for the analysis of endogenous markers involved in specific phases of cell cycle progression. For example, Cyclin A2 is the binding partner for cyclin-dependent kinase 2 (Cdk2), and plays a major role regulating progression through S-phase. Cyclin A2 and Cyclin B1 bind to and regulate Cdk1 involved in breakdown of the nuclear membrane and assembly of the mitotic spindle during entry into mitosis. At the beginning of mitosis, histone H3 becomes phosphorylated, and this distinguishes the mitotic cells in the G2/M peak of a DNA histogram. This can be combined with DNA content and/or thymidine analogue incorporation to build a highly detailed picture of cell cycle progression in cells [42, 44].

Metabolic regulation, maintenance of cellular antioxidant systems and epigenetic silencing by chromatin condensation which are popular research areas in cancer biology are applicable to flow cytometric analysis.

12.6.2.3. Immunologic Diseases

The flow cytometer is also used as a principal clinical laboratory platform for the analysis of immune status of patients, including determination of CD41 T cell blood counts after HIV infection [45]. The classification and prognosis of various primary immune deficiencies, such as autoimmune lymphoproliferative syndrome, severe combined immune deficiency, and Ab deficiencies can also be determined [46].

12.6.2.4. Cell-Based Therapy and Transplantation

The identification of CD341 as a marker for human hematopoietic stem and progenitor cells [17] led to interest in using flow cytometry to isolate and purify populations of phenotypically defined CD341 cells for transplantation. The rationale for this approach was that patients with cancer can be treated with high-intensity chemotherapy and radiation, followed by autologous stem cell rescue to achieve remission of malignancies that relapsed after conventional chemotherapy or for which conventional chemotherapy was inadequate for longterm remissions. The clinical use of cytometry-based FACS for isolation of autologous CD341 stem cells allowed the possibility of achieving a clinically significant depletion of contaminating tumor cells, reducing the risk for disease progression from the reinfusion of contaminating tumor cells in the stem cell graft.

Protocols for the study of autologous transplantation of patients with breast cancer, non-Hodgkin's lymphoma, and multiple myeloma using highly purified flow cytometry-sorted CD341 progenitors were initiated and the results of these studies were promising.

Another field of use for flow cytometry is the measurement of antibodies to HLA molecules that potentially developed after exposure through blood transfusions, or prior to transplantation, or pregnancy. It can be use to determine HLA genotypes for transplantation. Human histocompatibility testing has moved ahead in using multiplexing technology, particularly in providing detailed HLA antibody analysis, and molecular HLA typing.

The use of multiplexed bead assays and DNA ploidy assays in high-throughput diagnostic or prognostic evaluations, and studies involving detection of rare cell subsets can lead to the acquisition of extremely large data sets. Progress in data sharing is hampered by the lack of standardization in data analysis and reporting; as well as a means of efficiently sharing large data sets for independent evaluation and collaboration across different sites. To maximize the potential of flow cytometry for clinical use and investigation, informatics systems, such as those applied in analyses of massive genomics and proteomics datasets, will be needed as well as being integrated with other data-collection efforts. The emerging field of "cytomics" represents an integrated, whole-cell–based description of cellular physiology that includes aspects of genomics (genes and regulatory processes) and proteomics (protein abundance and structure) which endeavors to develop a more comprehensive picture of the biology of individual cells.

Automated systems for flow cytometry data analysis are being developed that use standardized ontologies and methods that treat such data sets as statistical objects embedded in dimensional space or clustered distributions that may be compared with one another to foster objective data capture, evaluation, and reporting [35].

These developments will advance the clinical use of flow cytometry and broaden its applications in the investigation of disease beyond hematologic malignancies and immune monitoring, into a new age of clinical discovery and focused pathophysiology.

REFERENCES

[1] Delude RL (2005) Flow Cytometry. *Crit. Care Med.,* 33:S426-428.

[2] Wilkerson MJ (2004) Principles of Flow Cytometry and Cell Sorting. In: 55th Annual Meeting of the American College of Veterinary Pathologists (ACVP) &39th Annual Meeting of the American Society of Clinical Pathology (ASVCP) pp. 1-4.

[3] Marie D, Brussard, CP D, Thyrhaug, R, et al (1999) Enumeration of marine viruses in culture and natural samples by flow cytometry. *Appl. Environ. Microbiol.,* 59: 905–911.

[4] Hedley DW (1989) Flow cytometry using paraffin-embedded tissue: Five years on. *Cytometry,* 10: 229–241.
[5] Gray JW. and Cram LS. (1990) Flow karyotyping and chromosome sorting, in Flow Cytometry and Sorting, 2nd edition (Melamed M R, Lindmo T, and Mendelsohn ML, eds.), Wiley-Liss, New York.
[6] Habbersett RC, Jett JH, and Keller RA. (2000) Single DNA fragment detection by flow cytometry, in Emerging Tools for Single-Cell Analysis (Durack G and Robinson JP, eds.), Wiley-Liss, New York, pp. 115–138.
[7] Givan AL (2011) Flow Cytometry: An Introduction, in Flow Cytometry Protocols, 3rd edition (Hawley TS, Hawley TG, eds.) Springer, New York, pp 1-7.
[8] Herzenberg LA, Parks D, Sahaf B et al (2002) The History and Future of the Fluorescence Activated Cell Sorter and Flow Cytometry:A View from Stanford. *Clinical Chemistry,* 48: 1819-1827.
[9] Ono S, Tsujimoto H, Matsumoto A et al (2004) Modulation of human leukocyte antigen-DR on monocytes and CD 16on granulocytes in patients with septic shock using hemoperfusion with polymyxin B-immobilized fiber. *Am. J. Surg.,* 188:150-156.
[10] Skrzeczynska J, Kobylarz K, Hartwich Z et al (2002) CD14 CD16 monocytes in the course of sepsis in neonates and small children: Monitoring and functional studies. *Scand. J. Immunol.,* 55: 629-638.
[11] Ibrahim SF, Engh GV (2007) Flow Cytometry and Cell Sorting. *Adv. Biochem. Engin/Biotechnol.,* 106: 19-39.
[12] Darzynkiewicz Z, Juan G, Li X, et al (1997) Cytometry in cell necrobiology: Analysis of apoptosis and accidental cell death (necrosis). *Cytometry,* 27, 1–20.
[13] Telford WG (2011) Lasers in Flow Cytometry. Methods Cell Biol 102: 375-409.
[14] Shapiro HM. (2004) Lasers for Flow Cytometry. *Curr. Protoc. Cytom. Chapter 1: Unit,* 1.9.1-1.9.16.
[15] Tamaki S, Eckert K, He D et al (2002) Engraftment of sorted/expanded human central nervous system stem cells from fetal brain. *J. Neurosci. Res.,* 69: 976–986.
[16] Huntly BJ, Gilliland DG (2005) Leukaemia stem cells and the evolution of cancer stem-cell research. *Nat. Rev. Cancer,* 5: 311–321.
[17] Ibrahim SF, van den Engh G (2003) High-speed cell sorting: fundamentals and recent advances. *Curr. Opin. Biotechnol.,* 14: 5–12.
[18] Chen W, Georgiou G (2002) Cell-surface display of heterologous proteins: from highthroughput screening to environmental applications. *Biotechnol. Bioeng.,* 79: 496–503.
[19] Daugherty PS, Iverson BL, Georgiou G (2000) Flow cytometric screening of cell-based libraries. *J. Immunol. Methods,* 243:211–227.
[20] Chalfie M, Tu Y, Euskirchen G, et al (1994) Green fluorescent protein as a marker for gene expression. *Science,* 263:802–805.
[21] Misteli T, Spector DL (1997) Applications of the green fluorescent protein in cell
[22] biology and biotechnology. *Nat. Biotechnol.,* 15: 961–964.
[23] Matz MV, Fradkov AF, Labas YA et al (1999) Fluorescent proteins from nonbioluminescent Anthozoa species. *Nat. Biotechnol.,* 17: 969–973.
[24] Arun KH, Kaul CL, Ramarao P (2005) Green fluorescent proteins in receptor research: an emerging tool for drug discovery. *J. Pharmacol. Toxicol. Meth.,* 51: 1–23.

[25] Waller A, Simons P, Prossnitz ER, Edwards BS, Sklar LA (2003) High throughput screening of G-protein coupled receptors via flow cytometry. *Comb. Chem. High Throughput Screen,* 6: 389–397.
[26] Krutzik PO, Nolan GP (2006) Fluorescent cell barcoding in flow cytometry allows high-throughput drug screening and signaling profiling. *Nat. Methods,* 3: 361–368.
[27] Farinas ET (2006) Fluorescence activated cell sorting for enzymatic activity. *Comb. Chem. High Throughput Screen*, 9: 321–328.
[28] Krutzik PO, Hale MB, Nolan GP (2005) Characterization of the murine immunological signaling network with phosphospecific flow cytometry. *J. Immunol.,* 175:2366–2373.
[29] Wood B (2006) 9-color and 10-color flow cytometry in the clinical laboratory. *Arch. Pathol. Lab. Med.,* 130:680–690.
[30] Pagnucco G, Vanelli L, Gervasi F (2002) Multidimensional flow cytometry immunophenotyping of hematologic malignancy. *Ann. NY Acad. Sci.,* 963:313–321.
[31] Jennings CD, Foon KA (1997) Recent advances in flow cytometry: application to the
[32] diagnosis of hematologic malignancy. *Blood,* 90: 2863–2892.
[33] Alvarez-Barrientos A, Arroyo J, Canton R, Nombela C, Sanchez-Perez M (2000) Applications of flow cytometry to clinical microbiology. *Clin. Microbiol. Rev.,* 13: 167–195.
[34] De Rosa SC, Roederer M (2001) Eleven-color flow cytometry. A powerful tool for
[35] elucidation of the complex immune system. *Clin. Lab. Med.,* 21: 697–712
[36] O'Gorman MR (2001) Clinically relevant functional flow cytometry assays. *Clin. Lab. Med.,* 21: 779–794
[37] Kohler G, and Milstein C. (1976) Derivation of specific antibody-producing tissue culture and tumor lines by cell fusion. *Eur. J. Immunol.,* 6: 511–519.
[38] Jaye DL, Bray RA, Gebel HM et al (2012) *Translational Applications of Flow Cytometry in Clinical Practice J. Immunol.,* 188:4715-4719.
[39] Purwosunu Y, Sekizawa A, Koide K, et al. (2006) Clinical potential for noninvasive prenatal diagnosis through detection of fetal cells in maternal blood. *Taiwan J. Obstet. Gynecol.,* 45:10–20.
[40] Graham JK (2001) Assessment of sperm quality: a flow cytometric approach. *Anim. Reprod. Sci.,* 68: 239–247.
[41] Wood BL, M Arroz D, Barnett J et al (2007). 2006 Bethesda International Consensus recommendations on the immunophenotypic analysis of hematolymphoid neoplasia by flow cytometry: optimal reagents and reporting for the flow cytometric diagnosis of hematopoietic neoplasia. Cytometry B Clin. Cytom. 72 (Suppl. 1): S14–22.
[42] Paiva B, MB Vidriales J, Cervero G et al (2008) GEM/PETHEMA Cooperative Study Groups. Multiparameter flow cytometric remission is the most relevant prognostic factor for multiple myeloma patients who undergo autologous stem cell transplantation. *Blood,* 112: 4017–4023.
[43] Kraus TS, CN Sillings DF Saxe et al (2010) The role of CD11c expression in the diagnosis of mantle cell lymphoma. *Am. J. Clin. Pathol.,* 134: 271–277.
[44] Moreton PB, Kennedy, G Lucas et al (2005). Eradication of minimal residual disease in B-cell chronic lymphocytic leukemia after alemtuzumab therapy is associated with prolonged survival. *J. Clin. Oncol.,* 23: 2971–2979.
[45] Darzynkiewicz Z, Crissman H, Jacobberger JW (2004) Cytometry of the cell cycle: cycling through history. *Cytometry A,* 58: 21-32.

[46] Chang Q, Hedley D (2012) Emerging applications of flow cytometry in solid tumor biology. *Methods,* 57: 359-367.

[47] Jacobberger JW, Sramkoski RM, Stefan T (2011) Multiparameter cell cycle analysis. *Methods Mol. Biol.,* 699:229-249.

[48] Barnett D, Walker B, Landay A et al (2008) CD4 immunophenotyping in HIV infection. *Nat. Rev. Microbiol.,* 6: S7–S15.

[49] Oliveira JB, Notarangelo LD, and Fleisher TA. (2008) Applications of flow cytometry for the study of primary immune deficiencies. *Curr. Opin. Allergy Clin. Immunol.,* 8: 499–509.

In: UV Radiation: Properties, Effects, and Applications
Editor: James A. Radosevich
ISBN: 978-1-63321-090-5

Chapter 13

APPLICATIONS OF ULTRAVIOLET RADIATION TO ANALYTICAL MASS SPECTROMETRY

Alexandre Giuliani*, Dr. Sc.
Synchrotron SOLEIL, L'Orme des Merisiers - Saint Aubin, France
CEPIA, Institut National de la Recherche Agronomique (INRA),
Rue de la Geraudiere, Nantes, France

ABSTRACT

Recently, ultraviolet (UV) radiation has found two promising applications in analytical mass spectrometry. UV light may be used to generate ions through photoionization of molecules of interest. The development of commercial atmospheric pressure photoionization (APPI) sources has triggered a growing use of this method. The large success of APPI comes from is compatibility with separative techniques and its large tolerance to salts, additives, and organic solvents. UV has also attracted attention as an activation method in tandem mass spectrometry. Indeed, photoabsorption in the UV range is a straightforward means of activation of ions. Although no commercial setups exist presently, UV activation of ions appears as an extremely promising method.

Keywords: Photon activation, Photoionization, structural analysis, mass spectrometry, lasers, synchrotron radiation, photochemistry

ABBREVIATIONS

APPI	Atmospheric Pressure Photoionization
CE	Capillary electrophoresis
CID	Collision induced dissociation
DEA	Dissociative Electron Attachment
DPI	Dissociative photoionization

* Email: giuliani@synchrotron-soleil.fr.

ECD	Electron capture Dissociation
EPD	Electron Photodetachment Dissociation
ESI	Electrospray ionization
ETD	Electron Transfer Dissociation
FT-ICR	Fourier Transform Ion Cyclotron Resonance
IE	Ionization Energy
LC	Liquid chromatography
MALDI	Matrix Assisted Laser Desorption Ionization
MS	Mass spectrometry
MS/MS	Tandem mass spectrometry
PA	Proton Affinity
PD	Photodissociation
SR	Synchrotron Radiation
ToF	Time of Flight
UV	Ultraviolet
VUV	Vacuum ultraviolet

13.0. INTRODUCTION

Mass spectrometry is recognized as a key technology in several general scientific domains and in analytical sciences specifically [1]. A mass spectrometer is composed of an ion source, a mass filter, and an ion detector. The ion source has the double task to place in the gas phase the sample and to provide it with an electrical charge to generate ions. The mass filter sorts out the ions from the source according to their mass to charge ratio (*m/z*). The abundance of the ions is measured by the detector. A mass spectrum is a representation of the ions abundance as a function of the *m/z* ratio. Although these pieces of information are very desirable data, more information can be obtained from tandem mass spectrometry. Especially, in a particular arrangement, an ion of interest can be selected, isolated either temporally or spatially in the mass filtering system, and activated. In the activation step, the internal energy of the ion is increased to produce fragmentation. Analysis of the fragments provides valuable data on the precursor ions, such as structural information.

Ultraviolet (UV) radiation may be used at two levels in the mass spectrometry workflow described above. First at all, UV may naturally be used in the ionization source to produce ions. For this, energetic photons in the vacuum-ultraviolet are needed to photoionize the species of interests. Although photoionization is a basic technique in molecular physics, which has been used for decades now [2], it has only recently found applications in analytical mass spectrometry with the advent of Atmospheric Pressure PhotoIonization (APPI) in the beginning of the 2000's [3,4]. APPI has been developed to fulfill the need of accessing some classes of molecules, which were difficult to analyze with the other ionization methods. In addition, APPI is compatible with liquid chromatography, allowing on-line analysis.

Another application of UV also lies in the activation step in tandem mass spectrometry. Usual activation methods rely on collisional activation [5] of the precursor ions of interest with inert gases such as noble gases or nitrogen. However, a straightforward mean of increasing internal energy of ions may be achieved through photoabsorption of energetic photons, such as in the UV wavelength range. This idea leaded to a new class of fragmentation techniques occuring, at least partly, on electronic excited potential energy surfaces of the ions.

In this chapter, atmospheric pressure photoionization and UV photon activation are discussed. We first address the technical aspects and the instrument arrangements before discussing the mechanistic features. Then, the domains of application of the techniques are briefly discussed. Finally, some perspectives of these two new techniques are given.

13.1. Atmospheric Pressure Photoionization

Direct sampling of a solution is often a necessity in analytical sciences, especially for the analysis of eluents from separation techniques such as liquid chromatography or capillary electrophoresis. However, the presence of solvents is a constraint, which complicates the necessary placement in the gas phase of the samples to be analyzed. Liquids evaporate under reduced pressure, which makes them difficult to handle under such conditions. It has therefore been necessary to invent desorption and ionization methods in which the molecules of interest are put in the gas phase at atmospheric pressure before entering the vacuum of the mass spectrometer. For this, it is mandatory to vaporize and to remove the solvents around the analytes (desolvatation) before providing it an electrical charge. Aside from electrospray ionization (ESI) [6], several other desorption-ionization methods exist. One of the most ancient and simple method is thermospray ionization. This technology is based on heating a spray of a solution containing the sample. Evaporation of the solvent may sometimes directly produce ions in the gas phase, which, in turn, may be analyzed and detected by mass spectrometry. However, the limitations of thermospray have led to use this method of desorption in conjunction with corrona discharge (atmospheric pressure chemical ionization, APCI) or UV irradiation. The later combination is referred to as Atmospheric Pressure Photoionization (APPI). APPI has been the last of the atmospheric pressure ionization method to appear in the beginning of the century. APPI completes advantageously the arsenal of the available techniques by providing access to molecular species of low polarity.

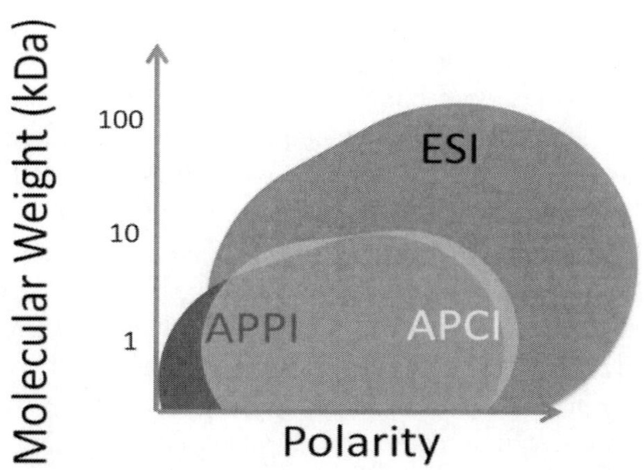

Figure 1. Schematic domain of application of atmospheric pressure ionization methods. APPI (dark gray), APCI (light gray) and ESI (grey).

13.1.1. Principles

APPI and its sister technique APCI both rely on thermal vaporization. It is then natural to address first thermospray before discussing the physical principle at work behind APPI.

13.1.1.1. Thermospray Ionization

Thermospray is a vaporization - desolvatation method that has been popular since the 1980s and 1990s. It was developed in 1983 by Vestal and coworkers [7]. Thermospray consists of spraying a solution under the combined action of a nebulizing gas and a vigorous heating. Initially, the liquid was sprayed under reduced pressure. The concept was later improved, and the solution was put in the gas phase at atmospheric pressure in a heated tube. The gas phase is composed of the analyte, the solvent, and the nebulizing gases.

For polar molecules, thermospray may release in the gas phase ions preexisting in solution. The addition of ammonium ions in the solvent increases the sensitivity. Under similar experimental conditions, thermospray and electrospray ionizations produce comparable ions, although both methods are not based on the same principles [8]. Thermospray creates charged droplets in the gas phase from an analyte in solution already in ionic form. It was proposed that thermospray ionization occurs according to the following processes: direct evaporation or ion chemical ionization by proton transfers [9]. However, the method itself does not lend well to the ionization of highly hydrophilic compounds [9,10]. This limitation surely comes from the difficulty to desolvate efficiently polar molecules. The molecular mass of the analyte also plays a role, as the method has been shown to offer good sensitivity for analytes of mass lower than 2,000 Da. Although based on a substantial heating of the solution (300-400°C), thermal fragmentations are hardly observed. Indeed, the analytes are placed in the gas phase without extensive pyrolysis owing to the rapid heating and to the protective effect of the solvent. As discussed below in more details, the commercial laboratory APPI sources are all based on thermal desolvation. Thus, most APPI sources can be operated in thermospray mode by turning off the UV radiation [11-13].

Thermospray is a robust mode of ionization, tolerant to salts and additives and compatible with all type of solvents (methanol, heptanes, dichloromethane...), even very non-polar ones. APPI has inherited these characteristics from thermospray.

13.1.1.2. Atmospheric Pressure Photoionization (APPI)

APPI is the latest of the atmospheric pressure ionization techniques compatible with liquid chromatographic separations. The technique was developed independently and almost simultaneously in the early 2000s by the groups of Bruins [4] and Syage [3]. The main motivation behind these developments was the need to expand the domain of application of the LC / MS to more apolar compounds, which were otherwise difficult to access by electrospray or other techniques [14], see Figure 1.

APPI sources are built on a heated nebulizer. The eluents from the LC are delivered though a metal capillary and sprayed by a flow of gas in the middle of a heated quartz tube. This aerosol is of the same nature as that produced by thermospray. When the efflux of the quartz tube reaches the interaction region, the gas phase is irradiated by the UV photons from the UV lamp.

Discharge lamps, commonly used in APPI, are made of a sealed chamber filled with an inert gas under reduced pressure. An electrical discharge is maintained between the electrodes of the lamp. The collisions of the discharge electrons with the medium produces electronically excited atoms, which relax by emitting a fluorescence radiation in the UV range. The wavelength of the emission is dependent on the nature of the gas. Discharge lamps in xenon (8.4 eV and 9.6 eV), krypton (10 eV and 10.6 eV) or argon (11.8 eV) are commercially available. The most commonly used lamps in APPI is the krypton one providing 10.6 eV photons.

Two ionization modes are possible under APPI conditions: direct photoionization or assisted photoionization. The operating conditions of the ion source will determine the physical state of the gas phase submitted to the radiation of the lamp, which affects ionization mechanisms. Note that the photoionization under APPI happens in a dense, heated, and complex medium (gas, solvents, analytes). These characteristics are responsible for the complexity of the reactions occurring in APPI sources.

13.2. Ionization Mechanisms

13.2.1. Direct Photoionization

If the ionization energy of the analyte is lower than the energy of the photons emitted by the lamp, a single photon photoionization phenomenon may occur:

$$M+h\nu \rightarrow M^{+\bullet}+e^-$$

Although the observation of the radical cation $M^{+\bullet}$ is expected under direct photoionization, the presence of this ion is often accompanied by the protonated molecules $[M+H]^+$. It has been proposed [15] the relative abundance of ions $M^{+\bullet}$ and $[M+H]^+$ were dependent on the solvent. For 1-aminonaphthalene, the products of APPI in methanol gave mainly an ion at *m/z* 144 corresponding to the protonated molecule. The same experiment conducted under identical conditions but in a totally aprotic solvent such as CCl_4 leaded to the formation of the radical cation $M^{+\bullet}$. Surprisingly, a mixture of both $M^{+\bullet}$ and $[M+H]^+$ species was observed in dichloromethane, which is not considered as a protic solvent [15]. In the model [15], the radical cation would react with the solvent by hydrogen transfer:

$$M^{+\bullet}+S \rightarrow [M+H]^+ +(S\text{-}H)^\bullet$$

Although this model qualitatively accounts for the observation, it should be further refined to achieve a more quantitative description. Some authors [16] have suggested that autoprotoniation phenomena might occur and lead to the formation of protonated molecules in totally aprotic solvents. Recently [17] there has been a review of these models and new mechanisms were proposed based on cluster chemistry.

Direct photoionization is however, not very effective. Indeed, whatever the photon energy of the photoionization source used, direct photoionization suffers from high UV absorption from the nebulizing gas and solvent vapors, which reduces the photon flux available for the ionization. In addition, the analyte is often present in low concentrations in the environment. These drawbacks explain the lack of sensitivity of APPI under direct photoionization conditions and have driven the appearance of a variant called assisted APPI.

13.2.2. Assisted Photoionization

If the ionization energy of the solvent is lower than the photon energy of the lamp, the solvent can be photoionized according to:

$$S + h\nu \rightarrow S^{+\bullet} + e^-$$

Here, we can consider that the solvent absorbs all the photon flux. The solvent radical cation $S^{+\bullet}$ can then react with the analyte by:

Charge transfer: $M + S^{+\bullet} \rightarrow M^{+\bullet} + S$, if the ionization energy of M is lower than that of S, and

Proton transfer: $M + S^{+\bullet} \rightarrow [M+H]^+ + (S\text{-}H)^\bullet$, if the proton affinity of $(S\text{-}H)^\bullet$ is lower than that of M.

These two chemical ionization reactions are very effective and increase substantially the sensitivity of APPI.

Sometimes, the solvent cannot be ionized by the UV photons of the lamp; that is, if the ionization energy of the solvent is greater than the photon energy. An additional molecule, called the dopant (D), ionizable by the UV photons of the lamp, may be introduced:

$$D + h\nu \rightarrow D^{+\bullet} + e^-$$

The ionization of the dopant initiates similar chemical ionization reactions:

Charge transfer: $M + D^{+\bullet} \rightarrow M^{+\bullet} + D$, if the ionization energy of M is lower than that of D or,

Proton transfer: $M + D^{+\bullet} \rightarrow [M+H]^+ + (D\text{-}H)^\bullet$, if the proton affinity of (D-H) is lower than that of M.

Some authors [4] have suggested a possible intervention of solvent clusters:

1) $D^{+\bullet} + nS \rightarrow (D\text{-}H)^\bullet + [S_n + H]^+$, if $AP(D\text{-}H)^\bullet < AP(S_n)$, where S_n represent a cluster of solvent. However, the presence of clusters of solvent under the experimental conditions found in APPI is subject to question. Indeed, clusters are non-covalent and fragile assemblies, whose existence at 400 ° C and at atmospheric pressure still need to be established.

13.2.3. Reactions Induced by Electron under Assisted APPI: The Negative Ionization Mode

Photoionization naturally generates positive ions following the absorption of ultraviolet radiation. However, photoionization of the solvent or the dopant also releases photoelectrons in the medium. These electrons induce reactions that lead to the creation of negative ions and can also be the cause of fragmentation reactions. Hence, in the negative ionization mode, the photoelectrons can attach to the analyte to produce an anion:

$$M + e^- \rightarrow M^{-\bullet}$$

This anion can also react in a dissociative electron attachment mechanism:

$$M^{-\bullet} \rightarrow [M\text{-}H]^- + H^\bullet$$

The atmospheric molecular oxygen, even in traces levels, can also attach an electron:

$$O_2 + e^- \rightarrow O_2^{-\bullet}$$

The superoxide anion can in turn react with the analyte by:

$$\text{Charge transfer: } O_2^{-\bullet} + M \rightarrow M^{-\bullet} + O_2 \text{ or}$$
$$\text{Proton transfer: } O_2^{-\bullet} + M \rightarrow [M\text{-}H]^- + HO_2^\bullet$$

Dissociative electron attachment reactions and dissociative electron transfer reactions have been reported for molecules containing halogens or NO_2 groups [18,19]. Some oxidation reactions have been reported, for example, with phenyl ethers [20].

13.2.4. Other Reactions Induced by Electron-Assisted Photoionization

Since APPI is based on thermospray mechanisms for desolvatation, positive ions may be released in the gas phase. Thus, the combination of thermospray and dopant-assisted APPI may produce two species in the gas phase prone to reaction: a positively charged ion and an electron. Sometimes, the thermospray ionization produces multiply protonated ions. These ions react according to a mechanism of dissociative electron capture (electron capture dissociation, ECD) or dissociative electron transfer (Electron Transfer Dissociation, ETD) [21]:

$$[M + nH]^{n+} + e^- \rightarrow [M + nH]^{(n-1)+} \rightarrow Fragments$$

$$[M + nH]^{n+} + T^{-} \rightarrow [M + nH]^{(n-1)+} + T \rightarrow Fragments$$

where T is an electron carrier like O_2.

This recombination mechanism has been evidenced by the observation of *c*- sequence ions during the ionization of peptides under assisted APPI [21,22]. A variant has been reported in negative ionization mode of hydrophobic peptides [23]. Following this observation, Robb and colleagues have modified an electrospray ion source to optimize the fragmentation induced by various sources [24].

13.3. Applications

The main application of APPI is centered on hydrophobic molecules, but the technique has been applied to the analysis of several other classes of analytes. APPI appears as one of the most versatile ionization techniques. Particularly, the method has been successfully used in the study of complex mixtures, such as the analysis of crude oils [25,26], lake sediments [27], characterization of dissolved organic matter (DOM) [28], etc. APPI generates a greater diversity of ion than other techniques because of the multitude of possible ionization mechanisms at work in the ions source. This specificity is interesting in the analysis of complex mixture.

13.4. Other Developments

13.4.1. Micro APPI

Commercial APPI sources require high solvent flow rates (typically a few tenth of μl / min to 1mL/min). Kostiainen and co-workers [29] have developed a chip based-APPI to reduce the solvent flow rates. APPI could then be coupled with gas chromatography or capillary liquid chromatography. The micro-APPI source (μAPPI) includes an entry for the LC solvent as well as nebulizing gas. Heating is done directly on the chip thanks to a pattern of aluminum elements in which an electrical current circulates. The chip accepts flow rates from 0.05 to 5 μl/min. The μAPPI source otherwise has the same characteristics as the conventional source.

13.4.2. Desorption APPI (DAPPI)

The emergence of ambient ionization methods (DART and DESI) has initiated a major activity in this area from which APPI has not escaped. Kostiainen and co-workers have adapted the μAPPI source for ambient analysis [30]. This technique is based on thermal desorption of the samples deposited on the surface to be analyzed, followed by conventional gas phase APPI ionization mechanisms.

13.4.3. Electron Capture Source at Atmospheric Pressure

The work of Laprevote and colleagues [21,23,31] has shown the possibility of forming *c*-type ions for polypeptides through mechanisms of electron capture or transfer at atmospheric pressure. Following this Robb and colleagues [24] have developed a new ion source to optimize these processes, which combines electrospray ionization and APPI.

13.4.4. Atmospheric Pressure Laser Ionization (APLI)

A variant to APPI has been introduced by Benter and collaborators [32], in which the UV lamp has been replaced by a UV laser, such as pulsed excimer laser at 248 nm, quadrupled Nd-YAG laser at 266 nm [17] or UV OPO lasers [33]. This technique, named Atmospheric Pressure Laser Ionization (APLI), is based on Resonance Enhanced Multi Photon Ionization (REMPI) and proceeds though sequential absorption of photons:

$$\mathrm{M} + \mathrm{n}h\nu \rightarrow \mathrm{M}^{*}$$

$$\mathrm{M}^{*} + \mathrm{m}h\nu \rightarrow \mathrm{M}^{+\bullet} + e^{-}$$

APLI is thus applicable to the analysis of chromophores showing absorption in the near UV region. This is a major restriction of APLI compared to APPI at 10 eV photons, for which virtually all analytes are subject to photoionization. However, since most solvents and gases do not absorb at the laser wavelength, APLI is very sensitive and produces radical cations with high abundance. However, the method suffers from a poorer duty cycle owing the pulsed nature of the lasers employed.

APLI has been applied to the analysis of poly-aromatic hydrocarbons, aromatic and highly-conjugated compounds. Crude oils have been successfully submitted to APLI-MS analysis [34]. Derivatization strategies have been used to access small molecules and polymers through APLI [33].

13.4.5. Wavelength Resolved APPI

APPI is based on the use of discharge lamps delivering discrete photon energy rays. If the photon energy has to be changed, then a discharge lamp in another gas has to installed. The use of the continuous light source capable of scanning a wide range of wavelength in the ultraviolet would bring an additional dimension to the method. Laprevote and collaborators have coupled an APPI source with a UV beamline at the SOLEIL synchrotron facility. This unique coupling was realized in 2011 [35] on a branch line of the DISCO beamline [36,37]. The tunability and the spectral purity of the radiation have been used to confirm the proposed mechanisms. In addition, wavelength resolved APPI offers the unique possibility to measure ionization energies of compounds for which the classical vaporization techniques failed [35].

13.5. UV ACTIVATION OF IONS

Since the introduction of modern ionization techniques (ESI and MALDI), mass spectrometry has been recognized as a powerful tool to study the structure of high molecular weight molecules, such as the amino acid sequence of proteins and co- and post-translational modifications [1,38]. Especially, tandem MS or MS/MS, a technique in which a particular ion of interest is isolated, fragmented and mass analyzed, has proven extremely useful for determination of primary structures of biopolymers. The usual activation methods are based on vibrational excitation of the target ions, such as low energy multiple collision-induced dissociation (CID), infrared multi photon dissociation (IRMPD) or black body infrared dissociation (BIRD) [39]. A technique based on the dissociative recombination of multiply protonated ions and referred to as Electron Capture Dissociation (ECD), has gained considerable attention in proteomics [40,41] owing to its unique properties. From a phenomenological point of view, ECD can fragment samples of high masses. It cleavages disulfide bridges and the protein backbones (mainly into *c*- and *z*- ions) over an unusually high proportion of the amino acids, while retaining the labile bonds from post-translational modifications. There has been much debate concerning the mechanism of ECD; but it is admitted that the recombination releases large amounts of internal energy into the ion (~6 eV) through the formation of charge reduced species $[M+nH]^{(n-1)\bullet +}$. However, because ECD cleaves mainly the N-C_α peptidic bond, prolines do not fragment into *c/z*- ions owing to their cyclic arrangement. Hence, there is a clear need for new MS/MS methods able to release large amount of internal energy into the ions on interests. A simple and straightforward mean to achieve this is though absorption of energetic photons by the target, typically in the ultra-violet range. This approach, however, is not trivial.

13.5.1. Experimental Set-Ups

Photon activation of ions requires the confinement of high densities of ions and photons at the same place and at the same moment. Since the first reports of photodissociation of ions in the 1950-1960 [42] and the pioneering work of R. C. Dunbar in the 1970's [43] in the visible, several experimental setups have been used to increase the photoproducts yields. As discussed by Ly and Julian [44] two main instrumental arrangements can be distinguished: the trapping and crossed-beam experiments. We have instead chosen to discuss separately the subject on the basis of the light sources.

13.5.1.1. Lasers

The photon activation field has fully benefitted of the appearance of lasers, as intense sources of low divergence became available. Numerous applications have appeared based on photodissociation and multi photon ionization of ions produced by ESI, MALDI, or other means [45-53].

The experimental arrangements either rely on ion traps or on crossed-beam setups. Ion traps have the advantages to allow the irradiation time to be adjusted [46]. McIver and coworkers have first implemented 193 nm excimer laser in Fourrier-Transform Ion Cyclotron Resonance mass spectrometers [45,46], followed by McLafferty and co-workers [47,48].

Radio frequency ion traps have been coupled by several groups with tunable lasers [49] and excimers lasers [51,54]. The orbitrap mass spectrometer [55] has been associated with 193 nm excimer lasers [56]. Recently, femto second lasers have been used successfully to irradiate ions in linear ion traps [57,58].

The crossed-beam arrangements have been used in tandem ToF experiments [54,59,60]. Fayeton and co-workers have developed a very powerful instrument in which the 263 nm beam from a Nd:YLF laser crosses the ion beam and both neutral and ionic photoproducts are measured in coincidence [61]. Crossed-beam experiments differ from the ion trapping one in the time scale. The delay between the excitation and detection is strongly reduced in cross beam experiments, which may affect the observed out comes. Indeed, some dissociation processes requiring longer time may not be observed [62].

13.5.1.2. Synchrotron Radiation Source

The high photon fluxes delivered by low divergent laser sources are invaluable for ion activation. Yet, the excitations are limited to a small portion of the UV range. Indeed, tunable lasers hardly exceed 6.2 eV (200 nm), which limits the accessible chromophores and only a few laser lines are available in the VUV range. A potential photon source, bearing the necessary high flux and low divergence, is synchrotron radiation. Moreover, SR is tunable in wide photon energy range, overlapping those covered by UV lasers [63]. Thissen et al., [64] has reported the first use of monochromatic SR for ion photoionization of xenon radical cation. Subsequent applications of SR for photon activation have followed shortly after [65-68].

In all these setups, SR is shined on ions trapped in FT-ICR or RF traps. Thissen et al., used a portable FT-ICR instrument based on permanent magnets [69]. While Bari et al. uses Paul trap for irradiation coupled with ToF detection [65], Hirsch et al. [66] and Giuliani et al. [68] have coupled linear ion traps with VUV beamlines. In these experiments, electromechanical photon shutters [70] are used to control irradiation times. Synchrotron radiation is monochromatized before entering the trap. However, the use of a gas filter [71] is clearly an advantage, since higher harmonic are removed, ensuring spectral purity [72,73]. It is worth mentioning that coupling with soft-X rays beamlines has been reported by Milosavljevic et al. [74].

13.6. Nature of the Photon Activation

Depending on the photon energy of the radiation and on the nature of the target ions, different excitation processes may occur, yielding different photoproducts. In the following, negative ions and positive ions will be distinguished.

13.6.1. Negative Ions

Dugourd and co-workers have extensively studied the outcomes of the irradiation of isolated anions in the gas phase [50] by UV photons down to 200 nm. Photodetachment appeared to be the main photoproduct:

$$[M\text{-}nH]^{n-} + h\nu \rightarrow [M\text{-}nH]^{(n-1)-\bullet} + e^-$$

where n is the number of deprotonated sites.

Two main mechanisms have been put forward to account for this phenomenon: direct or mediated photodetachment [75-77]. For mediate photodetachment, photoabsorption populates an excited state of the ion. This state may then cross with an autoionizing state with the emission of the electron in the vacuum, if the energy of the incoming radiation is above the sum the binding energy and the Coulomb barrier. The comparison of the photodetachment yields, measured as function of the photon energy with the photoabsorption spectra in solution, has reveled corresponding absorption bands [78,79]. However, this mechanism had been established for localized excitations in the near UV of proteins and nucleic acids.

Higher photon energy delivered by SR has allowed the validity of this model to be checked for non-localized absorption. The two steps model seemed to be valid in the VUV [80-82] up to 20 eV (60 nm). However, little fragmentation is observed over an extended photon energy range. Antoine and co-workers have combined photodetachment to collisional activation in the so-called activated EPD [83].

13.6.2. Positive Ions

The case of cations is slightly different from anions. Two photon energy regimes should be distinguished, since they rely on the different physics and produce different outcomes.

For photon energy below the ionization threshold, electronic excited states are populated upon photoabsorption:

$$[M+nH]^{n+} + h\nu \rightarrow [M+nH]^{n+*}$$

Dissociation may occur directly on the excited state manifolds, producing fragments of smaller masses. The total electrical charge is conserved during the process, meaning that no additional charge is created on the target. Otherwise, the system may relax though internal conversion to the ground electronic state, leading to a statistical redistribution of the energy into vibrational modes and eventually dissociation. The other relaxation pathways, which do not produce fragments, are transparent to mass spectrometric based experiments.

The photon energy also provides control on the chromophores involved in the process. PD can produce non-specific dissociations or specific ones at selected chromophores. The longer wavelength will target specifically localized chromophores (such as aromatic amino acids) [84] or to selected chromophore attached covalently or not to the target [85-87]. In contrast, radiation of shorter wavelength will be absorbed by delocalized chromophore (such as the peptidic backbone) [51,88]. These processes have been mainly probed using UV lasers and to a lesser extend using synchrotron radiation [65,88-90].

When the photon energy is increased above the ionization threshold, a radical cation may be produced:

$$[M+nH]^{n+} + h\nu \rightarrow [M+nH]^{(n+1)+\bullet} + e^-$$

A few eV above the ionization threshold, most of the photoabsorption cross section goes into photoionization [2]. Hence, although super excited states may be populate above the ionization threshold, most of the observed phenomena may be ascribed to the physics of the radical cation. This species is prone to dissociation in a so-called dissociative photoionization (DPI) scheme [88], which is similar in nature with electron ionization dissociation (EID) [91]. Photoionization of multiply charged cations requires energetic photons above 10 eV [92]. However, Kalcic and co-workers have reported abundant fragmentation upon femto second laser at 800 nm irradiation of trapped protonated peptides cations [57]. Although not emitting in the VUV, these fs laser activations produce ionization though electron tunneling [58] and thus lead to similar outcomes as DPI and EID activations.

13.7. Applications

Ion photon activation techniques have been used to address analytical challenges for the main molecular families, such as peptides, and proteins, nucleic acids, sugars... Our aim is not to be exhaustive in listing all the applications, but rather to analyze the literature to identify the most interesting features of this large panel of methods.

Under electrospray ionization conditions, the nature of the sample guides the selection of the ionization mode. For example, nucleic acids are most of the time studied in the negative ion mode. Thus, electron photodetachment dissociation (EPD) techniques have been successfully applied to structural analysis of this class of compound [76,77,93,94]. However, at least doubly deprotonated ions are required for EPD. Activated EPD has been also applied with success to top-down sequencing of proteins polyanions [83]. The method has undoubtedly some potential in top-down mass spectrometry; it is handicapped by several of its features. First, activated EPD requires an additional activation step after the formation of the radical cation, which complicates the analytical sequence. More important, proteins and peptides are very often studied under position ionization, mainly for traditional reasons. Consequently, the vast majority of the tools, such as *in silico* prediction of the fragmentations, and mass spectra databases are built for the positive ion mode. This fact certainly prevents the establishment of EPD as a routine method in proteomics. In contrast, photon activation at 193 nm and 157 nm of positive ion doesn't suffer from these drawbacks. The utility of these methods has been demonstrated in proteomic analysis [52,95], and their complementary nature with the standard activation methods has been shown. However, dissociative photoionization (DPI) appears to provide higher sequence coverage than 193 and 157 nm PD. The fragmentations produce under DPI activation conditions are spread over all the peptidic backbone, producing sequence ions of all nature [88]. Interestingly, DPI preserves non-covalent association during the fragmentation process, thereby allowing identification of the non-covalent binding sites [88]. Use of a tunable photon source, such as synchrotron radiation appears extremely interesting for *de novo* sequencing, as it allows spanning from the PD regime to DPI. The photon energy tunability offers a remarkable control over the nature of the fragments [88]. However, coupling of SR with liquid chromatography, which is a prerequisite for further uses in proteomics, has still to be realized.

Oligosaccharides have been submitted to EPD, with some success [87,96]. However, these compounds are essentially transparent in the near UV, which necessitates the grafting of a chromophore, either covalently or not. Again, the technique is restricted to acidic sugar. PD in the VUV range has been demonstrated for sugar [51,97]. DPI at 16 eV has shown the greatest promise for structural analysis of oligosaccharides [98]. However, a proper comparison of the different photon activation method is lacking.

SUMMARY AND CONCLUSION

In the workflow of analytical mass spectrometry, two applications of UV radiation have appeared. UV may be used to generate ions, which will be subsequently analyzed by mass spectrometry. However, this natural application of this portion of the electromagnetic spectrum has really become of practical interest with the development of atmospheric pressure photoionization. These ion sources are compatible with the liquid chromatographic methods and extend the range of molecules amenable to mass spectrometric analysis. The possibility of interrogating a system at atmospheric pressure by photoionization is still in its infancy, and other opportunities should appear in the near future. Especially, analysis of gas mixture is easy by photoionization in real conditions should develop later.

Another use of UV has been found in ion activation in tandem mass spectrometry. Most of the structural information is gained from tandem mass spectrometry. In this technique, the internal energy of an ion of interest is increased until this ion fragment. The analysis of the products of the fragmentation is information about the structure and the energetic of the precursor. A direct mean to increase the internal energy of an ion is through absorption of light with enough energy, typically in the UV range. Aside from the experimental difficulties of photon activation of ions, several implementations of light sources on mass spectrometers have been demonstrated. UV photon activation of ions shows great promise, but still lack from a commercially available instrument.

REFERENCES

[1] Yates JR III. (2011) A century of mass spectrometry: from atoms to proteomes. *Nature Methods 8*, 633–637.

[2] Berkowitz J. (1979) *Photoabsorption, photoionization, and photoelectron spectroscopy*; Academic Pr.

[3] Syage J, Evans MD, Hanold K. (2000) Photoionization mass spectrometry. *American laboratory 32*, 24.

[4] Robb D, Covey T, Bruins A. (2000) Atmospheric Pressure Photoionization: An Ionization Method for Liquid Chromatography− Mass Spectrometry. *Analytical chemistry 72*, 3653–3659.

[5] Futrell J. (2000) Development of tandem mass spectrometry: one perspective. *International journal of mass spectrometry 200*, 495–508.

[6] Fenn J, Mann M, Meng C, Wong S, Whitehouse C. (1990) Electrospray Ionization-Principles and Practice. *Mass spectrometry reviews 9*, 37–70.

[7] Blakley CR, Vestal ML, (1983) Thermospray interface for liquid chromatography/mass spectrometry. *Analytical chemistry 55*, 750–754.

[8] Covey T, Bruins A. (1988) Comparison of thermospray and ion spray mass spectrometry in an atmospheric pressure ion source. *Organic mass spectrometry 23,* 178-186.

[9] Vestal ML. (1983) Studies of ionization mechanisms involved in thermospray LC-MS. *International journal of mass spectrometry and ion physics 46*, 193–196.

[10] Bagag A, Giuliani A, Laprevote O. (2008) Atmospheric pressure photoionization mass spectrometry of oligodeoxyribonucleotides. *European journal of mass spectrometry 14*, 71–80.

[11] Turnipseed SB, Roybal JE, Andersen WC, Kuck LR. (2005) Analysis of avermectin and moxidectin residues in milk by liquid chromatography–tandem mass spectrometry using an atmospheric pressure chemical ionization/atmospheric pressure photoionization source. *Analytica chimica acta 529*, 159–165.

[12] Turnipseed SB, Andersen WC, Karbiwnyk CM, Roybal JE, Miller KE. (2006) No-discharge atmospheric pressure chemical ionization: evaluation and application to the analysis of animal drug residues in complex matrices. *Rapid communications in mass spectrometry 20*, 1231–1239.

[13] Robb D, Blades M. (2008) State-of-the-art in atmospheric pressure photoionization for LC/MS. *Analytica chimica acta 627*, 34–49.

[14] Hayen H. (2003) Strategies for the liquid chromatographic–mass spectrometric analysis of non-polar compounds. *Journal of chromatography A 1000*, 549–565.

[15] Syage J. (2004) Mechanism of [M+ H]+ formation in photoionization mass spectrometry. *Journal of the American society for mass spectrometry 15*, 1521–1533.

[16] Marotta E, Seraglia R, Fabris F, Traldi P. (2003) Atmospheric pressure photoionization mechanisms. *International journal of mass spectrometry 228*, 841–849.

[17] Klee S, Albrecht S, Derpmann V, Kersten H, Benter T. (2013) Generation of ion-bound solvent clusters as reactant ions in dopant-assisted APPI and APLI. *Analytical and bioanalytical chemistry.*

[18] Song L, Wellman AD, Yao H, Adcock J. (2007) Electron capture atmospheric pressure photoionization mass spectrometry: analysis of fullerenes, perfluorinated compounds, and pentafluorobenzyl derivatives. *Rapid communications in mass spectrometry 21*, 1343–1351.

[19] Song L, Wellman AD, Yao H, Bartmess JE. (2007) Negative ion-atmospheric pressure photoionization: Electron capture, dissociative electron capture, proton transfer, and anion attachment. *Journal of the American society for mass spectrometry 18*, 1789–1798.

[20] Debrauwer L, Riu A, Jouahri M, Rathahao E, Jouanin I, Antignac J, Cariou R, Lebizec B, Zalko D. (2005) Probing new approaches using atmospheric pressure photo ionization for the analysis of brominated flame retardants and their related degradation products by liquid chromatography–mass spectrometry. *Journal of chromatography A 1082*, 98–109.

[21] Debois D, Giuliani A, Laprevote O. (2006) Fragmentation induced in atmospheric pressure photoionization of peptides. *Journal of mass spectrometry 41*, 1554–1560.

[22] Delobel A, Halgand F, Laffranchise-Gosse B, Snijders H, Laprevote O. (2003) Characterization of Hydrophobic Peptides by Atmospheric Pressure Photoionization-Mass Spectrometry and Tandem Mass Spectrometry. *Analytical chemistry 75*, 5961–5968.

[23] Bagag A, Giuliani A, Laprevote O. (2011) Atmospheric pressure photoionization of peptides. *International journal of mass spectrometry 299*, 1–4.

[24] Robb DB, Rogalski JC, Kast J, Blades MW. (2011) A New Ion Source and Procedures for Atmospheric Pressure-Electron Capture Dissociation of Peptides. *Journal of the American society for mass spectrometry 22*, 1699–1706.

[25] Purcell JM, Hendrickson CL, Rodgers RP, Marshall AG. (2006) Atmospheric Pressure Photoionization Fourier Transform Ion Cyclotron Resonance Mass Spectrometry for Complex Mixture Analysis. *Analytical chemistry 78*, 5906–5912.

[26] Purcell JM, Juyal P, Kim DG, Rodgers RP, Hendrickson CL, Marshall AG. (2007) Sulfur Speciation in Petroleum: Atmospheric Pressure Photoionization or Chemical Derivatization and Electrospray Ionization Fourier Transform Ion Cyclotron Resonance Mass Spectrometry. *Energy fuels 21*, 2869–2874.

[27] Chiaia-Hernandez AC, Krauss M, Hollender J. (2013) Screening of Lake Sediments for Emerging Contaminants by Liquid Chromatography Atmospheric Pressure Photoionization and Electrospray Ionization Coupled to High Resolution Mass Spectrometry. *Environmental science & technology 47*, 976–986.

[28] Mopper K, Stubbins A, Ritchie JD, Bialk HM, Hatcher PG. (2007) Advanced Instrumental Approaches for Characterization of Marine Dissolved Organic Matter: Extraction Techniques, Mass Spectrometry, and Nuclear Magnetic Resonance Spectroscopy. *Chemical reviews 107*, 419–442.

[29] Kauppila, T. J.; OEstman, P.; Marttila, S.; Ketola, R. A.; Kotiaho, T.; Franssila, S.; Kostiainen, R. (2004) Atmospheric Pressure Photoionization-Mass Spectrometry with a Microchip Heated Nebulizer. *Analytical chemistry 76*, 6797–6801.

[30] Haapala M, Pol J, Saarela V, Arvola V, Kotiaho T, Ketola RA, Franssila S, Kauppila TJ, Kostiainen R. (2007) Desorption Atmospheric Pressure Photoionization. *Analytical chemistry 79*, 7867–7872.

[31] Delobel A, Roy S, Touboul D, Gaudin K, Germain DP, Baillet A, Brion F, Prognon P, Chaminade P, Laprevote O. (2006) Atmospheric pressure photoionization coupled to porous graphitic carbon liquid chromatography for the analysis of globotriaosylceramides. Application to Fabry disease. *Journal of mass spectrometry 41*, 50–58.

[32] Constapel M, Schellentrager M, Schmitz OJ, Gab S, Brockmann KJ, Giese R, Benter T. (2005) Atmospheric-pressure laser ionization: a novel ionization method for liquid chromatography/mass spectrometry. *Rapid communications in mass spectrometry 19*, 326–336.

[33] Schiewek R, Monnikes R, Wulf V, Gab S, Brockmann KJ, Benter T, Schmitz OJ. (2008) A Universal Ionization Label for the APLI-(TOF)MS Analysis of Small Molecules and Polymers. *Angewandte chemie international edition 47*, 9989–9992.

[34] Schrader W, Panda SK, Brockmann KJ, Benter T. (2008) Characterization of non-polar aromatic hydrocarbons in crude oil using atmospheric pressure laser ionization and Fourier transform ion cyclotron resonance mass spectrometry (APLI FT-ICR MS). *Analyst 133*, 867.

[35] Giuliani A, Giorgetta JL, Ricaud JP, Jamme F, Rouam V, Wien F, Laprevote O, Refregiers M. (2012) Atmospheric pressure photoionization using tunable VUV synchrotron radiation. *Nuclear instruments and methods in physics B 279,* 114-117

[36] Giuliani A, Yao I, Lagarde B, Rey S, Duval JP, Rommeluere P, Jamme F, Rouam V, Wein F, De Oliveira C, Ros M, Lestrade A, Desjardins K, Giorgetta JL, Laprevote O, Herbaux C, Refregiers M. (2011) A differential pumping system to deliver windowless VUV photons at atmospheric pressure. *Journal of synchrotron radiation 18*, 546–549.

[37] Giuliani A, Jamme F, Rouam V, Wien F, Giorgetta JL, Lagarde B, Chubar O, Bac S, Yao I, Rey S, Herbeaux C, Marlats JL, Zerbib D, Polack F, Refregiers, M. (2009) DISCO: a low-energy multipurpose beamline at synchrotron SOLEIL. *Journal of synchrotron radiation 16*, 835–841.

[38] Biemann K. (2002) Four decades of structure determination by mass spectrometry: from alkaloids to heparin. *Journal of the American society for mass spectrometry 13*, 1254–1272.

[39] McLuckey SA, Goeringer DE. (1997) Slow Heating Methods in Tandem Mass Spectrometry. *Journal of mass spectrometry 32*, 461–474.

[40] Zubarev RA, Kelleher N, McLafferty F. (1998) Electron capture dissociation of multiply charged protein cations. A nonergodic process. *Journal of the American chemical society 120*, 3265–3266.

[41] Zubarev RA. (2003) Reactions of polypeptide ions with electrons in the gas phase. *Mass spectrometry reviews 22*, 57–77.

[42] Dehmelt H, Jefferts K. (1962) Alignment of the H 2 Molecular Ion by Selective Photodissociation. I. *Physical review 125*, 1318.

[43] Dunbar R. (1971) Photodissociation of CH3Cl+ and N2O+ Cations. *Journal of the American chemical society 93*, 4354–4358.

[44] Ly T, Julian RR. (2009) Ultraviolet Photodissociation: Developments towards Applications for Mass-Spectrometry-Based Proteomics. *Angewandte chemie international edition 48*, 7130–7137.

[45] Lebrilla CB, Wang D, Mizoguchi TJ, McIver RT. (1989) Comparison of the Fragmentation Produced by Fast Atom Bombardment and Photodissociation of Peptides. *Journal of the American chemical society 111*, 8593–8598.

[46] Bowers WD, Delbert SS, Hunter RL, McIver RT. (1984) Fragmentation of Oligopeptide Ions Using Ultraviolet-Laser Radiation and Fourier-Transform Mass-Spectrometry. *Journal of the American chemical society 106*, 7288–7289.

[47] Williams ER, McLafferty FW. (1990) 193 nm Laser photoionization and photodissociation for isomer differentiation in Fourier-transform mass spectrometry. *Journal of the American society for mass spectrometry 1*, 361–365.

[48] Guan Z, Kelleher N, O'Connor P, Aaserud D, Little D, McLafferty F. (1996) 193 nm photodissociation of larger multiply-charged biomolecules. *International journal of mass spectrometry and ion processes 157*, 357–364.

[49] Tabarin, T.; Antoine, R.; Broyer, M.; Dugourd, P. (2005) Specific photodissociation of peptides with multi-stage mass spectrometry. *Rapid communications in mass spectrometry 19*, 2883–2892.

[50] Antoine R, Dugourd P. (2011) Visible and ultraviolet spectroscopy of gas phase protein ions. *Physical chemistry chemical physics 13*, 16494.

[51] Reilly JP. (2009) Ultraviolet photofragmentation of biomolecular ions. *Mass spectrometry reviews 28*, 425–447.

[52] Madsen JA, Boutz DR, Brodbelt JS. (2010) Ultrafast Ultraviolet Photodissociation at 193 nm and its Applicability to Proteomic Workflows. *Journal of proteome research 9*, 4205–4214.

[53] Boyarkin O, Mercier S, Kamariotis A, Rizzo T. (2006) Electronic spectroscopy of cold, protonated tryptophan and tyrosine. *Journal of the American chemical society 128*, 2816–2817.

[54] Thompson M, Cui W, Reilly J. (2004) Fragmentation of singly charged peptide ions by photodissociation at λ= 157 nm. *Angewandte chemie international edition 43*, 4791–4794.

[55] Hu Q, Noll RJ, Li H, Makarov A, Hardman M, Cooks GR (2005) The Orbitrap: a new mass spectrometer. *Journal of mass spectrometry 40*, 430–443.

[56] Vasicek LA, Ledvina AR, Shaw J, Griep-Raming J, Westphall MS, Coon JJ, Brodbelt JS. (2011) Implementing Photodissociation in an Orbitrap Mass Spectrometer. *Journal of the American society for mass spectrometry 22*, 1105–1108.

[57] Kalcic C, Gunaratne T, Jones A, Dantus M, Reid G. (2009) Femtosecond Laser-Induced Ionization/Dissociation of Protonated Peptides. *Journal of the Americna chemical society 131*, 940–942.

[58] Kalcic CL, Reid GE, Lozovoy VV, Dantus M. (2012) Mechanism Elucidation for Nonstochastic Femtosecond Laser-Induced Ionization/Dissociation: From Amino Acids to Peptides. *Journal of physical chemistry A 116*, 2764–2774.

[59] Barbacci D, Russell D. (1999) Sequence and side-chain specific photofragment (193 nm) ions from protonated substance P by matrix-assisted laser desorption ionization time-of-flight mass spectrometry. *Journal of the American society for mass spectrometry 10*, 1038–1040.

[60] Moon JH, Yoon SH, Kim MS. (2005) Photodissociation of singly protonated peptides at 193 nm investigated with tandem time-of-flight mass spectrometry. *Rapid communications in mass spectrometry 19*, 3248–3252.

[61] Lepere V, Lucas B, Barat M, Fayeton JA, Picard VJ, Jouvet C, Carcabal P, Nielsen I, Dedonder-Lardeux C, Gregoire G, Fujii A. (2007) Comprehensive characterization of the photodissociation pathways of protonated tryptophan. *The journal of chemical physics 127*, 134313.

[62] Gabelica V, Pauw ED. (2005) Internal energy and fragmentation of ions produced in electrospray sources. *Mass spectrometry reviews 24*, 566–587.

[63] Hatano Y. (1999) Interaction of vacuum ultraviolet photons with molecules. Formation and dissociation dynamics of molecular superexcited states. *Physics reports 313*, 109–169.

[64] Thissen R, Bizau JM, Blancard C, Coreno M, Dehon C, Franceschi P, Giuliani A, Lemaire J, Nicolas C. (2008) Photoionization cross section of Xe+ ion in the pure 5p5 2P3/2 ground level. *Physical review letters 100*, 223001–.

[65] Bari S, Gonzalez-Magana O, Reitsma G, Werner J, Schippers S, Hoekstra R, Schlathoelter T. (2011) Photodissociation of protonated leucine-enkephalin in the VUV range of 8–40 eV. *The journal of chemical physics 134*, 024314.

[66] Hirsch K, Lau J, Klar P, Langenberg A, Probst J, Rittmann J, Vogel M, Zamudio-Bayer V, Moeller T, Issendorff B. (2009) X-ray spectroscopy on size-selected clusters in an

ion trap: from the molecular limit to bulk properties. *Journal of physics B: atomic, molecular and optical physics 42*, 154029.

[67] Milosavljevic AR, Nicolas C, Lemaire J, Dehon C, Thissen R, Bizau JM, Refregiers M, Nahon L, Giuliani A. (2011) Photoionization of a protein isolated in vacuo. *Physical chemistry chemical physics 13*, 15432–15436.

[68] Milosavljevic AR, Giuliani A, Nicolas C, Gil JF, Lemaire J, Refregiers M, Nahon L. (2010) Gas-phase spectroscopy of a protein. *Journal of physics: conference series 257*, 012006.

[69] Mauclaire G, Lemaire J, Boissel P, Bellec G, Heninger M. (2004) MICRA: a compact permanent magnet Fourier transform ion cyclotron resonance mass spectrometer. *European journal of mass spectrometry 10*, 155–162.

[70] Milosavljevic AR, Nicolas C, Gil JF, Canon F, Refregiers M, Nahon L, Giuliani, A. (2012) Fast in vacuo photon shutter for synchrotron radiation quadrupole ion trap tandem mass spectrometry. *Nuclear instruments and methods in physics research section B: beam interactions with materials and atoms 279*, 34–36.

[71] Mercier B, Compin M, Prevost C, Bellec G, Thissen R, Dutuit O, Nahon L. (2000) Experimental and theoretical study of a differentially pumped absorption gas cell used as a low energy-pass filter in the vacuum ultraviolet photon energy range. *Journal of vacuum science & technology A: vacuum, surfaces, and films 18*, 2533.

[72] Nahon L, de Oliveira N, Garcia GA, Gil JF, Pilette B, Marcouille O, Lagarde B, Polack F. (2012) DESIRS: a state-of-the-art VUV beamline featuring high resolution and variable polarization for spectroscopy and dichroism at SOLEIL. *Journal of synchrotron radiation (2012). 19, 508-520.*

[73] Milosavljevic AR, Nicolas C, Gil JF, Canon F, Refregiers M, Nahon L, Giuliani A. (2012) VUV synchrotron radiation: a new activation technique for tandem mass spectrometry. *Journal of synchrotron radiation 19*, 174–178.

[74] Milosavljevic AR, Canon F, Nicolas C, Miron C, Nahon L, Giuliani A. (2012) Gas-Phase Protein Inner-Shell Spectroscopy by Coupling an Ion Trap with a Soft X-ray Beamline. *The Journal of physical chemistry letters 3*, 1191–1196.

[75] Matheis K, Joly L, Antoine R, Lepine F, Bordas C, Ehrler OT, Allouche AR, Kappes, MM, Dugourd P. (2008) Photoelectron Spectroscopy of Gramicidin Polyanions: Competition between Delayed and Direct Emission. *Journal of the American chemical society 130*, 15903–15906.

[76] Gabelica V, Rosu F, Tabarin T, Kinet C, Antoine R, Broyer M, De Pauw E, Dugourd, P. (2007) Base-Dependent Electron Photodetachment from Negatively Charged DNA Strands upon 260-nm Laser Irradiation. *Journal of the American chemical society 129*, 4706–4713.

[77] Gabelica V, Rosu F, De Pauw E, Antoine R, Tabarin T, Broyer M, Dugourd P. (2007) Electron photodetachment dissociation of DNA anions with covalently or noncovalently bound chromophores. *Journal of the American society for mass spectrometry 18*, 1990–2000.

[78] Joly L, Antoine R, Broyer M, Lemoine J, Dugourd P. (2008) Electron Photodetachment from Gas Phase Peptide Dianions. Relation with Optical Absorption Properties. *Journal of physical chemistry A 112*, 898–903.

[79] Joly L, Antoine R, Allouche AR, Broyer M, Lemoine J, Dugourd P. (2007) Ultraviolet Spectroscopy of Peptide and Protein Polyanions in Vacuo: Signature of the Ionization State of Tyrosine. *Journal of the American chemical society 129*, 8428–8429.

[80] Brunet C, Antoine R, Dugourd P, Canon F, Giuliani A, Nahon L. (2012) Formation and fragmentation of radical peptide anions: insights from vacuum ultra violet spectroscopy. *Journal of the American society for mass spectrometry 23*, 274–281.

[81] Brunet C, Antoine R, Allouche AR, Dugourd P, Canon F, Giuliani A, Nahon L.(2011) Gas phase Photo-Formation and Vacuum UV Photofragmentation Spectroscopy of Tryptophan and Tyrosine Radical-Containing Peptides. *Journal of physical chemistry A 115*, 8933–8939.

[82] Brunet C, Antoine R, Dugourd P, Canon F, Giuliani A, Nahon L. (2013) Photo-induced electron detachment of protein polyanions in the VUV range. *The journal of chemical physics 138*, 064301.

[83] Larraillet V, Antoine R, Dugourd P, Lemoine J. (2009) Activated-Electron Photodetachment Dissociation for the Structural Characterization of Protein Polyanions. *Analytical chemistry 81*, 8410–8416.

[84] Antoine R, Broyer M, Chamot-Rooke J, Dedonder C, Desfrancois C, Dugourd P, Gregoire G, Jouvet C, Onidas D, Poulain P, Tabarin T, van der Rest G. (2006) Comparison of the fragmentation pattern induced by collisions, laser excitation and electron capture. Influence of the initial excitation. *Rapid communications in mass spectrometry 20*, 1648–1652.

[85] Diedrich JK, Julian RR. (2008) Site-Specific Radical Directed Dissociation of Peptides at Phosphorylated Residues. *Journal of the American chemical society 130*, 12212–12213.

[86] Wilson JJ, Kirkovits GJ, Sessler JL, Brodbelt JS. (2008) Photodissociation of non-covalent peptide-crown ether complexes. *Journal of the American society for mass spectrometry 19*, 257–260.

[87] Enjalbert Q, Racaud A, Lemoine J, Redon S, Ayhan MM, Andraud C, Chambert S, Bretonniere Y, Loison C, Antoine R, Dugourd P. (2012) Optical Properties of a Visible Push–Pull Chromophore Covalently Bound to Carbohydrates: Solution and Gas-Phase Spectroscopy Combined to Theoretical Investigations. *Journal of physical chemistry B 116*, 841–851.

[88] Canon F, Milosavljevic AR, van der Rest G, Refregiers M, Nahon L, Sarni-Manchado P, Cheynier V, Giuliani A. (2013) Photodissociation and Dissociative Photoionization Mass Spectrometry of Proteins and Noncovalent Protein-Ligand Complexes. *Angewandte chemie international edition 52*, 8377–8381.

[89] Milosavljevic AR, Cerovski VZ, Canon F, Nahon L, Giuliani A. (2013) Nanosolvation-Induced Stabilization of a Protonated Peptide Dimer Isolated in the Gas Phase. *Angewandte chemie international edition 52*, 7286–7290.

[90] Gonzalez-Magana O, Reitsma G, Bari S, Hoekstra R, Schlatholter T. (2012) Length effects in VUV photofragmentation of protonated peptides. *Physical chemistry chemical physics 14*, 4351.

[91] Fung YME, Adams CM, Zubarev RA. (2009) Electron Ionization Dissociation of Singly and Multiply Charged Peptides. *Journal of the American chemical society 131*, 9977–9985.

[92] Giuliani A, Milosavljevic AR, Hinsen K, Canon F, Nicolas C, Refregiers M, Nahon L. (2012) Structure and Charge-State Dependence of the Gas-Phase Ionization Energy of Proteins. *Angewandte chemie international edition 51*, 9552–9556.

[93] Gabelica V, Tabarin T, Antoine R, Rosu F, Compagnon I, Broyer M, De Pauw E, Dugourd P. (2006) Electron Photodetachment Dissociation of DNA Polyanions in a Quadrupole Ion Trap Mass Spectrometer. *Analytical chemistry 78*, 6564–6572.

[94] Silvestri C, Brodbelt JS. Tandem mass spectrometry for characterization of covalent adducts of DNA with anticancer therapeutics. *Mass spectrometry reviews* 2012, *32*, 247–266.

[95] Zhang L, Reilly JP. (2010) De NovoSequencing of Tryptic Peptides Derived from Deinococcus radioduransRibosomal Proteins Using 157 nm Photodissociation MALDI TOF/TOF Mass Spectrometry. *Journal of proteome research 9*, 3025–3034.

[96] Racaud A, Antoine R, Joly L, Mesplet N, Dugourd P, Lemoine J. (2009) Wavelength-tunable ultraviolet photodissociation (UVPD) of heparin-derived disaccharides in a linear ion trap. *Journal of the American Society for mass spectrometry 20*, 1645–1651.

[97] Devakumar A, Thompson M, Reilly J. (2005) Fragmentation of oligosaccharide ions with 157 nm vacuum ultraviolet light. *Rapid communications in mass spectrometry, 19*, 2313–2320.

[98] Ropartz D, Lemoine J, Giuliani A, Bittebiere Y, Enjalbert Q, Antoine R, Dugourd P, Ralet MC, Rogniaux H. (2014) Deciphering the structure of isomeric oligosaccharides in a complex mixture by tandem mass spectrometry: photon activation with vacuum ultra-violet brings unique information and enables definitive structure assignment. *Analytica chimica acta. 807: 84-95.*

In: UV Radiation: Properties, Effects, and Applications
Editor: James A. Radosevich
ISBN: 978-1-63321-090-5

Chapter 14

EFFECT OF ULTRAVIOLET RADIATION INDUCED CROSSLINKING ON NATURAL FIBER AND BIOMATERIALS

M. Fizur Rahman, Jahid M. M. Islam and Mubarak A. Khan
Institute of Radiation and Polymer Technology
Bangladesh Atomic Energy Commission, Bangladesh

ABSTRACT

Irradiation of natural fiber and biomaterials with photo initiator by ultraviolet (UV) radiation leads to the formation of very reactive intermediates in the forms of excited states, ions and free radicals which are almost instantaneously used up in several reaction pathways and consequently result in the arrangement or formation of new bonds structures. These reactions lead to grafting, crosslinking, curing and degradation of biomaterials. Use of non-ionizing radiation (UV) has been employed successfully for significant physical and chemical changes as well as changes in surface structure and surface energy of the fibers. It is possible to replace synthetic polymer by irradiated biomaterials having better mechanical and swelling property. UV radiation induced photo curing provides many advantages, such as: low energy consumption, ambient temperature operation, no emission of solvents, reduces reaction time, enhanced product quality, thus reducing costs. Radiation processing offers a clean and additive-free method for preparation of value-added novel materials based on renewable, non-toxic and biodegradable natural polymers. Effect of UV radiation stimulate grafting, crosslinking, curing and degradation on various biomaterials like chitosan, calcium alginate, gelatin and natural fiber will be discussed in this chapter.

Keywords: Crosslinking, natural fibers, biomaterials

ABBREVIATIONS

BA	butyl acrylate
Eb	Elongation at break
EHA	2-ethyl hexyl acrylate
HDDA	1,6-hexanediol diacrylate
HEA	2-hydroxy ethylacrylate
HEMA	2-hydroxyethylmethylacrylate
JPC	Jute polymer composite
MeOH	methanol
MMA	methyl methacrylate
NaOH	sodium hydroxide
NVP	1-vinyl 2-pyrrolidone
PE	polyethylene
PL	polymer loading
PP	polypropylene
ST	styrene
TS	Tensile Strength
TSPMA	3-triethyloxyl silyl propylmethacrylate
U	urea
UV	Ultraviolet

14.0. INTRODUCTION

The awareness in natural fiber-reinforced polymer composite materials is rapidly increasing both in terms of their industrial applications and fundamental research. Natural fibers have low abrasion multi-functionality, low density, low cost, availability, high toughness, acceptable specific strength properties, good thermal properties, enhanced energy recovery, completely or partially recyclable, and biodegradability [1]. Plants, such as flax, cotton, hemp, jute, sisal, kenaf, pineapple, ramie, bamboo, banana, etc., as well as wood, used from time immemorial as a resource of lignocellulosic fibers, are more and more often applied as the reinforcement of composites. Their availability, renewability, low density, and price as well as satisfactory mechanical properties make them an attractive ecological substitute to glass, carbon and man-made fibers used for the manufacturing of composites. The natural fiber containing composites are more environmentally friendly, and are used in transportation (automobiles, railway coaches, aerospace), military applications, building and construction industries (ceiling paneling, partition boards), packaging, consumer products, etc. because of their outstanding thermo-mechanical properties [2].

Jute is a stiff fiber, lustrous in appearance, and generally has a rough texture. Jute has been widely regarded as a versatile, helpful and valuable agro-fiber. It is difficult to propose another natural fiber possessing such diverse properties and basic utilities as jute in solicitous use in carpets, ropes, sacks, Hessian clothes, etc. With the discovery of different types of new synthetic substances, their vast use has adversely affected jute products in the world market. Marketing of traditional jute products like bags and Hessian clothes is facing serious challenge, causing shrinkage of jute market. But the synthetic packing materials are non biodegradable and cause serious environmental pollution, the possibility for increased use of jute and jute products that are biodegradable is getting brighter. Jute has been used as a reinforcing element of polymer composites. Jute contains relatively hard cellulose fibers; it has high tensile modulus and low elongation at break (Eb) [3]. Jute is biodegradable and easily decomposable in the environment and is thus environmentally friendly compared to the synthetic polymers that are not biodegradable and cause pollution of the environment. One of the drawbacks of jute fiber is its high hydrophilicity. It is therefore necessary to improve the characteristic properties of jute chemically so that, the resulting fiber will behave like synthetic polymers with respect to durability, sustainability and mechanical strength but at the same time will be degradable and eco-friendly [4].

The most serious concern with natural fibers is their hydrophilic nature due to the presence of pendant hydroxyl and polar groups in various constituents. These groups are also responsible for poor adhesion between fibers and hydrophobic matrix polymers [5, 6]. The hydrophilic nature of the fiber surface also leads to high moisture uptake for the natural fibers which seriously lower the mechanical properties of the fibers themselves. For the same reason, the natural fibers are inherently incompatible with nonpolar-hydrophobic thermoplastics, such as polyolefins. Moreover, difficulty in mixing because of poor wetting of the fibers with the matrix is another problem that leads to composites with weak interface [7].

Chemical treatments can clean the fiber surface, modify the chemistry on the surface, lower the moisture uptake and increase the surface roughness. As natural fibers bear hydroxyl groups from cellulose and lignin; they are amenable to modification. The hydroxyl groups may be involved in the hydrogen bonding within the cellulose molecules thereby reducing the activity towards the matrix. Surface modifications may activate these groups or can introduce new moieties that can effectively lead to chemical interlock with the matrix. Mercerization, isocyanate treatment, acrylation, permanganate treatment, acetylation, silane treatment and peroxide treatment with various coupling agents and other pretreatments (i.e., UV radiation) of natural fibers have achieved various levels of success for improving fiber strength, fiber fitness and fiber-matrix adhesion. [8-11]

The surface modification can be achieved by-

1. Light (generally UV light) induced modification
2. Ionizing radiation (γ-rays, electron beam) induced modification
3. Plasma induced modification.

These three methods share the same objectives i.e., creating free radicals onto the polymeric substrates and then these radical sites are used as initiators for copolymerization reactions with monomers present in the grafting solution. Various parameters such as concentration of the monomer, time of treatment, radiation dose, type and concentration of catalyst used, if any, could greatly affect the grafting efficiency and need to be optimized.

Radiation is a very convenient method for surface treatment of jute fiber. Techniques concerning UV are particularly good because UV sources are relatively flexible, low-priced and easy to establish. UV curing technology offers the advantages of curing at ambient temperatures, faster processing, low energy consumption, i.e., energy savings, greater reaction control, no emission of solvent, safety, cost effectiveness, savings on plants floor space, and the production of polymer materials having tailor-made properties. Radiation curing processes become more important and useful because this process reduces or eliminates the need of a volatile solvent and the energy required to cure, are substantially less than that required in conventional methods [12]. Other major advantages of radiation curing process are:

- It generates little or no heat
- It is space saving and one pack system.

A UV curing coating contains a mixture of liquid pre-polymers, monomers, pigments and extenders, together with a photo-active compound (photoinitiator). When UV light penetrates the coating, the photoinitiator breaks down into free radicals. The free radicals are then able to initiate a rapid addition polymerization reaction, with unsaturated groups which are present in the oligomer and monomers [13]. *Figure 1* shows the mechanism of the free radical formation due to UV radiation.

On the other hand, the history of man is strongly linked to a wide variety of natural occurring polymers. One of the current priorities in polymer chemistry is to explore bio-based polymers from renewable resources. The development of innovative biopolymer materials has been underway for a number of years, and continues to be an area of interest for many scientists. Biopolymers have been considered as alternatives for non-biodegradable petroleum-based polymers because of their abundance, renewability, low cost, and good biodegradability. Biopolymers can be sustainable, carbon neutral and are always renewable because they are made from plant materials or animal which can be grown year on year indefinitely. Most of these plant materials come from agricultural non food crops. The environmental problems generated by disposal of conventional plastics and other related polymers have intensified researches with polymers that have their origin from renewable resources such as prawn shell, animal bone, brown seaweeds, plant resin etc. Therefore, the use of biopolymers would create a sustainable industry. In addition, biopolymers have the potential to cut carbon emissions and reduce CO_2 quantities in the atmosphere: this is because the CO_2 released when they degrade can be reabsorbed by crops grown to replace them; this makes them close to carbon neutral.

Biopolymers have great commercial potential for bio-plastic, but some of the properties such as brittleness, low heat distortion temperature, high gas permeability, low melt viscosity for further processing etc. restrict their use in a wide-range of applications. Therefore, modification of the biopolymers through innovative technology is a formidable task for scientists. Although there are various modification methods for biopolymers, photo curing by UV radiation is one of the best techniques for its diversified merits as described earlier. In this chapter we will discuss on various approaches of UV modification to natural fibers (especially jute) as well as other biopolymers like chitosan, gelatin, calcium alginate, plant resin, and potato starch.

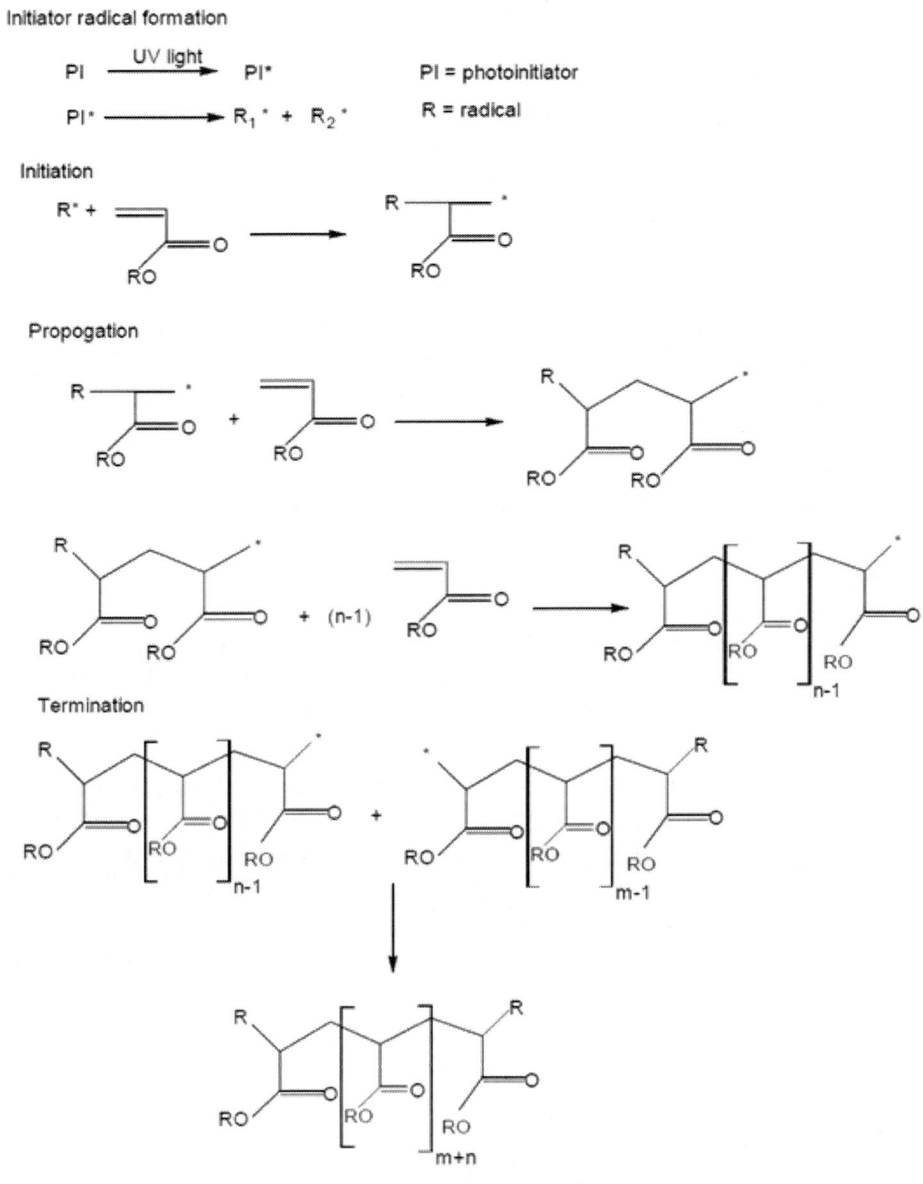

Figure 1. Mechanism path of free radical polymerization.

14.1. Modification of Natural Fiber

14.1.1. Modification with 2-Hydroxy Ethylacrylate

Because of the universal concern about environmental pollution, scientists are becoming more and more interested in natural fibers. Jute is a complex polymeric substance and is composed chiefly of cellulose, hemicellulose, and lignin [14, 15]. Curing or grafting of monomer on cellulose and cellulosic fibers by using a radiation technique is one of the most modern and effective methods to improve the mechanical properties, and reduce the water uptake tendency. Curing of cellulosic materials by ultraviolet radiation offers advantages such as reduced reaction time, reaction occurs at ambient temperature, improved monomer stability, continuous process, eco-friendly, and increased design flexibility through process control.

The radiation curing can markedly influence the properties of the resulting products. Moisture absorption of jute fiber could be reduced significantly by monomer treatment co-polymerization method. Several acrylate monomers have been used and it is reported that treatment of jute fiber with various oligomers and monomers using a UV radiation technique showed improved mechanical properties and biodegradability under certain environmental conditions [16-18].

Current processes of grafting monomers onto fibers and yarns, along with the determination of the resultant electrical properties of jute fibers, have also been receiving considerable attention by the scientists. This is particularly true for Indian and Bangladeshi scientists because of the availability of raw jute produced in this region. Khan et al. studied the dielectrical properties of photocured jute with 2-Hydroxyethyl Methacrylate (HEMA) and has observed that dielectric constant and loss tangent (tanΩ) increased with increasing temperature up to the transition temperature [19].

In our laboratory, we improved the mechanical and electrical properties of jute fabrics by means of grafting with monomer 2-hydroxy ethylacrylate (HEA) and aliphatic urethane diacrylate oligomer (M-1200) using UV radiation. For this purpose, jute fabric (Hessian cloth) was grafted with HEA and aliphatic urethane diacrylate oligomer (M-1200) under UV radiation at different intensities in order to improve the mechanical and electrical properties. The concentrations of the monomer (HEA), and oligomer (M1200), radiation dose, and soaking time were optimized with respect to mechanical property such as tensile strength (TS) of the treated and untreated Hessian cloth. The 73 % HEA, 25 % oligomer (M-1200) and 2 % photoinitiator (Darocur-1116) in a 1 minute soaking time showed the highest TS at 50th UV pass [20]. The electrical properties such as dielectric constant and loss tangent of both treated and untreated sample were measured at frequency 10 kHz with the variation of temperature. It was observed that dielectric constant and loss tangent increased with increasing temperature up to the transition temperature and then decreased.

14.1.2. Jute Fabrics Reinforced Urethane Based Thermoset Composites

A rapid growth has occurred in the consumption of fiber reinforced polymer composites, yielding a unique combination of high performance, great versatility, and processing advantages at favorable costs by permutation and combination of various fibers and polymers for the last three decades. Composites having unique properties for versatile applications as alternatives to conventional materials like metals, woods, etc. have been prepared [21, 22]. Glass fiber reinforced thermoplastic materials based on commodity resins have proven to be very successful in high volume markets because of their excellent price performance ratio, although there are some disadvantages such as abrasiveness, they are non-recyclable, non-renewable and skin irritating. Thus, scientists are looking to find other alternatives [23].

Among commodities, thermosetting urethane acrylate possesses outstanding properties like good surface hardness, scratch resistance, and excellent electrical properties. As a material of construction, jute urethane acrylate composite can play a vital role all over the world. Jute urethane acrylate composite can be used as particle board, ceilings, blocks for building construction, and furniture.

By the commercial production of jute urethane acrylate composite, it is possible to reduce the use of particle board and cement. Moreover, most of the nature fibers like sisal, flax, hemp, etc are more expensive than jute. In order to compete with the synthetic fiber reinforced composite, the mechanical properties of jute urethane acrylate composite should be improved. Some researchers prepared jute urethane acrylate composite with the improved mechanical properties by coating the synthetic urethane acrylate polymers over the natural fiber, jute. To enhance the use of jute fibers, a broader systematic investigation is necessary to search for chemical modifiers and physical treatments. Several processes such as chemical treatments [24] and photochemical treatments [25-29] have been developed to modify the fiber surfaces.

Jute fabrics reinforced urethane acrylate based thermoset composites were prepared by compression molding. The jute fabrics were modified by special formulated solutions and curing under UV radiation. Here potassium permanganate (a strong oxidizing agent) is used to modify jute surface. Previously, pineapple leaf fiber reinforced polyethylene composites were treated with $KMnO_4$ as an oxidizing agent and improved properties were achieved. During $KMnO_4$ treatment, organic constituents of jute (lignin, pectin, impurities, traces of metals, etc) might be dissolved and the surface of the cellulose in jute became smooth, which can enhanced fiber matrix adhesion (Figure 2). Physico-mechanical properties of potassium permanganates pretreatment of Hessian cloth reinforced polymer matrix composite were investigated and subjected to water uptake, degradation studies such as soil degradation, and simulating weathering of the untreated and treated composites for diversified industrial applications.

Jute content in the composites were optimized with regards to the extent of mechanical properties and 55wt% jute content in the composite showed the highest mechanical properties. Irradiated jute-urethane acrylate thermoset composite at 25th pass of UV radiation performed the best mechanical properties compared to the control composite [9]. Again, jute fabrics were treated with $KMnO_4$ solution and cured under UV radiation, and then their mechanical properties were measured.

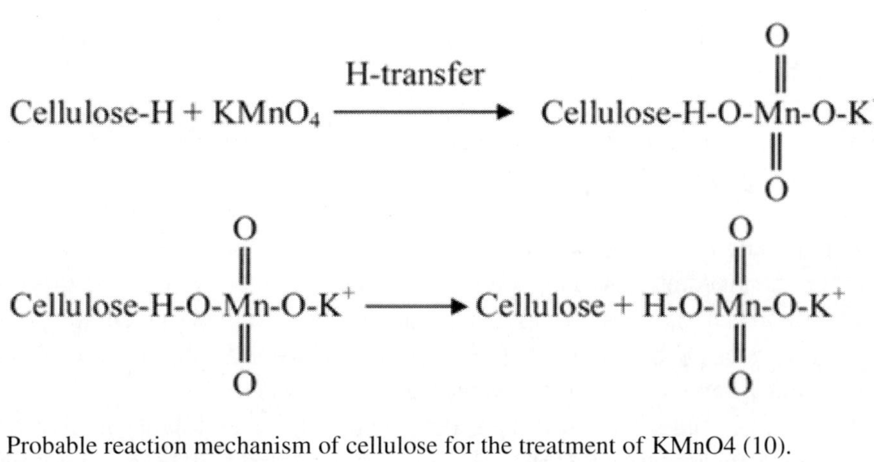

Figure 2. Probable reaction mechanism of cellulose for the treatment of KMnO4 (10).

But mechanical properties of the composites were found lower than that of UV treated urethane acrylate-based thermoset composites. SEM supported the idea that the improvement in interfacial bonding was most likely responsible for those improvements. Water uptake behavior of treated composite showed a significantly lower trend relative to untreated composite. Soil degradation studies demonstrated that TS and TM of the treated sample were higher than that of the untreated sample with respect to their degradation time which indicates higher environmental stability of the developed composite.

14.1.3. UV Radiation on Methyl Methacrylate Treated Jute and Flax Fibers

Products of jute and flex are not very durable and are rather easily destroyed and decomposed. Various treatments have been used to improve properties of the natural fibers [30, 31]. Low grade jute and its cuttings have been improved by microbial activity [32]. Combining synthetic polymers with the natural polymers under chemical or radiation treatment brings about a composite that has less affinity toward water and is thus hydrophobic in nature. Such modification provides relatively strong, durable, and sustainable materials. Many researchers have prepared such composites by treating jute with monomers [33, 34] under γ radiation or with urethane oligomer [25, 18] under UV radiation. In all these studies, there was a large uptake of the impregnating solutions. The initial increase of the mechanical strength (tenacity) was followed by decreased tenacity against higher polymer loadings. This was particularly true with the γ radiation system [35]. This trend was improved under UV irradiation. The tenacity was enhanced by 200% with increased elongation up to 60% when jute was treated with a urethane acrylate under UV radiation in the presence of certain additives (36). In these cases, there was a large uptake of the impregnating materials.

In our research laboratory, the mechanical strength (tenacity) of jute and flax fibers was enhanced by 110% and 50%, respectively, when these fibers were treated with MMA in methanol solutions under UV radiation [27], which is a significant achievement. The extent of grafting MMA onto these fibers varied between 3 and 14%. The treated fibers also gained stretching ability up to 45% for flax and 25% for jute.

14.1.4. Grafting of Different Monomers onto Jute Yarn by In Situ UV-Radiation Method

Jute polymer composite (JPC) was prepared where a single impregnating monomer 2-hydroxyethylmethacrylate (HEMA) was used and polymer loading (PL) was kept to a minimum level (26). In our study, three single impregnating monomers methyl methacrylate (MMA), butyl acrylate (BA), and styrene (ST) were used. The PL was kept at a minimum level to maintain the inherent degradable nature of the jute, as well as to reduce the amount of impregnating monomer. The physico-mechanical properties of jute yarn using different types of monomer throughout the preparation of JPC were compared and the presence of different additives used for the improvement of the properties of JPC was studied.

Jute yarn was improved through in situ grafting under UV radiation technique using three different types of bulk monomers: MMA, BA and ST. For almost all cases, bulk monomer was mixed with methanol to swell the cellulose backbone of jute for better impregnation of monomer solution. The effects of variation in irradiation time, monomer concentration, and additives on the extent of grafting were studied. Various physico-mechanical properties like grafting water uptake, tensile strength, and elongation at break for both treated and untreated jute yarns were performed. The properties of the treated jute yarn with different monomers were also compared. The highest polymer loading was found for the monomer BA but MMA showed better tensile properties amongst the monomers investigated. Five types of additives (1%) of different chemical nature, such as 2-hydroxyethylmethacrylate, ethylhexylacrylate, acrylamide, N-vinylpyrrolidine, and urea (U) were added into the bulk monomer solution to study the effects on grafting and mechanical properties of jute yarn. Among all additives, only U showed better properties in the MMA+MeOH system. Water uptake of the treated and untreated samples was also performed and BA-grafted samples showed lower water uptake to that of virgin as well as other treated samples [3].

14.1.5. Effect of Alkali and UV Radiation Pretreatment on Jute Yarn

Jute is a natural polymer composed mainly of cellulose, hemicellulose, and lignin. Jute fiber contains 82-85% of holocellulose of which 58-63% is α-cellulose. The latter is composed of 55-59% glucosan, 1.8–3.0% xylan, and 0.8–1.2% polyuronide [37]. It also contains trace quantities of galactan, araban, mannan, and rhammosan. Due to the building blocks, especially for hemicellulosic constituents jute is more susceptible to alkali, particularly NaOH. The action of aqueous alkali solutions on jute is generally associated with a loss in weight, attributable mainly to some loss of its hemicellulosic component, and it depends on temperature, the concentration of the alkali, and the duration of treatment [38]. The treatment is commonly known as "Mercerization," which is one of the oldest methods of treating textile fibers. Mercerization is not a fiber-purification process, but it induces desirable changes in yarn and fabric properties (e.g., mechanical properties, dyeing properties, and luster). The treatment involves soaking the fibers in a dilute solution of sodium hydroxide. In the presence of less-concentrated sodium hydroxide (NaOH) solutions, various ternary complexes can form between cellulose, sodium hydroxide, and water. In these complexes called soda celluloses (Na–cell), a certain number of hydrating water molecules of the sodium hydroxide hydrates are replaced by hydroxyl groups of cellulose. The formation of each type of Na–cell occurs at well-defined NaOH concentrations and temperatures [39].

Because Mercerization is one of the effective methods of treating fibers and because radiation is a very convenient method for graft initiation, which can induce desirable changes in fibers, the sum of these two effects can produce an outstanding result. Techniques involving UV radiation are particularly attractive, given that UV sources are relatively cheap, flexible, and easy to install. Because natural fiber (jute) is more sensible and environmentally friendly, this treatment may promote innovative and diverse applications of jute and jute products. M. A. Khan et al. [8] upgrade the properties by maintaining the inherent degradable nature of jute and at the same time by keeping polymer loading at a minimum level. Therefore, the amount of impregnating monomer can be reduced and the production cost can be minimized, while substantially enhancing the tensile properties.

Jute yarns were grafted with 1,6-hexanediol diacrylate (HDDA) monomer and its solution in methanol (MeOH). Solutions containing different HDDA concentrations were prepared in methanol along with a photoinitiator. The concentration of monomer, soaking time in the monomer solutions, and intensity of UV radiation were optimized and coupled with the optimized mechanical properties. Optimum conditions to carry out further treatment were established at 5% HDDA concentration, 5 min soaking time, and the 4th pass of UV radiation. Enhanced tensile strength (67%), elongation at break (14%), modulus (108%), and polymer loading (11%) were achieved after treating the jute yarns with HDDA/MeOH and curing under UV radiation. To further improve the tensile properties, the surface of jute yarns was pretreated by alkali along with non-ionizing UV radiation. Alkali and UV pretreatment promoted more enhancements in tensile properties than did the simple alkali treatment. The best performance was achieved with alkali and UV-pretreated and HDDA-grafted jute samples, enhancing tensile strength (103%), elongation at break (46%), and modulus (114%) [8]. It was found that tensile properties could be increased by simple Mercerization, although the present investigation reveals that if the Mercerized samples are UV pretreated and grafted with monomer, the tensile properties increase significantly, however, the enhancement in polymer loading was not satisfactory.

14.1.6. Effect on Urethane Acrylate-based Composites

Surface modification by UV radiation also leads to the improvement of mechanical properties of jute/resin composite. The high strength of jute fibers was not been extensively exploited in reinforced composites. This was, at least partially, because they were hygroscopic and due to their low wettability by resins. These characteristics result mainly due to the presence of hydroxyl groups and other polar groups in several constituents of the fiber, mainly cellulose and lignin. Therefore, to enhance the use of jute and other vegetal fibers, a broader system of investigations was necessary to search for chemical modifiers and physical treatments that would alter these undesirable characteristics. Several processes, such as chemical treatments [40], photochemical treatments [35, 25] have been developed to modify fiber surfaces. Improved physicomechanical properties of natural fibers with different impregnating solutions under UV radiation have been reported [8]. Others [41] have studied the effect of surface modification by ultraviolet radiation on the mechanical properties of jute/glass fiber-reinforced unsaturated polyester hybrid composite made by a hand layup technique. Better mechanical properties and enhanced interfacial adhesion were found in this research.

Surface modification of jute fabrics by UV treatment usually increases the polarities of fiber surface, which increased the fiber wettability as well as composite strength. An intense UV radiation results in a decrease of mechanical properties, a reduced degree of polymerization is observed, and two opposing phenomena such as photo-crosslinking and photo-degradation take place simultaneously when UV radiation is used as a method of surface pretreatment [10, 42-45].

Coupling agents are effective in modifying the natural fiber/matrix interface. Khan et al. prepare jute fabric-reinforced urethane acrylate-based composites by using urethane acrylate oligomer (M-1200) in methanol solvent along with photoinitiator (Darocur-1173) followed by ultraviolet (UV) radiation. Different formulations were prepared by varying oligomer in the solvent by 50–90% (by weight). Jute fabrics treated with 70% oligomer, 28% methanol, and 2% photoinitiator followed by UV radiation showed the best mechanical properties of the resulting composites. For further improvement of the mechanical properties of the composites, jute fabrics were first treated with different percentages of methylmethacrylate (MMA) in methanol with photoinitiator and cured under UV radiation and then again the composites were fabricated by the oligomer. Urea was incorporated into the MMA formulation and was found to have the potential to improve the mechanical properties of the composites [46].

14.1.7. Surface Modified Jute/polypropylene Composite

The effects of surface modification on the functional properties of fiber and polymeric materials are given in *Scheme 1.*

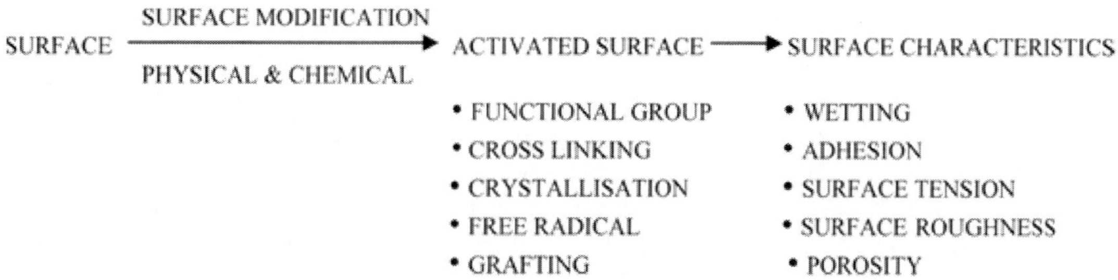

Scheme 1. Effect of surface modification on functionality.

14.1.8. Effect of Acrylic Monomers

Thermoplastics can be prepared by using synthetic polymers as coating over the natural polymer. Synthetic polymers like polypropylene (PP), polyethylene, polyester, etc. are already proved to be applicable, advantageous, and successful in a wide variety of applications and high volume market because of their low density, high softening point, good surface hardness, scratch resistance, and excellent price performance ratio [28, 43, 47, 48]. PP is used as the polymer matrix and jute is used as the reinforcing agent to reduce the content of synthetic PP, keeping environmental consciousness in mind. Different chemical treatments and radiation induced surface modification of reinforcing jute were performed to get improved mechanical properties and good weathering resistance, water uptake, interfacial adhesion, etc. [41, 49]. UV has been selected as the radiation source for pretreatment and curing instead of gamma radiation.

Jute fabrics reinforced with PP composites (40% fiber by weight) were prepared by compression molding using the surface modified Hessian cloth and PP sheet. The surface of jute is modified by photo grafting with acrylic monomers and oligomer to study the effect of mechanical properties on these resulting composites. Different formulations using oligomer urethane diacrylate (M-1100), monomers 2-ethyl hexyl acrylate (EHA), 1-vinyl 2-pyrrolidone (NVP), and photo-initiator (Irgacure-369) in methanol were prepared in order to modify the surface of jute fabrics. Jute fabrics are soaked in the prepared formulations and then cured by UV radiation source of varying intensities. The formulation containing 5% NVP, 5% EHA, 2.5% oligomer, and 2% photo-initiator and cured at 15 UV passes perform the highest mechanical properties. Tensile strength (TS), bending strength (BS), tensile modulus (TM), and bending modulus (BM) of the composites were found to be 54, 62, 915, and 2030 MPa, respectively. To investigate the effect of thermal initiator, benzyl peroxide is used in the optimized formulation instead of the Irgacure-369 photo-initiator. The TS, BS, TM, and BM of the composites were found to be 54, 63, 998, and 2150 MPa, respectively by treating jute with 2% benzoyl peroxide as the thermal initiator [11]. It was found that jute fabrics treated with formulations containing oligomer-monomer in methanol along with both thermal and photo-initiator performed higher mechanical properties of PP matrix composites than the untreated jute composites.

14.1.9. Effect of Starch and Coupling Agents

Since the 17th century, starch has been considered a good candidate in thermoplastics technology for the production of biodegradable plastics. Starch is the most abundant, naturally occurring polymer, a truly renewable resource, and it is a biodegradable polymer because it can readily be metabolized by a wide range of microorganisms. Unlike other synthetic thermoplastic polymers, starch can be processed into a thermoplastic material in the presence of plasticizers and under the influence of heat. The most important sources of starch are wheat, potato, rice, tapioca, and corn. Starch is a polysaccharide produced by many plants as a storage polymer. It is composed of glucose monomers joined by 2-(1-4) linkages. The use of granular starch as filler in plastics has advantages. Starch is biodegradable, abundant, low cost, easily available and a renewable resource. Many researchers reported higher mechanical properties of biodegradable composites based on starch and natural fibers [50-54]. Coupling agents are frequently used to improve the interaction and adhesion between the fiber and polymer matrix. The coupling agents used hereby contain chemical groups where one group can react with the fiber and the other one with the polymer. The bonds are covalent and hydrogen bonds as they improve the interfacial adhesion. The aim of this study was to modify the surface of jute by photocuring with monomer and starch, then finding out the effect of mechanical, thermal, and electrical properties on the resulting composites.

Composites based on PP matrix and both, untreated HEMA and starch treated jute fabrics were prepared by compression molding. Fiber content in the composites was optimized (45% by weight) with the extent of mechanical properties. HEMA solutions of different concentrations in methanol (MeOH) along with the photoinitiator Darocur-1173 were prepared. Jute fabrics treated with 15% HEMA at 20th UV pass for 15 min showed better performance. For the improvement of the properties, jute fabrics were treated with starch solutions of different concentrations for different soaking times. Improved mechanical properties, such as tensile strength (19%), tensile modulus (29%), bending strength (27%), bending modulus (12%), and impact strength (60%), were found in 5% starch treated jute fabrics/PP composites as compared to the untreated composite. HEMA treated jute fabrics were again treated with starch (5%) solution and found to deliver the best results, such as tensile strength (23%), bending strength (35%), tensile modulus (33%), bending modulus (15%), and impact strength (71%), over the untreated composite [55].

14.1.10. Effect of Green Dye and Development of Hybrid Composite

Many researchers are researching on various aspects of jute fibers and fiber-reinforced composites. Khan et al. [56] has produced degradable jute reinforced polymer composites using vinyl and silicon monomers as coupling agents. Also, several reports have documented that jute fibers were used as reinforcement in thermoplastics like polyethylene (PE) and polypropylene (PP) and thermosets such as unsaturated polyester and epoxy resin [57]. In all the cases the mechanical properties increased due to the reinforcement with jute fibers.

To improve the fiber-matrix adhesion, a pretreatment of the fiber surface or the incorporation of surface modifier during processing is required. Natural fibers also were improved through genetic mutation, to modify some characteristic properties. The main objectives of the genetic modifications were to introduce extra group(s) to the fiber surface which resulted in better bonding/adhesion to the reinforcing polymer. [47, 58, 59]. Green dye fiber treatments have also been found to improve the strength of jute based composites. Green dye treatment of jute fabrics has been successfully improved the fiber surface adhesion characteristic by removing natural and artificial impurities from the fiber, thus increasing the number of reactive hydroxyl groups on the fiber surface available for chemical bonding, which increases tensile properties, elasticity and dimensional stability [60]. The increase in the percentage crystallinity index of green dye-treated fibers occurs because of the removal of the cementing materials, which leads to better packing of the cellulose chains. Green dye treatment of jute fibers also leads to fiber-bundle fibrillation; which is the breaking down of the composite fiber bundle into smaller fibers. This increases the effective surface area available for contact with the polymer-matrix.

Hayder et al. prepared jute-reinforced PE, PP and mixture of PP/PE composites. They investigated the effect of UV radiation and green dye as a method of surface treatment of the jute fabrics/polymer matrix composites. Water and soil degradation study of the resulting composites were also performed. From this study, it was found that 90% PP and 10% PE matrices based jute reinforced composites performed the best. UV radiation at different intensities was applied both on matrices and jute. Mechanical properties of the irradiated jute and matrices based composites were found to increase significantly. The increase of TS and BS of jute polymer composite increasing UV radiation dose may be due to the inter-crosslinking between the neighboring cellulose molecules, which resulted in the strength of natural/synthetic fiber. It was observed that tensile properties increase with UV pretreatment up to a certain limit and then decrease due to the two opposing phenomena, namely, photo cross-linking and photo degradation that take place simultaneously under UV radiation [61]. At lower doses, free radicals are stabilized by a combination reaction and, as a result, photo cross-linking occurs. The higher the number of active sites generated on the polymeric substrate, the greater the grafting efficiency. But at higher radiation, the main chain may be broken down and polymer may degrade into fragments and, as a result, tensile properties will decrease after certain UV doses. An intense radiation results in a loss of tensile strength, and a reduced degree of polymerization is observed [62]. Optimized jute fabrics were also treated with different concentrations of green dye (0.1–1%, w/w) with 2% $K_2O_8S_2$ in methanol solution for 2–8 min. A composite made of 0.5% green dye jute (5 min soaking time) and an irradiated matrix showed the best mechanical properties [60].

14.1.11. Influences on the Electrical Property

Improvement of jute fiber through UV cured film of urethane acrylate has been studied [36]. This group investigated physical and mechanical properties of the UV-cured film, the role of plasticizers, and water uptake were also studied. Though plasticizers decrease some textile properties of the UV cured films, they substantially enhance the tensile strength of jute by 300%. This group [28] also prepared a number of formulations in the presence of plasticizers and monomers of different characteristics using urethane acrylate prepolymer with an aliphatic chain. Hessian cloth was treated with these formulations and the tensile properties of these jute plastic composites were found to increase. The degradation character of the composite was also studied. Modification of jute by grafting was reported [63]. Moisture absorption of jute fibers could be reduced significantly by treatment with graft copolymerization using MMA. The tensile strength and elongation at the break of jute fiber decreased on grafting whereas the initial modulus increased with increasing percentage of grafting. Another report outlined [34] the grafting copolymerization of acrylonitrile onto jute fiber using the lippuric acid redox initiator system.

In our laboratory Hessian cloth was treated with HEMA under ultraviolet radiation in order to improve the mechanical and electrical properties. The concentration of the monomer HEMA, radiation dose, and soaking time were optimized with respect to mechanical properties such as tensile strength and elongation at break of the treated and untreated Hessian cloth. The 10% HEMA, 15 min soaking time, and 15th pass of radiation rendered the best tensile properties [15]. The variations of dielectric properties with temperature were measured at 10 kHz frequency. It was observed that the dielectric constant and loss tangent increased with increasing temperature up to the transition temperature and then decreased, and at the end become almost constant.

14.1.12. A Comparative Study between Gamma and UV Radiation Based Modification

The treatment of jute fibers with high-energy radiation (gamma) and nonionizing radiation (UV) have been employed successfully for significant physical and chemical changes as well as changes in surface structure and surface energy of the fibers. One group [64] studied the effect of corona discharge and UV radiation on the properties of jute fiber epoxy composites. Some jute-reinforced composites with urethane oligomer under gamma radiation were prepared and characterized [65]. The influence of ionizing radiation on polymers has been studied quite extensively over the past few decades. Gamma radiation is known to deposit energy in solid cellulose by Compton scattering and the rapid localization of energy within molecules which produces trapped macrocellulosic radicals. The radicals thus generated are responsible for changing the physical, chemical, and biological properties of cellulose fibers [29].

The effects of ^{60}Co γ-radiation and UV radiation on the jute fabrics/PP composites and the effect of starch as a method of surface treatment were investigated. Samples were fabricated by compression molding and the mechanical properties were evaluated. For UV treated jute fabrics/PP composites, TS, TM, BS, BM, and IS of the composites were found to be 59.69 MPa, 1.32 GPa, 65.37 MPa, 1.62 Gpa, and 26.14 kJ/m^2, respectively. On the other hand, for gamma treated jute fabrics/PP composites, TS, TM, BS, BM, and IS of the composites were found to be 57.95 MPa, 1.24 GPa, 62.45 MPa, 1.58 Gpa, and 24.25 kJ/m^2, respectively. From this investigation it was found that the mechanical properties of UV treated jute fabrics/PP composites had increased TS, TM, BS, BM, and IS compared to that of the gamma treated jute fabrics/PP composites. SEM images of the fracture sides of the composites were supportive of the fact that gamma treated jute fabrics/PP composites had poorer fiber matrix adhesion than that of the UV treated jute fabrics/PP composites [66].

14.1.13. Jute Composite with MMA by Gamma and UV Radiations in the Presence of Additives

Jute yarns were also modified by methyl MMA in the presence of various additives and co-additives under ^{60}Co gamma and ultraviolet radiations. Jute yarns treated with MMA+MeOH solutions were irradiated either with ^{60}Co gamma source or with UV radiation. In gamma radiation, polymer loading of MMA onto jute increased quite substantially, but the strength of the composite decreases sharply after 15% polymer loading. Besides, gamma-treated jute samples were very brittle. On the other hand, jute yarns irradiated *in situ* under UV radiation was found to be grafted with MMA. The tensile strength of the UV-cured jute yarn composite increases with an increase of grafting level, in contrast to the behavior observed with the gamma-irradiated jute composite samples. The tensile properties of the composites can be further enhanced by the incorporation of certain additives and co-additives into MMA+MeOH solutions. This opens diverse applications for jute materials [43].

14.2. MODIFICATION OF BIOMATERIALS

14.2.1. Chitosan

Chitosan (Figure 3) is a linear polysaccharide prepared by the alkaline or enzymatic N-deacetylation of chitin which is found in the outer shells of crabs, prawn, shrimp, lobsters, etc. The repeating units of chitosan are β-(1→4)-linked glucosamines. Thus, chitosan contains a large number of hydroxyl and amino groups. These two functional groups provide several possibilities for grafting of desirable bioactive groups [67-69]. Since chitosan is nontoxic and biocompatible with the human physiological system, it has been investigated as biomaterials in the fields such as biomedicine, pharmacology and biotechnology. Chitin and chitosan have already been used in agricultural, food, industrial and medical fields [70-72].

Chitosan can be photo grafted using various types of monomers like silane monomer, acrylic monomers, ethylene glycol, ethylene glycol dimethacrylate and is reported to have higher mechanical properties while keeping its inherent biodegradable properties [73-75]. Silanes monomer, which is actually a monomeric silicon, is one of the most promising monomer to modify both organic and inorganic compounds. This belongs to a class of organosilane compounds having at least two reactive groups. It works at an interface between an inorganic substrate (such as glass, metal or mineral) and an organic material (such as an organic polymer, coating or adhesive) to bond, or couple, the two dissimilar materials. The organo-functional portion of the silane can react, and bond to the polymer backbone. These versatile materials are used as promoters, dispersing agents, crosslinking agents, surface modifiers etc. As a crosslinking agent, it is used in various polymers like polyethers, polyurethanes, polyesters, etc. to improve properties such as tear resistance, elongation at break, abrasion resistance, etc. A silane coupling agent can be made to be compatible with almost every type of organic polymer, and thermosetting resins through elastomers. It may be applied to the substrate as a pretreatment [76]. It has the ability to improve adhesion between organic and inorganic phases [77].

Khan et al. [78] prepare a chitosan film which was extracted from prawn shell waste, and characterized of thin films of chitosan. They observe the effect of the silane monomer named 3-trimethoxyl silyl propylmethacrylate (TSPMA) on the physico-mechanical properties of the chitosan films using a photocuring technique. It was found that, Chitosan films treated (4 min soaking time) with 30% TSPMA performed with better properties at the 24th UV pass and were 4 times higher TS value compared to raw chitosan film. It was also observed that optimized film showed lower water uptake. The SEM study showed homogenous surfaces, without separation of phases between the two polymers, and a compact structure in the film formation. The FTIR study revealed the intermolecular interaction among chitosan polysaccharide chain, TSPMA, and water molecule in the film. Finally, it can be said that silane monomer improved the mechanical properties of the chitosan films [78].

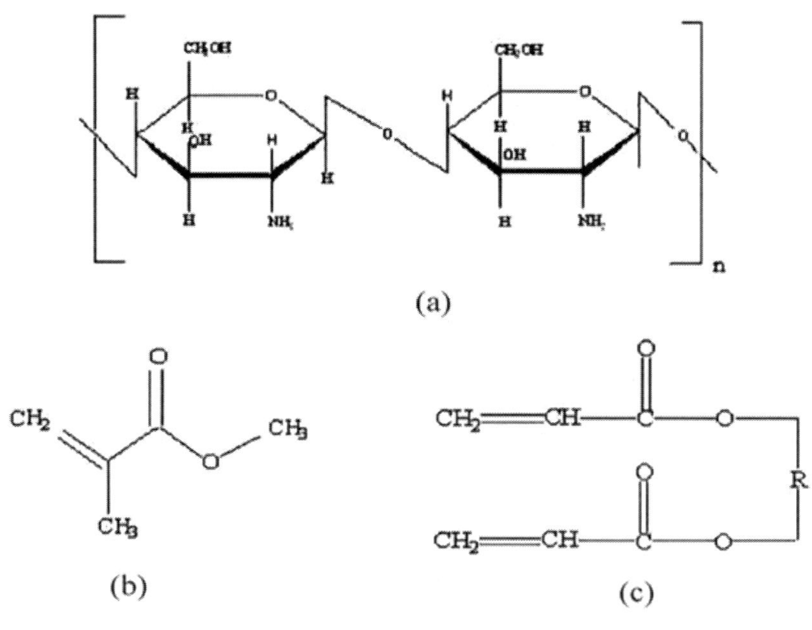

Figure 3. Chemical structure of Chitosan (a), MMA (b), and aliphatic urethane diacrylate (c).

A multifunctional vinyl monomer promotes rapid free-radical propagation reaction leading to network (crosslinking) polymer structures through grafting via their double bonds. Multifunctional unsaturated acrylic monomer is used in the treatment of the polymer film to reduce the UV-radiation intensities to obtain optimum properties [79]. The extent of grafting and homo-polymer formation was found to depend on chitosan, amount of initiator, monomer concentration, reaction temperature and reaction time [80]. Others [81], prepared various formulations of MMA and aliphatic urethane diacrylate oligomer (Figure 3) cured chitosan films with dissimilar soaking times and varying UV radiation intensities. The cured films were characterized by measuring TS, Eb, PL, water absorption and gel content properties. The formulation containing 43% MMA and 15% oligomer in methanol solution showed the best performance at 20th UV pass for 4 min soaking time [81]. For all cases the mechanical properties of the treated films were higher than that of the untreated raw chitosan films. This creates a new field of research on chitosan using acrylic monomers and UV radiation to improve the inherent chitosan films properties.

Blending of chitosan with other synthetic hydrophilic polymers such as polyvinyl pyrrolidone [82, 83], polyethyleneoxide [83], or natural polymers: alginate [84], and starch [85] has been suggested as a promising approach to producing polymers for specific uses especially drug delivery systems. Again, chitosan dissolved in weak organic acids such as acetic acid can be cast into membranes, fibers, and sponges. Chitosan and its oligomers are well known for their interesting biological properties, which have led to various applications.

Urea has considerable effect on a polysaccharide solution. In particular, the effect of urea on the structure of biopolymers and bioaggregates is not well understood at the molecular level, despite the expressive number of papers published in this area. Urea is the most widely used denaturating agent, being extensively applied in biochemistry, not only to denature proteins at high urea concentration, but also to promote controlled folding. Whether urea interacts directly in the biopolymer, promoting a better solvation of the polar residues, or indirectly by decreasing the water structure is the main topic of discussion in this research area [85, 86]. Other mechanisms have also been proposed to explain the effect of urea in biopolymers and bioaggregates [87, 88]. Since chitosan is a brittle material and absorbs moisture; so, improvement of its toughness and water resistance is necessary. Urea was supposed to reduce the water absorption capability of the polysaccharide film.

Khan et al., prepared chitosan/urea biodegradable film by solution casting. The film was photocured with 1-vinyl-2-pyrrolidone using different formulations. Photocuring and grafting of the chitosan/urea film improved the physico-mechanical properties. Monomer formulation containing 30% NVP and 25th UV pass were most effective in enhancing the mechanical properties of the films. The TS of NVP grafted photocured chitosan/urea film increased by about 258% and 152.5% compared to chitosan/urea and chitosan films, respectively [89].

Starch (Figure 4) and chitosan are hydrocolloids with well-known good film forming capacities. Both have been reported as potential materials for food packaging, especially as edible films and coatings [90]. However, the use of a mixture of chitosan and corn starch to obtain biodegradable films has not been extensively studied [90].

Khan et al. [91], prepared chitosan/starch blend films of 2:1 ratio by solution casting. They employed UV radiation to improve the chitosan/starch film properties. The mechanical properties such as tensile strength, elongation at break and the gel content and swelling ratio of the photocured films were studied. Tensile strength (TS) and elongation at break (Eb) of chitosan/starch blend film were found to be 9.33MPa and 3.6%, respectively. The blend films (2% Chitosan) were soaked in six different formulations of MMA monomer and oligomer (M-1200) at different time intervals and irradiated under UV radiation at different intensities. Formulations containing 43% MMA and 15% oligomer in methanol solution showed 209% increase in TS at 25th UV pass for 4 min soaking time [91]. SEM and FTIR were used to investigate the molecular interaction and morphology of the blend film, respectively.

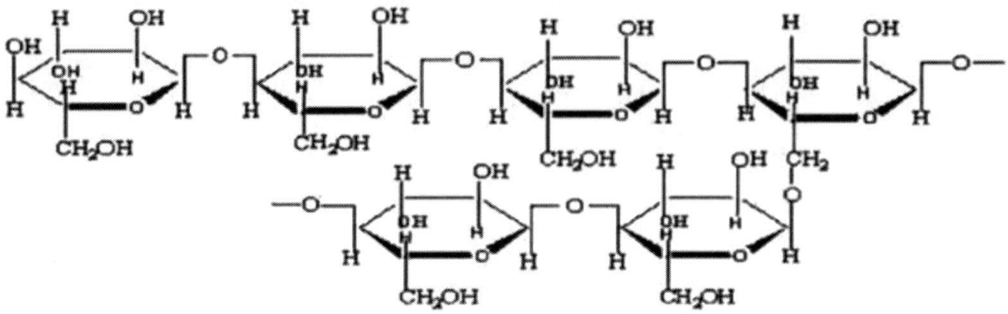

Figure 4. Chemical structure of Starch.

14.2.2. Natural Resin Based Composites

Making bio-degradable polymer from natural renewable sources is very much familiar and ancient work. But further work may be carried out for making bio-degradable polymer by using raw materials, which are locally available, cheap and quite friendly to the environment. Moreover, an additional demand is now widely proposed that polymeric materials should come from renewable resources instead of from petrochemical sources. From that point of view Jiga gum has been used as a raw material for making polymer film. This Jiga gum is found from a south Asian plant named Jibal or Kaphila, scientific name of the plant is *Lannea Grandis Dennst.* A yellowish-white thick gum secretes from the damaged part of the tree when it is infected by bacteria. Fresh gum is soluble in water, forming rather thin mucilage with good adhesive properties. From elementary organic analysis, it is found that the gum contains D-Galactose, L-Arabinose, D-galacturonic acid and Aldobiuronic acid [92, 93]. One group [94] described the thermogram of Jiga (*Lannea Grandis*) gum. Here for grafting, UV radiation has been used because UV curing technology offers great advantages.

Alkoxysilane fictionalization has been used in the production of inorganic-organic hybrid materials, because alkoxysilanes can provide covalent bonding between inorganic and organic components [95-97]. Alkoxysilanes also have been used in the production of biodegradable cross-linked materials (98). Radiation technology has been used frequently to improve the plastic products due to the chemical reaction among polymer molecules under irradiation [99, 100]. To enhance the rapid free-radical propagation reaction, leading to network (cross-linking) polymer structures through grafting via their double bonds, multifunctional vinyl monomer were used.

Khan et al. successfully extracted hard Jiga resin from the natural gum from that secretion by Jiga plants using the solvent mixture of ethanol and benzene. Thin polymer films of the bio-blends of Jiga resin and starch were prepared in aqueous medium by casting. A number of solutions using vinyltrimethoxy silane of varying concentrations in methanol along with a photo initiator were prepared. The prepared films (Jiga/starch) were soaked in the solution at varied soaking time and cured under UV radiation. The highest polymer loading (PL), tensile strength (TS) and elongation at break (Eb) were found with a formulation of 5% monomer + 93% methanol + 2% photoinitiator at 3 min soaking time under 15th pass of UV radiation for 1:1 Jiga/starch film [101]. Jiga/starch blend films can be prepared by changing the composition. If the percentage of Jiga resin is increased in the composition, the film does not form, but with the optimum starch content, it enables the formation of the degradable bio-blended films.

14.2.3. Modification of Alginic Acid

Alginates are linear copolymers of β-(1-4)-linked D-mannuronic acid and α-(1-4)-linked L-guluronic acid units, which exist widely in many species of brown seaweeds. Alginate has been used in a wide range of industries, such as food, textile printing, paper, and pharmaceuticals, and for many other novel end uses [102]. In recent years, alginate has been widely used in the wound management industry as a novel material for the manufacture of 'moist healing' products such as gels, foams, and fibrous nonwoven dressings that are used to cover wounds [103-106]. But one of the major problems associated with the use of natural fibers in composites is their high moisture sensitivity, leading to severe reduction of mechanical properties and delamination. The reduction in mechanical properties may be due to poor interfacial bonding between resin matrices and fibers. It is therefore necessary to modify the fiber surface to render it more hydrophobic and also more compatible with resin matrices [56, 107, 108].

An effective method of chemical modification of natural fibers is by a chemical coupling method. Silane chemical coupling presents three main advantages such as, they are commercially available in large scale, at one end, they bear alkoxysilane groups capable of reacting with OH-rich surface, and at the other end, they have a large number of functional groups, which can be tailored as a function of the matrix to be used [1, 108, 109-111]. Some acrylic monomers like 2-ethylhexyl acrylate and 2-hydroxyethyl methacrylate were found to act as effective coupling agents for increasing the mechanical properties of calcium alginate-plastic composite prepared under radiation technique.

Coupling agents such as MMA usually improve the degree of crosslinking in the interface region and offer a perfect bonding result. Coupling agents are effective in modifying the natural fiber/matrix interface [112]. Khan A. et al. used methyl methacrylate (MMA) as the coupling agent. The silane and MMA-treated calcium alginate fibers reinforced PP composites were successfully prepared by compression molding method followed by UV curing. The mechanical properties of the composites revealed promising values. The changes in tensile and bending properties as a function of fiber content were investigated. The mechanical performance of the composites was maximum at 20% (by weight) fiber loading. From this investigation it was evident that the MMA-treated composites showed best mechanical properties at 30% MMA concentration at 30th UV pass, whereas the silane-treated composites exhibited best mechanical properties at 5% silane concentration at 30th UV pass [113].

14.2.4. Modification of Gelatin

Gelatin is a protein derivative obtained by boiling collagen. It is reported that to overcome the poor mechanical properties of gelatin, laminates were prepared using gelatin and starch reinforced with fabrics [114]. Gelatin is biodegradable and has good biocompatibility, which makes it to use as tissue replacement material [115]. In order to improve the property and to develop a new material chitosan, starch, silk etc. are used with gelatin [116]. Chitosan is graft copolymerized to enhance the property and also combined with gelatin to use as wound dressing materials [74, 75,115]. Crosslinked gelatin sponges are investigated for their potential application [117]. Photocrosslinkable polymers possess functional groups, which can undergo light induced reactions to form directly a cross-linked polymer. Photoinitiator performs widespread application in case of this photoinduced polymerization [118]. Use of additives in radiation crosslinking is common. For example: acrylic monomers, urea, glycerin etc. used for the modification of natural polymers. Research had also carried out on the polymerization of acrylic monomer in polymer matrix [119, 120]. Urea as additive influences the processing condition and the property of the finished product during curing reactions. Another important thing is that, presence of additive in curing reactions reduces homopolymer formation and incorporates specific properties [121].

Thin films of gelatin were prepared by casting. Then the films were photocured and the mechanical properties were studied. Cross-linking and chain scission occurs when polymers are irradiated under UV radiation. The TS of the UV cured films increases with the increase of radiation intensities because at low radiation intensities cross-linking is dominating. Higher radiation causes degradation of the polymers, which is responsible for the decrease of TS values [75]. Natural polymer, when exposed to radiation degrades due to breaking of glycosidic linkage [122]. Accordingly, gelatin which is a natural biopolymer has shown the same effect when exposed to higher UV radiation.

The tensile strength of UV cured gelatin films showed about 10% enhancement than that of raw gelatin films. Minor amount of urea was used as additive in aqueous gelatin solution and films were prepared using same technique. Four formulations were prepared in methanol with 2-ethylhexyl acrylate in the presence of photoinitiator (darocur-1664). The films were soaked in the prepared formulations and then cured under UV radiation at different intensities. Percentage of urea, monomer concentration, soaking time and radiation intensities were optimized with the extent of polymer loading, TS and elongation at break of the photocured film. The films containing 2% urea, cured with 3% EHA for 3 min at 15th UV pass showed the highest mechanical properties. A significant improvement of TS (31%) occurred when EHA (3%) was incorporated [123].

In our laboratory flexible gelatin-based films were also fabricated using HEMA monomer followed by UV radiation which induces poly-HEMA formation (Figure 5). Poly-HEMA is an important hydrophilic polymer, with current uses in contact lenses, biocompatibilization and drug delivery [124, 125]. Poly-HEMA is also known as a hydrogel with high hydrophilicity [126]. It is reported that Poly-HEMA has adequate biocompatibility and has many potential applications in medical sciences [127, 128]. Some researchers successfully implanted Poly-HEMA-collagen (fibrillar) composites into bone as medical implants [129, 130]. HEMA, a vinyl monomer, is used as a coupling agent for natural fibers like jute, cotton, silk etc. [26, 131,132].

When UV light penetrates the polymer, the photoinitiator breaks down into free radicals which are able to initiate a rapid addition polymerization reaction, with unsaturated groups which are present in the oligomer and monomer. So, upon absorption of UV light, the molecules can be excited to a higher electron energy level and dissociated to form free radicals. UV radiation can only initiate surface reaction [133]. The acrylic monomer is used in the polymer film to reduce the UV radiation intensities to obtain optimum properties or to achieve an increased cure state at the same UV radiation intensities [81].

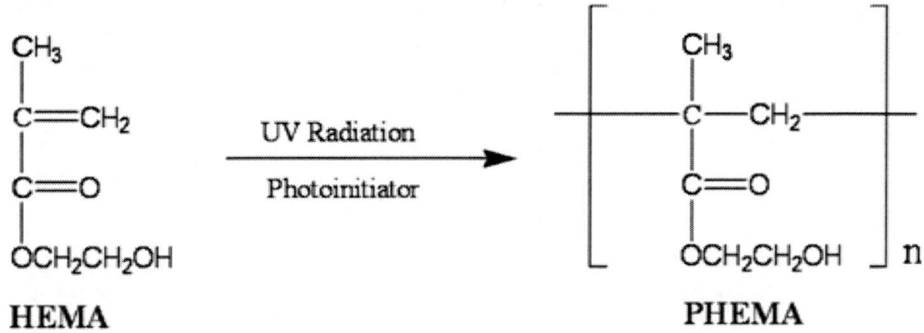

Figure 5. Formation of poly (HEMA) from HEMA during UV curing.

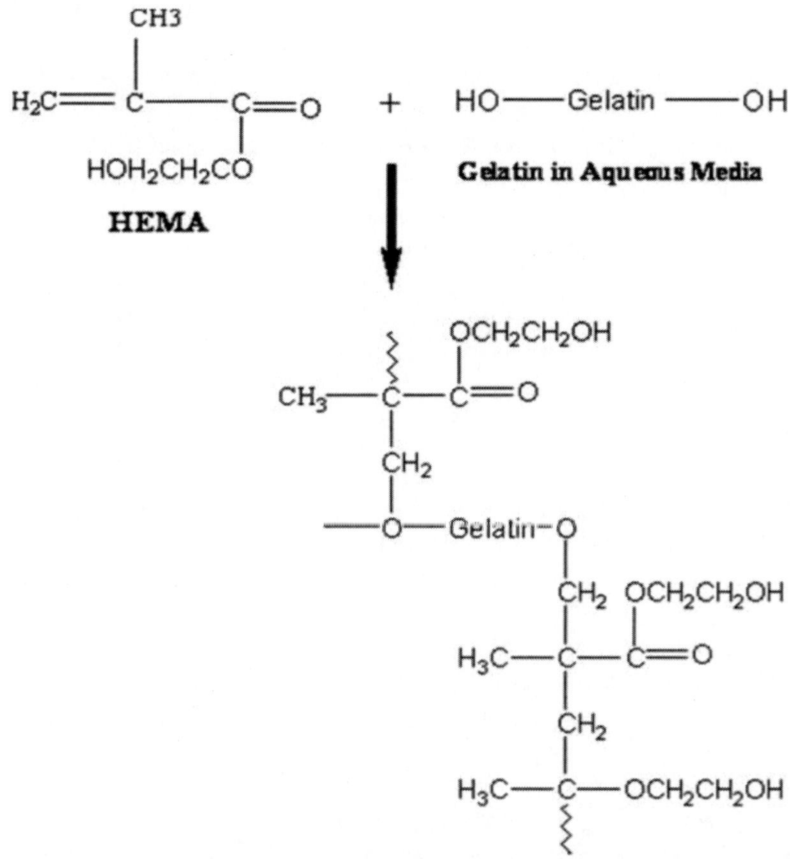

Figure 6. Reaction mechanism between HEMA and gelatin.

Gelatin films soaked in 5 different formulations containing HEMA (10–50%, by wt), methanol and photo-initiator and treated under UV radiation elevated mechanical properties at 20th UV pass for 30% HEMA solution. Besides, thin films prepared from gelatin/HEMA blend solution shows enhanced tensile strength and tensile modulus with respect to UV radiation in the presence of photo initiator. The reaction between HEMA and gelatin is attributed from the improvement of the mechanical properties of the HEMA/Gelatin films compared to pure gelatin films. With the increase of number of UV pass, the TS of the gelatin films increases, which indicates the crosslinking of acrylic group of HEMA with hydroxyl group of gelatin, as shown in Figure 6 [134].

CONCLUSION

The modern world is developing in accelerated approaches. The immerging advancement of technology also induces the utilization of finite resources of Earth leading to a vulnerable state. This is why development is not only a concern, it has to be sustainable, promoting utilization of renewable resources, minimizing waste through reduce, recycle and reuse, maximizing utilization of raw material, cutting the CO_2 emission etc. This will ultimately lead to zero waste production facilities without affecting the ecosystem. In this contrast, application of natural fiber is very promising for its natural origin, diversified uses, and high strength. Application of UV radiation for the surface modification leads to induce hydrophobicity as well as stability. Thus limitations associated with uses of natural fiber minimize what has been created to date but offers a new dimension in the composite sector. Similarly, UV induced crosslinking significantly improve mechanical strength of biomaterials having tissue engineering and biocomposite purposes. So, by all contrasts, UV radiation has been proven to be a low cost and clean technique for the sustainable development new composites.

REFERENCES

[1] Saheb, D. N., Jog, J. P., (1999), Natural fiber *polymer* composites: a review, *Adv. Polym. Technol.*, 18 (4):351–363.

[2] Khan, R. A., Khan, M. A., Sultana, S., Khan, M. N., Shubhra, Q. T. H. and Noor, F. G., (2010), Mechanical, Degradation, and Interfacial Properties of Synthetic Degradable Fiber Reinforced Polypropylene Composites, *J. of Reinforced Plastics and Composites,* 29 (3):466-476.

[3] Khan, M. A., Rahman, M. A., Akhunzada, K. S., (2002), Grafting of Different Monomers Onto Jute Yarn by In Situ UV-Radiation Method: Effect of Additives, *J. Polym. Plast. Technol. Eng.,* 41 (4):677–689.

[4] Shah, A. N., Lakkad, S. C., (1981), Mechanical properties of jute-reinforced plastics, *Fiber Sci. Technol.*, 15:41.

[5] Rong, M. Z., Zhang, M. Q., Liu, Y., Yang, G. C. and Zeng, H.M., (2001), The effect of fiber treatment on the mechanical properties of unidirectional sisal-reinforced epoxy composites, *Compos. Sci. Technolo.*, 61:1437.

[6] Bledzki, A. K. and Gassan, J., (1996) Natural fiber reinforced plastics. Kassel, Germany: University of Kassel.

[7] John, M. J., Rajesh, D., and Anandjiwala, (2008), Recent developments in chemical modification and characterization of natural fiber-reinforced composites, *Polym. Compos*., 29 (2):187-207.

[8] Khan, M. A., Shehrzade, S. and Hassan, M. M., (2004), Effect of alkali and ultraviolet (UV) radiation pretreatment on physical and mechanical properties of 1,6-hexanediol diacrylate–grafted jute yarn by UV radiation, *J. Appl. Polym. Sci.*, 92:18.

[9] Zaman, H. U., Khan, A., Khan, R. A., Huq, T. Khan, M. A., Shahruzzaman, M., Rahman, M. M., Al-Mamun, M. and Poddar, P., (2010), Preparation and Characterization of Jute Fabric Reinforced Urethane Thermoplastic Composites: Effect of UV radiation, *Fibers and Polymers*, 11 (2):258.

[10] Hassan, M. M., Khan, M. A., Islam, M. R., (2003), Influence of Mercerization and UV Radiation treatment on the Improvement of Mechanical Properties of Photo-grafted Jute Yarn with Silane and Acrylamide, *Poly. Plas. Tec. Eng*., 42 (4):515–531.

[11] Khan, M. A., Khan, R. A., Zaman, H. U., Noor-A Alam, M., and Hoque, M. A., (2010), Effect of Surface Modification of Jute with Acrylic Monomers on the Performance of Polypropylene Composites, *J. of Reinforced Plastics and Composites*, 29 (8):1195-1205.

[12] Roffey, C. G., In Photopolymerization of Surface Coatings, *Wiley,* New York, 1982.

[13] Woods, J. G., In Radiation Curing of Polymeric Materials, *ACS Symposium Series, American Chemical Society*, Washington, DC, 417, 1990, p. 333.

[14] Hassan, M. M., Islam, M. R., Sawpan, M. A. and Khan, M. A., (2003), Effect of Silane on Mechanical and Degradable Properties of Photografted Jute Yarn with Acrylamide, *J. Appl. Polym. Sci.*, 89:3530.

[15] Khan, M. A., Das, L. R., Khan, A. H., Ahmed, F. and Rahman, S. M. B., (2003), Study on the Mechanical and Dielectric Properties of Photo-cured Jute Fabrics with 2-Hydroxyethyl Methacrylate (HEMA), J. *Appl. Polym. Sci*., 89:656.

[16] Hassan, M. M., Islam, M. R., and Khan, M. A., (2003), Improvement of Physico-mechanical Properties of Jute yarn by Photografting with 3-(trimethoxysilyl)-propyl methacrylate, *J. Adhesion Sci. and Technology*, 17 (5):737.

[17] Khan, M. A., Hinrichsen, G. and Drzal, L. T., (2001), Influence of Novel Coupling Agents on Mechanical of Jute Reinforced Polypropylene Composite, *J. Mater. Sci. Lett*., 20:1711.

[18] Khan, M. A., Uddin, M. K., Islam, M. N. and Ali, K. M. I., (1995), Degradable Property of UV Cured Hessian (Jute) Cloth, *J. Appl. Polym. Sci.*, 58:31.

[19] Khan, M. A., Khan, R. A., Zaman, H. U. and Khan, A. H., (2009), Effect of Gamma Radiation on the Physico-Mechanical and Electrical Properties of Jute Fiber-Reinforced Polypropylene Composites, *J. Reinforced Plastics and Composites*, 28 (13):1651-1660.

[20] Khan, M. A., Majumder, S. C., Rahman, M. A., Noor, F. G., Zaman, H. U., Mollah, A. J. I., Khan, R. A. and Das, L. R., (2010), Mechanical and Electrical Properties of Photocured Jute Fabric with 2-hydroxy ethyl acrylate, *Fibers and Polymers*, 11 (3):391-397.

[21] Amash, A. and Zugenmaier, P., (2000), Morphology and properties of isotropic and oriented samples of cellulose fibre-polypropylene composites, *Polymer*, 41:1589.

[22] Shah, N. and Banerjee, A. N., (1996), Dynamic mechanical study on unidirectional polyethylene–glass fibers: PMMA hybrid composite laminates, *J. Appl. Polym. Sci*., 62:1199.

[23] Jenkings, A. D. and Ledwith, A., Structure, Reactivity and Mechanism in Polymer Chemistry, *Wiley*, London, UK, 1974, pp.1-25.

[24] Benedetto, A. T. D., Haddad, G., Schilling, C. and Osterholtz, F. in Interfacial Phenomena in Composite Materials '89: *Proceedings of an International Conference Sponsored by the Journals Composites and International*, (Jones, F. R. Ed.), Butterworth-Heinemann, 1989, pp.12-24.

[25] Khan, M. A., Hossian, M. M., Hossian, M. A. and Ali, K. M. A., (1997), Improvement of Jute yarn through *in-situ* Copolymerization by UV radiation, *Polym. Plast. Technol. Eng.*, 36:285.

[26] Ali, K. M. I., Khan, M. A. and Akhunzada, K. S., (1999), *In situ* grafting of jute yarn with HEMA by UV radiation., *J. Appl. Polym. Sci.*, 77:841.

[27] Ali, M. A., Khan, M. A., Ali, K. M. I. and Hinrichsen, G., (1998), Mechanical Studies of MMA treated Jute and Flax Fibers under UV radiation, *J. Appl. Polym. Sci.*, 70:843.

[28] Uddin, M. K., Khan, M. A. and Ali, K. M. I., (1997), Degradable Jute Plastic Composite (JPC*), Polym. Degrad. Stabil.*, 55:1-7.

[29] Ali, K. M. I., Khan, M. A. and Ali, M. A., (1997), Study on Jute materials with Urethane acrylate by UV Curing, *Radiat. Phys. Chem.*, 49 (3):383.

[30] Semsarzadeh, M. A., (1985), Low Cost, High Strength, Hard Cellulosic Fiber Reinforced Polyester Structures and Units, *Polym. Tech. & Eng.*, 24 (4):323-334.

[31] Sanadi, A. R., Prasad, S. V. and Rohatgi, P. K., (1996), Sunhemp fibre-reinforced polyester, *J. Mater. Sci.*, 21:4299.

[32] Mohiuddin, G., (1992), Upgrading of Low-grade Jute and Cuttings. Part I: Microbial Activity and its Enhancement by Using Chemicals, *J. Text. Inst.*, 83 (4):527.

[33] Ghosh, P., Bandyopadhyay, A. R. and Das, S., (1983), Radiation (^{60}Co)-Induced Graft Copolymerization of Acrylamide on Jute Fiber, *J. Macromol. Sci. Chem. A*, 19:1165.

[34] Huque, M. M., Habibuddowla, M. D., Mahmood, A. J. and Mian, A. J., (1980), Graft copolymerization onto jute fiber: Ceric ion-initiated graft copolymerization of methyl methacrylate, *J. Appl. Polym. Sci. Polym. Chem. Ed.*, 18:1447.

[35] Khan, M. A., Ali, K. M. I., Al Imam, M. R. and Mannan, R. J., (1997), Radiation Induced Jute Plastic Composite (JPC) and Its IR Studies, *Polym. Tech. & Eng.*, 36:89.

[36] Ali, K. M. I., Uddin, M. K., Bhuiyan, M. I. U. and Khan, M. A., (1994), Improvement of Jute fibers Through UV Cured Films of Urethane acrylate, *J. Appl. Polym. Sci.*, 54:303.

[37] Mohanty, A. K., Misra, M., Hinrischen, G., (2000), Biofibres, biodegradable polymers and biocomposites: An overview, *Macromol Mater Eng.*, 276/277:1.

[38] Rowell, R. M., Stout, H. P., In: Jute and Kenaf, Lewin, M., Pearce, E. M. Eds., *Handbook of Fiber Chemistry, 2nd ed.*, Marcel Dekker: New York, 1998.

[39] Sobue, H., Kiessig, H., Hess, K., (1939), The cellulose-sodium hydroxide-water system as a function of the temperature, *J. Phys. Chem. B*, 43:309.

[40] Khan, M. A. and Hassan, M. M., Surface Modification of Natural Fibers, In: K. I. Mital, (eds.), *Polymer Surface Modification: Relevance to Adhesion*, Ch. 3, pp. 263–271, VSP Publications, Netherlands, 2004.

[41] Kafi, A. A., Abedin, M. Z., Beg, M. D. H., Pickering K. L. and Khan, M. A., (2006), Study on the Mechanical Properties of Jute/Glass hybrid composites, Effect of UV radiation, *Journal of Reinforced Plastics and Composites*, 25 (6):575–588.

[42] Ali, K. M. I., Khan, M. A. and Islam, M. N., (1996), Improvement of Physico-mechanical Properties of Hessian Cloth (Jute), by Urethane acrylate, with Ultraviolet Radiation, *Poly. Plas. Tec. Eng.*, 35 (1):53.

[43] Khan, M. A., Hossain, M. and Ali, K. M. I., (1999), Jute composite with MMA by gamma and UV radiations in the presence of additives, *J. Applied Poly. Sci.*, 74:900.

[44] Ali, K. M. I., Khan, M. A. and Husain, M. M., (1994), Study of Wood Plastic Composites in Presence of Nitrogen Containing Additives, *Radiation Physics and Chemistry*, 44 (4):427–429.

[45] Khan, M. A., Khan, R. A., Zaman, H., Hossain, A. and Khan, A. H., (2009), Effect of Gamma Radiation on the Physico-Mechanical and Electrical Properties of Jute Fiber-Reinforced Polypropylene Composites, *J. Reinf. Plastic Composites*, 28 (13):1651-1660.

[46] Zaman, H. U., Khan, A. H., Hossain, M. A., Khan, M. A. and Khan, R. A., (2010), Effect of Chemical Modification of Jute Fabrics on the Performance of Urethane Acrylate-based Composites, *J. Reinforced Plastics and Composites*, 29 (13):2027-2036.

[47] Karmaker, C. and Hinrichen, G., (1991), Processing and Characterization of Jute Fiber Reinforced Thermoplastic Polymers, *Polymer Plastic Tec. Eng.* 30 (5–6):609–629.

[48] Roe, P. J. and Ansell, M. P., (1985), Jute Reinforced Polyester Composites, *Journal of Material Science*, 20 (11):4015–4020.

[49] Khan, M. A., Shehrzade, S., Chowdhury, A. S. M. and Rahman, M. M., (2002), Effect of Pre-treatment with UV-radiation on Physical and Mechanical Properties of Photocured Jute Yarn with 1, 6, Hexanediol diacrylate (HDDA), *Journal of Polymers and the Environment*, 9 (3):115–124.

[50] Ramaraj, B., (2007), Crosslinked poly(vinyl alcohol) and starch composite films. II. Physicomechanical, thermal properties and swelling studies, *J. Applied Polymer Science*, 103:1127-1132.

[51] Ramaraj, B., (2007), Crosslinked poly(vinyl alcohol) and starch composite films. II. Physicomechanical, thermal properties and swelling studies, *Journal of Applied Polymer Science*, 103:909-916.

[52] Rosa, M. F., Chiou, B. S., Eliton, S. and Imam, S. H., (2009), Biodegradable composites based on starch/EVOH/glycerol blends and coconut fibers, *J. Appl. Poly. Sci.*, 111:612.

[53] Aji, P. M., Thielemans, W. and Dufresne, A., (2008), Mechanical properties of nanocomposites from sorbitol plasticized starch and tunicin whiskers, *J. Appl. Poly. Sci.*, 109:4065-4074.

[54] Cao, X., Chen, Y., Chang, P. R., Stumborg, M. and Huneault, M. A., (2008), Green composites reinforced with hemp nanocrystals in plasticized starch, *J. Appl. Poly. Sci.*, 109:3804-3810.

[55] Zaman, H. U., Khan, A. H., Hossain, M. A., Khan, M. A. and Khan, R. A., (2010), Fabrication and Characterization of Jute Reinforced Polypropylene Composite: Effectiveness of Coupling Agents, *Journal of Composite Materials*, 44 (16):1945-1963.

[56] Khan, M. A., Kopp, C., Hinrichsen, G., (2001), Effect of Vinyl and Silicon Monomer on Mechanical and Biodegradable Properties of Biodegradable Jute-Biopol Composite, *J. Reinf. Plast. Comp.*, 20:1414–1429.

[57] Mohanty, A. K., Misra, M., (1995), Studies on jute composites - a literature review, *Polym. Plast. Technol. Eng.*, 34 (5):729.

[58] Maldas, D., Kokta, B. V., (1990), Effect of Extreme Conditions on the Mechanical Properties of Wood Fiber-Polystyrene Composites. Ii. Sawdust as Reinforcing Filler, *Polym. Plast. Technol. Eng.*, 29 (1&2):119.

[59] Raj, R. J., Kokta, B. V., Grouleau, G., Daneault, C., (1990), The Influence of Coupling Agents on Mechanical Properties of Composites Containing Cellulosic Fillers, *Polym. Plast. Tec. Eng.*, 29 (4):339.

[60] Zaman, H. U., Khan, M. A. and Khan, R. A.,(2009), Improvement of Mechanical Properties of Jute Fibers-Polyethylene/Polypropylene Composites: Effect of Green Dye and UV Radiation, *Poly. Plas. Tec. Eng.*, 48:1130–1138.

[61] Marcovich, N. F., Ostrovsky, A. N., Aranguren,M. I., Reboredo, M. M., *Proceedings from the 3rd International symposium on Natural Polymers and Composites*-IS NaPOl, Sao Pedro, SP, Brazil, May, 2000, pp 14–17.

[62] Wunderlich, B. *Thermal Analysis, Academic Press*: Boston, 1990.

[63] Patnaik, S., Sarangi, S., Mohanty, A. K., Singh, B. C., (1989), Graft copolymerization of acrylonitrile onto jute fibers (studies on Ce(IV)-hippuric acid redox system), *J. Appl. Polym. Sci.*, 37:2099.

[64] Gassan, J. and Gutowski, V. S., (2000), Effects of corona discharge and UV treatment on the properties of jute/epoxy composites, *J. Compos. Sci. Technol.*, 60:2857-2863.

[65] Khan, M. A., Haque, N., Kafi, A. A., Alam, M. N. and Abedin, M. Z., (2006), Jute Reinforced Polymer Composite by Gamma Radiation: Effect of Surface Treatment with UV Radiation, *Polym. Plast. Technol. Eng.*, 45 (5):607-613.

[66] Zaman, H. U., Khan, M. A., Khan, R. A., Mollah, A. Z. I., Pervin, S. and Al-Mamun, M., (2010), A Comparative Study between Gamma and UV Radiation of Jute fabrics/Polypropylene Composites: Effect of Starch, *Journal of Reinforced Plastics and Composites*, 29 (13):1930-1939.

[67] Muzzarelli, R. A. A., *Chitin, Pergamon Press*, New York, USA, 1977, pp 1–3.

[68] Roberts, G. A. F., *Chitin chemistry, Mac Millan Press*, Houndmills, USA, 1992, p 1–10.

[69] Sugimoto, M., Morimoto, M., Sashiwa, H., Saimoto, H., Shigemasa, Y., (1998), Preparation and Characterization of Water-Soluble Chitin and Chitosan derivatives, *Carbohydr. Polym.*, 36:49–59.

[70] Crescenzi, V., (1994), Polysaccharide science and technology: development and trends, *Trend. Polym. Sci.*, 3 (2):104–109.

[71] Miskiel, F. J., Pazur, J. H., (1991), The preparation and characterization of antibodies with specificity for the carbohydrate units of gum arabic and gum mesquite, *Carbohydr. Polym.*, 16 (1):17–35.

[72] Stanford, P. A. *Front carbohydrate research, Elsevier*, Amsterdam, 1992, pp-250.

[73] Khan, M. A., Ferdous, S., Mustafa, A. I., (2005), Improvement of Physico-mechanical Properties Chitosan Films by Photo-curing with Acrylic monomers, *J. Polym. Environ.*, 13 (2):193–201.

[74] Ferdous, S., Mustafa, A., Khan, M. A., (2003), Study on Photo-cured Bio-blend Films of Chitosan/PVA and PEO/PVA with Acrylic monomers, *J. Macromol. Sci. Pure. Appl. Chem. A*, 40 (8):817–832.

[75] Haque, P., Mustafa, A. I., Khan, M. A., (2007), Effect of crosslinking monomers on the physico-mechanical and degradation properties of photografted chitosan film, *Carbohydr. Polym.*, 68 (1):109–115.

[76] Sterman, S., and Marsden, J. G., (2004), The effect of silane coupling agents in improving the properties of filled or reinforced thermoplastics, *Polym. Eng. Sci.*, 6 (2):97–112.

[77] Plueddemann, E. P. *Silane Coupling Agent, Plenum Press*: New York, 1982, 25–33.

[78] Khan, M. A., Rashidul A., Farhana G. N., Khan, R. A. and Arifur M. R.,(2009), Modification and Characterization of Chitosan Films Using 3-trimethoxyl silyl propylmethacrylate, *Journal of Macromolecular Science, Part A: Pure and Applied Chemistry*, 46:751–758.

[79] Caner, C. and Cansiz, O., (2007), Effectiveness of chitosan based coating in improving shelf-life of eggs, *Journal of the Sci. of Food and Agriculture*, 87:227.

[80] Renbutsu, E., Okabe, S., Omura, Y., Nakatsubo, F., Minami, S., Saimoto, H. and Shigemasa,Y., (2007), Synthesis of UV-curable chitosan derivatives and palladium (II) adsorption behavior on their UV-exposed films, *Carbohydrate Poly.*, 69 (4):697–706.

[81] Alam, R., Khan, M. A., Khan, R. A., Ghoshal, S., Mondal, M. I. H., (2008), Study on the physico-mechanical properties of photo-cured chitosan films with oligomer and acrylate monomer, *J. Polym. Environ.*, 16 (3):213-219.

[82] Vasudev, S. C., Chandy, T., Sharma, C. P., (1997), Development of chitosan/polyethylene vinyl acetate co-matrix: controlled release of aspirin-heparin for preventing cardiovascular thrombosis, *Biomaterials,* 18:375.

[83] Khoo, C. G. L., Frantzich, S., Rosinski, A., Sjostrom, M., Hoogstraate, J., (2003), Oral gingival delivery systems from chitosan blends with hydrophilic polymers, *Eur. J. Pharm. Biopharm.*, 55:47.

[84] Zhang, L., Guo, J., Peng, X., Jin, Y., (2004), Preparation and release behavior of carboxymethylated chitosan/alginate microspheres encapsulating bovine serum albumin, *J. Appl. Polym. Sci.*, 92:878.

[85] Enea, O., Jollcoeur, C., (1982), Heat capacities and volumes of several oligopeptides in urea-water mixtures at 25^{0}C Some implications for protein unfolding, *J. Phys. Chem. B.*, 86:3870.

[86] Liepinsh, E., Otting, G., (1994), Specificity of urea binding to proteins, *J. Am. Chem. Soc.*, 116:9670.

[87] Dias,L. G., Florenzano, F. H., Reed, W. F., Baptista, M. S., Souza, S. M. B., Alvarez, E. B., Chaimovich, H., Cuccovia, I. M., Amaral, C. L. C., Brasil, C. R., Romsted, L. S., Politi, M., (2002), Effect of Urea on Biomimetic Systems: Neither Water 3-D Structure Rupture nor Direct Mechanism, Simply a More "Polar Water", *J. Langmuir*, 18:319–324.

[88] Gradolnik, J., Mar'echal, Y., (2002), Urea and urea/water solutions: an infrared study, *J. Mol. Struct.*, 615:177.

[89] Khan, M. A., Alam, R., Rahman, M. R., Noor, F. G. and Khan, R. A., (2009), Physico-Mechanical and Degradation Properties of Urea-Modified Chitosan Film Photocured with 1-Vinyl-2 Pyrrolidone, *Poly. Plastics Technology and Engineering*, 48:1206–1215.

[90] Garcia, N. L., Ribba, L., Dufresne, A., Aranguren, M. I., Goyanes, S., (2009), Physico-Mechanical Properties of Biodegradable Starch Nanocomposites, *Macromol. Mater. Eng.*, 294 (3):169–177.

[91] Khan, M. A., Rahman, M. A., Khan, R. A., Rahman, N., Islam, J. M. M., Alam, R. and Mondal, M. I. H., (2010), Preparation and Characterization of the Mechanical

Properties of the Photocured Chitosan/Starch Blend Film, *Polymer Plastics Technology and Engineering*, 49:748–756.

[92] Twilley J. W., (1984), The Analysis of Exudates Plant Gums in Their Artistic Applications: An Interim Report, *Advances in Chemistry*, 205:357-394.

[93] Larson, K. C., An Investigation of Pectic Substance in the Developing Xylem of *Populus Tremuloides*, The Institute of Paper Chemistry, Appleton, Wisconsin, 1967.

[94] Kulshrestha, V. K., (1962), Differential thermal studies on plant gums, *J. Pol. Sci.*, 58:791–808.

[95] Huang, H. H., Wilkes, G. L., Carlson, J. G., (1989), Structure-property behaviour of hybrid materials incorporating tetraethoxysilane with multifunctional poly-(tetramethylene oxide), *J. Polymer*, 30:2001–2012.

[96] Kioul, A., Mascia, L., (1994), Compatibility of polyimide-silicate ceramers induced by alkoxysilane silane coupling agents, *J. Non-Cryst. Solids*, 175:169–186.

[97] Mascia, L., Tang, T., (1998), Polyperfluoroether-silica hybrids, *J. Polym.*, 39 (14): 3045–3057.

[98] Helminen, A., Korhonen, H., Seppala, J. V., (2001), Biodegradable crosslinked polymers based on triethoxysilane terminated polylactide oligomers, *J. Polym.*, 42:3345–3353.

[99] Bhattacharya, A., (2000), Radiation and industrial polymers, *Prog. Poly. Sci.*, 25:371–401.

[100] Gehring, J., (2000) With *radiation* crosslinking of engineering plastics into the next millennium, *Rad. Phys. Chem.*, 57 (3–6):361–365.

[101] Khan, M. A., Khan, R. A., Rashid, M. H. and Kabir, A. N. M. H., (2009), Modification and characterization of bio-blend of jiga resin/starch with vinyltrimethoxy silane by UV-radiation, *Polymer-Plastics Technology and Engineering*, 48:786–792.

[102] Fan, L., Du, Y., Huang, R., Wang, Q. and Zhang, L., (2005), Preparation and characterization of alginate/gelatin blend fibers, *J. Appl. Polym. Sci.*, 96:1625–1629.

[103] Fan, L., Zhu, H., Zheng, H., Xu, Y. and Zhang, C., (2007), Structure and properties of blend fibers prepared from alginate and konjac glucomannan, *J. Appl. Polym. Sci.*, 106:3903.

[104] Lee, Y. J., Shin, D. S., Kwon, O. W., Park, W. H., Choi, H. G., Lee, Y. R., et al., (2007), Preparation of atactic poly(vinyl alcohol)/sodium alginate blend nanowebs by electrospinning, *J. Appl. Polym. Sci.*, 106:1337–1342.

[105] Qin, Y., (2006), The characterization of alginate wound dressings with different fiber and textile structures, *J. Appl. Polym. Sci.*, 100:2516–2520.

[106] Qin, Y., (2004), Gel swelling properties of alginate fibers, *J. Appl. Polym. Sci.* 91:1641–1645.

[107] Idicula, M., Joseph, K. and Thomas, S., (2010), Mechanical Performance of Short Banana/Sisal Hybrid Fiber Reinforced Polyester Composites, *J. Reinf. Plast. Compos.*, 29:12–29.

[108] Khan, R. A., Khan, M. A., Sultana S., Khan, M. N., Shubhra Q. T. H. and Noor, F. G., (2010), Mechanical, Degradation, and Interfacial Properties of Synthetic Degradable Fiber Reinforced Polypropylene Composites, *J. Reinf. Plast. Compos.*, 29:466–476.

[109] Kuswaha, P. K. and Kumar, R., (2010), Effect of Silanes on Mechanical Properties of Bamboo Fiber-epoxy Composites, *J. Reinf. Plast. Compos.*, 29:718.

[110] Khan, R. A., Parsons, A. J. Jones, I. A., Walker, G. S. and Rudd, C. D., (2010), Preparation and Characterization of Phosphate Glass Fibers and Fabrication of Poly(caprolactone) Matrix Resorbable Composites, *J. Reinf. Plast. Compos.*, 29 (12):1838-1850.

[111] Ismail, H., Shuhelmy, S., Edyham, M. R., (2002), The effects of a silane coupling agent on curing characteristics and mechanical properties of bamboo fibre filled natural rubber composites, *European Polymer Journal,* 38 (1):39.

[112] Rahman, S. M. M. A., Mustafa, A. I. and Khan, M. A., (2009), Jute reinforced polypropylene composite: Effect of surface pretreatment by photocuring with acrylic monomers, *J. Reinf. Plast. Compos.*, 28:1733–1745.

[113] Khan, A., Huq, T., Saha, M., Khan, R. A. and Khan, M. A., (2010), Surface modification of calcium alginate fibers with silane and methyl methacrylate monomers, *Journal of Reinforced Plastics and Composites*, 29 (20):3125-3132.

[114] Boyanova, M., Stambolieva, N., Apostolov, A. A., Fakirov, S., (2003), Dissolution and Enzymatic Degradation Studies Before and After Artificial Ageing of Silk- or Linen-Reinforced Gelatin Laminates, *Macromol. Mater. Eng.*, 288:399.

[115] Li, J., Chen, Y., Yin, Y., Yao, F., Yao, K., (2007), Modulation of nano-hydroxyapatite size via formation on chitosan-gelatin network film in situ, *Biomaterials,* 28:781.

[116] Sivakumar, M., Radhakrishnan, P. R. G., Kothandaraman, H., (1991), Grafting of glycidyl methacrylate onto gelatin, J. *Appl. Poly. Sci.*, 43:1789.

[117] Sarkar, S., Chourasia, A., Maji, S., Sadhukhan, S., Kumar, S., Adhikari, B., (2006), Synthesis and characterization of gelatin based polyester urethane scaffold, *Indian Acad. Sci.* 29:475.

[118] Salamone, J. C. *Polymeric material encyclopedia, CRC Press*, New York, 1996 pp 7.

[119] Kronggauz, V. V., yohannan, R. M., (1990), Photopolymerization kinetics and monomer diffusion in polymer matrix, *Polymers*, 31:1131.

[120] Fitzgerald, J. J., Landry, C. J. T., (1990), Vitrification and curing studies of a photopolymerizable semi-interpenetrating polymer network, *J. Appl. Poly. Sci.*, 40:1727.

[121] Dworjanyn, P. A., Garnett, J. L., Khan, M. A., Maojun, X., Meng-Ping, Q., Nho, Y. C., (1993), Novel Additives for Accelerating Radiation Grafting and Curing Reactions, *Radiat. Phys. Chem.*, 42:31.

[122] Song, C. L., Yoshii, F., Kume, T. J., (2001), Radiation crosslinking of biodegradable poly(butylene succinate) at high temperature, *J. Macromol. Sci. Pure Appl. Chem.*, 38:961.

[123] Noor, F. G., Khan, M. A., Ghoshal, S., Khan, R. A., Mazid, R. A., Chowdhury, A. M. S., (2010), Effect of Urea on the Mechanical Properties of Gelatin Films Photocured with 2-Ethylhexyl Acrylate, *J. Poly. Environ.*, 18(3):224-230.

[124] Mirzadeh, H., Katbab, A. A., Gorgin, E., Golestani, A., (1995), Cell attachment to laser-induced AAm-and HEMA-grafted ethylenepropylene rubber as biomaterial: in vivo study, *Biomaterials*, 16:641.

[125] Guice, K. B., Loo, Y. L., (2006), Azeotropic Atom Transfer Radical Polymerization of Hydroxyethyl Methacrylate and (Dimethylamino)ethyl Methacrylate Statistical Copolymers and Block Copolymers with Polystyrene, *Macromol.,* 39:2474–2480.

[126] Khorasani, M. T., Mirzadeh, H., Sammes, P. G., (1999), Laser surface modification of polymers to improve biocompatibility: HEMA grafted PDMS, in vitro assay—III, *Radiat. Phys. Chem.*, 55:685–689.

[127] Hoffman, A. S., Cohn, D., Hanson, S. R., Ratenar, B. D., Reynolds, L. O., (1983), Application of radiation-grafted hydrogels as blood-contacting biomaterials, *Radiat. Phys. Chem.*, 22:267–283.

[128] Kejlova, K., Labsky, J., Jirova, D., Bendova, H., (2005), Hydrophilic polymers: biocompatibility testing in vitro, *Toxicol. in Vitro*, 19:957–962.

[129] Cifkova, I., Stolt, M., Holuza, R., Adam, M., (1987), Calcification of poly(2-hydroxyethyl methacrylate)-collagen composites implanted in rats, *Biomaterials*, 8:30-34.

[130] Smetana, K., Stol, M., Novak, M., Adam, M., (1992), Implantation of p(HEMA)-collagen composite into bone, *Biomaterials*, 13:639–642.

[131] Khan, M. A., Hassan, M. M., Drazel, L. T., (2005), Effect of 2-Hydroxyethyl Methacrylate on the Mechanical and Thermal Properties of Jute-Polycarbonate Composite, *Composites Pt. A*, 36:71–81.

[132] Shukla, S. R., Athalye, A. R., (1992), Ultraviolet-radiation induced graft-copolymerization of hydroxyethyl methacrylate onto cotton cellulose, *J. Appl. Polym. Sci.*, 44:435.

[133] Khan, M.A., Khan, R. A., Aliya, B. S., Nasreen, Z., (2006), Effect of the Pretreatment with UV and Gamma Radiations on the Modification of Plywood Surface by Photocuring with Epoxy Acrylate, *J. Polym. Environ.*, 14 (1):111-118.

[134] Sabrina, S., Khan, R. A., Khan, M. A., Zaman, H. U., Shahruzzaman, M., Banu, P., Khan, M. N. and Mustafa, A. I., (2010), Preparation and Mechanical Characterization of Gelatin-Based Films Using 2-Hydroxyethyl Methacrylate Cured by UV Radiation, *Polymer Plastics Technology and Engineering*, 49(6):560-566.

INDEX

#

A

B

C

D

E

F

G

J

K

L

M

N

O

P

Q

R

T

U

V

W

X

Y

Z